THE RAY SOCIETY

INSTITUTED 1844

This Volume is No. 147 of the series

LONDON 1973

BRITISH TORTRICOID MOTHS

COCHYLIDAE AND TORTRICIDAE: TORTRICINAE

J. D. BRADLEY, PH.D., F.R.E.S.
Commonwealth Institute of Entomology, London

W. G. TREMEWAN, M.I.BIOL.
British Museum (Natural History), London

&

ARTHUR SMITH, A.R.C.A.
British Museum (Natural History), London

With additional colour illustrations by

BRIAN HARGREAVES, N.D.D., M.G.L.C., M.F.P.S.

Published September 1973

ISBN 0 903874 01 6

Sold by: Johnson Reprint Co., 24–28 Oval Road,
London, England, NW1 7DX
    &
E. W. Classey Ltd, 353 Hanworth Road,
Hampton, Middlesex, England TW12 3EN

For the Ray Society
c/o British Museum (Natural History)
Cromwell Road, London, England SW7 5BD

The text has been set in Monotype Ehrhardt and
printed on Curwen Wove specially made for
The Curwen Press Ltd by William Sommerville.
The colour plates have been reproduced in
four colours and printed by offset lithography
on Paladin coated cartridge.

Designed by James Shurmer at The Curwen Press Ltd.
Colour reproduction, composition and printing
by The Curwen Press Ltd, London, England.

Binding by Webb Son and Company, Ferndale, Wales.

To L. T. FORD 1881 – 1961
who did much to inspire others to find interest and pleasure in
collecting and rearing Microlepidoptera in the British Isles

The L. T. Ford Collection of British Microlepidoptera
is preserved in the British Museum (Natural History)

PREFACE

This work was started more than ten years ago, after we had become aware of the growing demand for a new illustrated book on the British Tortricoidea. The kind of book most needed seemed to be one not primarily for the specialist, but one readily usable by anyone interested in Lepidoptera, and requiring minimal knowledge of taxonomy. It seemed specially desirable that it should not be solely an aid to the identification of the moths or imagines, but should collate available data on the bionomy of the various species and, so far as possible, illustrate larval feeding habits, an aspect which hitherto had not been attempted for the Tortricoidea. We have endeavoured to fulfil these requirements in preparing this volume for the Ray Society.

Initially it was expected that the whole of the British Tortricoidea, comprising the two families Cochylidae and Tortricidae, could be included in a single volume. This unfortunately proved impracticable and it was necessary to segregate the subfamily Olethreutinae. The present volume thus contains the Cochylidae and the sub-family Tortricinae.

Like most other Lepidoptera, the majority of Cochylids and Tortricids can be readily identified by the colour-pattern of the forewings. Meticulous attention was therefore paid to detail when preparing the colour paintings of the moths and, if used in conjunction with the written descriptions, these should enable most specimens to be determined to species. Inevitably there will be some doubtful determinations, particularly among the more variable species, and it may then be necessary to examine the genitalia of the specimens.

The authors and artists have been fortunate in having ready access to the collections of Microlepidoptera in the British Museum (Natural History), including the extensive British collections of E. R. Bankes, W. G. Sheldon, and L. T. Ford, which together contain examples of nearly all the known species of British Tortricoidea and their varieties.

Although this work has been a joint effort on the part of the three authors, it should be stated that the taxonomic part and the biological descriptions are the responsibility of the two senior authors and any errors must be attributed to them. The biological illustrations of larval spinnings and foodplants have all been drawn by Arthur Smith. Originally it was envisaged that he would also undertake the colour paintings of the moths, but that task proved so prodigious and time-consuming that it became necessary to seek assistance. We were therefore fortunate that Mr. Brian Hargreaves could be commissioned to paint the colour plates of the Tortricinae, with the exception of the plate depicting *Acleris literana* (Linnaeus) which, together with the colour plates of the Cochylidae, are the work of Arthur Smith.

The work has been accomplished with the help and advice of many individuals. We wish to express our indebtedness to the Trustees of the British Museum (Natural History) and the Director of the Commonwealth Institute of Entomology for the many facilities provided, and in particular to the successive Keepers of Entomology, Dr. W. E. China, C.B.E., Mr. J. P. Doncaster, C.B.E., and Dr. Paul Freeman. From the outset we were fortunate in having the co-operation of Messrs. H. C. Huggins, H. N. Michaelis, E. C. Pelham-Clinton (Royal Scottish Museum, Edinburgh) and S. Wakely, all of whom generously contributed numerous notes and observations drawn from knowledge accrued during many years of practical experience in the field, and who provided some of the larval spinnings used for illustration and description. We are also grateful to the following who have helped in various ways: Messrs. R. G. Adams (Pest Infestation Control Laboratory, Slough), C. R. B. Baker (Plant Pathology Laboratory, Harpenden), J. M. Chalmers-Hunt, D. L. Coates, Dr. G. H. L. Dicker (East Malling Research Station), the late Capt. J. Ellerton, D.S.C., Lt.-Col. A. M. Emmet, M.B.E., D. W. H. Ffennell, R. L. E. Ford, J. Heath (Biological Records Centre, Abbots Ripton), S. N. A. Jacobs, Lt.-Col. W. B. L. Manley, the late R. M. Mere, V.R.D., Mrs. M. J. Morgan (University College, Bangor), F. C. Nutbeam (Head Gardener, Buckingham Palace), Dr. J. R. G. Turner, R. W. J. Uffen, J. D. R. Vernon (Agricultural Development & Advisory Service, Bristol), R. G. Warren, T. G. Winter (Forest Research Station, Alice Holt Lodge, Farnham), and to our colleagues in the British Museum (Natural History): D. J. Carter, B. J. Clifton, D. S. Fletcher, E. W. Groves, M. A. Kirby, Dr. J. F. Perkins, Dr. K. Sattler and W. H. T. Tams and others who have contributed to this work. We are especially indebted to Mrs. J. D. Bradley, Lt.-Col. A. M. Emmet, Mr. E. C. Pelham-Clinton, Dr. J. F. Perkins and Mr. W. H. T. Tams for reading the text and making constructive criticisms and suggestions, and to Dr. W. S. Bristowe who initially proposed that the work should be published by the Ray Society and gave much encouragement.

The publication of this work under the auspices of the Ray Society has largely been achieved through the generous support of the John Spedan Lewis Trust for the Advancement of Natural Sciences, and we wish to acknowledge the efforts of Mr. G. H. Locket in acting as intermediary. We are also indebted to Imperial Chemical Industries Ltd, and the Worshipful Company of Armourers and Brasiers, who have contributed towards the cost of printing this and other Ray Society publications, and to the Honorary Secretary of the Ray Society, Mr. D. Macfarlane, for his sympathetic and unfailing response to our numerous requests.

It is a pleasure to acknowledge the great interest and care taken by Mr. Basil Harley and the staff of The Curwen Press in the production of this book. Their expertise in the field of illustrated works on natural history follows in the best tradition.

John David Bradley
Walter Gerald Tremewan

J. D. Bradley and W. G. Tremewan have been responsible for the text and have worked in close collaboration with Arthur Smith who has been responsible for the biological drawings (monochrome plates 1–21), including much of the field work in quest of larval material from which the drawings have been made, and also for the paintings of some of the imagines (colour plates 22–25, 47).

The Ray Society has always prided itself on the scientific contribution to its publications made by the illustrator and has therefore gladly included Arthur Smith's name as an author.

Council, The Ray Society

INTRODUCTION

In the British Isles the large cosmopolitan group comprising the superfamily Tortricoidea is represented by two families, following current concepts of classification of the Lepidoptera. These are the Cochylidae (Phaloniidae), of which 48 species appear in the British list, and the Tortricidae, of which 321 species are listed. Among these species are a number which are not indigenous and some of uncertain taxonomic status. The family Tortricidae is divided into two subfamilies, the Tortricinae comprising 94 species, and the Olethreutinae (Eucosminae) comprising 227 species. These two subfamilies may be subdivided into tribes: the Tortricinae into the Archipini, Sparganothini, Cnephasiini and Tortricini; and the Olethreutinae into the Olethreutini, Eucosmini and Laspeyresiini. The Cochylidae is a more homogeneous family and does not show such strong tribal tendencies as the Tortricidae. However, the status of this group as a separate family is by no means assured, since the imagines and larvae show many of the morphological characteristics and behavioural patterns of the Tortricidae, which have led some authors to treat the group as no more than a subfamily of the Tortricidae, or even a tribe of the Tortricinae.

The limitations of families and genera are to a certain extent a matter of personal opinion. In the present work the narrower concept of the family and other suprageneric categories and the broader concept of the genus have been adopted. The species thus form large but reasonably homogeneous generic groups which are more convenient for practical purposes but, if desired, can be divided into subgeneric groups, as shown in the check-list.

Although very varied in wing shape and colour pattern, Tortricoid moths have a characteristic general appearance. In the subfamily Tortricinae, the species in the tribe Archipini are among the most distinctive. The Archipines are predominantly brown in coloration and have the apices of the comparatively broad forewings strongly produced, the moth resembling a bell in outline when at rest, the folded wings being flattened against the substrate. In contrast, species of the tribe Cnephasiini generally have elongate-ovate forewings with predominantly grey and black coloration, while those of the tribe Tortricini have characteristic rectangular-shaped forewings, often with roughened scaling or scale-tufts. Many species of the subfamily Olethreutinae may at once be recognized by the presence of a specialized marking known as the ocellus, situated in the tornal area of the forewing and often with metallic or iridescent scaling; the development of this marking is generally considered to indicate an advanced wing pattern, often associated with diurnal habits. The Olethreutines usually rest with their wings folded close to the abdomen in a somewhat ridge-like attitude. A number of species, particularly in the genera *Pardia*, *Apotomis* and *Hedya*, have cryptic black and white coloration and resemble bird-droppings when at rest. Others, such as those of the genera *Rhyacionia* and *Blastesthia*, when at rest resemble the buds of the conifers in which the larvae feed.

The Cochylids show less diversity in wing shape; the comparatively narrow forewings are more triangular and are folded against the abdomen in a steep ridge-like position when the moth is at rest. In many species the thickening of the principal veins, leaving hollows between them, gives the forewings a rather characteristic convoluted appearance.

Tortricoid moths, commonly known as leaf-rollers because of the leaf-rolling habit of the larvae of some species, are familiar to most collectors of Lepidoptera, even if they do not collect the "Micros". The imagines of many species are frequently to be seen on the wing or disturbed from rest during the day, and at night they often come to light, especially ultra-violet light.

Many collectors, otherwise interested only in the "Macros", have been induced to collect Tortricoid moths because of their comparatively robust build. Provided a minuten-pin of suitable gauge and a setting board of appropriate size are used, the procedure for preparing specimens is much the same as with the Macrolepidoptera and with a little practice perfect cabinet specimens can be obtained.

The inadequacy of the literature available to the general collector and the consequent difficulties in determination of specimens has undoubtedly deterred many from pursuing the study of this well represented and attractive group of moths. Apart from the illustrated edition of volumes 10 and 11 of *The Lepidoptera of the British Islands* by Barrett (1904–07), the few works dealing with the British Tortricoidea are either not adequately illustrated or are out of date or too technical for the non-specialist. The one most widely used is *A Revised Handbook of the British Lepidoptera* by Meyrick (1928). Bradley & Martin (1956) and Bradley (1959) published illustrated lists with revised nomenclature in the *Entomologist's Gazette*.

An important work which began a new era in the classification of the Tortricoidea was that of Pierce & Metcalfe (1922) on the structure of the genitalia of the British species. This coincided with a similar study by Heinrich (1923; 1926) of the North American Olethreutinae. The classification in these works was based on genitalic structures, which were shown to be of value in assessing relationships, in addition to providing characters for the identification of species. Prior to their publication the classification of the Tortricoidea was based mainly on superficial and external characters, such as wing shape and scaling, labial palpi, wing venation and secondary sexual characters, such as the presence of a costal fold in the forewing of the male. These characters are now known to be of limited value in the differentiation of taxa at generic level since in some cases they show parallel development or convergence in unrelated groups and can be misleading. It is now generally accepted that the structure of the genitalia offers a sounder basis for classifying the Lepidoptera, that of the male usually showing the best characters. Modern classification of the Tortricoidea is therefore based on genitalic morphology together with venational, palpal and other external characters. This is well demonstrated in the important work by Obraztsov (1954–57) on the genera of the Palaearctic Tortricidae and in the subsequent works of Hannemann (1961; 1964) on the German Tortricidae and Cochylidae, Bentinck & Diakonoff (1968) on the Tortricidae of the Netherlands and, more recently, Razowski (1970) on the Palaearctic Cochylidae.

This extensive use of genitalic characters for classification has made it virtually impossible to construct reliable keys to the Tortricoidea on other external characters alone. The key to the families, subfamilies and tribes which is provided below is therefore of limited application, since it is based on venational, palpal and superficial characters.

The aim of the present work is to help the non-specialist to identify British Tortricoid moths and learn something about their biology, distribution and variation. Rearing of the immature stages and study of the geographical distribution and variation of species offer scope for original research and field work. For example, the specific status and distribution of *Aphelia unitana* (Hübner), the status in the British Isles of *Paramesia gnomana* (Clerck) and *Acleris abietana* (Hübner), and the genetics of polymorphic species, particularly of the genus *Acleris*, need fuller investigation. The subfamily Olethreutinae includes a greater number of species about which little is known.

Descriptions and Illustrations

In the descriptions of the imagines, the measurements in millimetres (mm) denote the expanse of the fully spread forewings, including the cilia, and are taken from the smallest and largest specimens examined. When the sexes show a marked difference in size, the measurements for each are given separately, although there usually is an overlap. Sexual dimorphism is described, including antennal differences in some species-groups, and the presence of secondary sexual characters, such as a costal fold on the forewing of the male, is mentioned.

The illustrations of the moths in the colour plates are approximately $\times 2 \cdot 4$. The descriptions of the wing patterns are intended to supplement the colour illustrations and emphasize the diagnostic characters of each species. They are also necessary because variation in the colour pattern of some species is so extensive that it has not been possible to illustrate the full range. The head and thorax may generally be assumed to be of the same colour as the markings of the forewing, and the abdomen like that of the hindwing.

The colour terms used are necessarily somewhat general because coloration in the Lepidoptera, besides being naturally variable, often varies with the age and condition of both live and dead specimens. Many pigments are liable to fade or change through exposure to light or with the passage of time. In particular, the rich grey coloration characteristic of fresh specimens of many species may become distinctly brownish. Also, the chemical action of killing agents may affect certain pigments, especially if the agent is used in excess or the specimens are left too long in the killing jar.

Colour is also subject to different interpretations depending upon the light source and specimens should be examined under good daylight conditions, not in artificial light. Ground colour is usually produced by pale unicolorous scales, and the colour patterns or markings are nearly always produced by scales which are darker-coloured medially. A low power lens of about $\times 5$ should be used to examine specimens. Higher magnifications tend to disrupt the actual colour pattern, since the colour demarcations of individual scales become discernible.

In the descriptions of the larvae, the known foodplants are cited for monophagous and oligophagous species. Only a selection of the foodplants is given for polyphagous species.

Wing Pattern

In the forewing the Tortricoidea retain vestiges of the transverse fasciate pattern of the primordial Lepidoptera. This appears essentially as basal, sub-basal, median, postmedian and subterminal fasciae. The fasciae show varying degrees of reduction or modification within the different generic groups; in some they may be obsolescent or obsolete, while in others they either form or are replaced by specialized patches or longitudinal streaks. The hindwing is usually more or less unicolorous, sometimes with darker strigulation.

In the Cochylidae (text-figs. 1, 2) the assumed basic pattern consists of basal, sub-basal, median, postmedian and subterminal fasciae, and a tornal marking. In some species the basal, sub-basal and subterminal fasciae and the tornal marking may be obsolescent or obsolete.

In the Tortricinae (text-figs. 3–5) the assumed basic pattern consists of poorly differentiated basal and sub-basal fasciae, a well-developed, outward-oblique median fascia from the middle of the costa to near the tornus, and a relatively small pre-apical spot, or a large costal blotch. The latter marking is formed by the costal part of the median fascia, which is obsolete dorsally, coalescing with the pre-apical spot.

In the Olethreutinae the wing pattern of species in the Olethreutini, in which the pattern appears to be most primitive, is basically similar to that of the Archipini in the subfamily Tortricinae. In the Eucosmini and Laspeyresiini (text-fig. 6) the transverse fasciate pattern is often obsolescent and reduced to numerous strigulae along the costa, whilst a complex marking known as the ocellus is developed in the tornal area, and a semi-elliptical or falcate patch of ground colour is often present on the dorsum.

A structure of minor importance sometimes mentioned in the descriptions is the dorsal tuft, which is situated near the base of the forewing on the dorsum and consists of appressed or semi-erect scales, and often projects slightly beyond the margin. A small marking in the discal area, known as the discal or discocellular spot, is present in many species. More important is the fold running from the base to the tornus of the forewing, following the course of either the cubital vein, or that of vein 1c (CuP), and referred to in the descriptions as either the median fold (usually in the Cochylidae) or the submedian fold (usually in the Tortricidae), according to its position on the wing.

Variation

The majority of Tortricoids represented in the British fauna are widely distributed in the Holarctic region. A number of these species show insular subspeciation, an aspect of polytypy that appears to have been little studied within the group and is beyond the scope of the present work. Among the British Tortricoidea a few species show relatively marked geographical variation or subspeciation within the British Isles, notably *Eupoecilia angustana* (Hübner), *Syndemis musculana* (Hübner), *Lozotaenia forsterana* (Fabricius) and *Eana penziana* (Thunberg & Becklin). Ecological variation, such as that found in *E. angustana*, also appears to have been little studied.

A number of species are polymorphic and have distinct genetical forms. Certain of these recurrent forms are more frequent than others, sometimes varying in frequency according to locality. Polymorphism is especially frequent in species of the genus

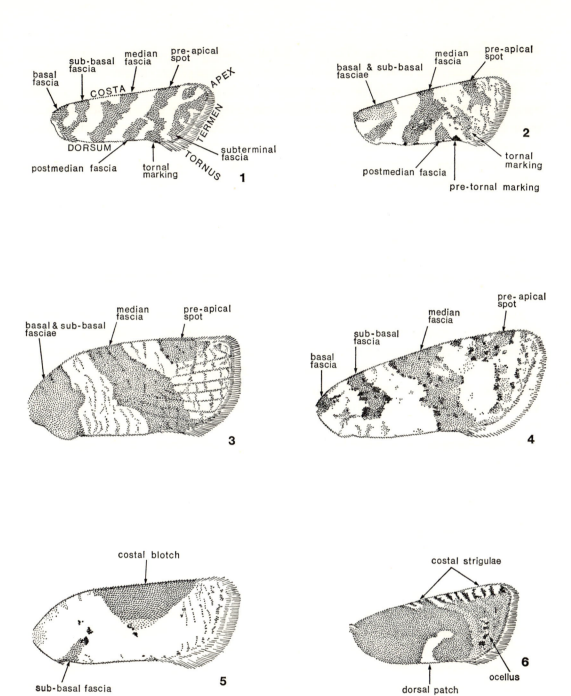

Wing patterns of Tortricoid moths

Figs. 1, 2, Cochylidae. Figs. 3–5, Tortricidae: Tortricinae. Fig. 6, Tortricidae: Olethreutinae

Acleris, especially *A. cristana* (Denis & Schiffermüller) and *A. hastiana* (Linnaeus), which are among the most strikingly polymorphic species of Lepidoptera.

The significance of this polymorphism is unknown. In some species, isolated populations produce forms which are either found only in a particular population or are more frequent in it than in others. This suggests the influence of some selective agency or mechanism, possibly connected with cryptic coloration.

Many of the more distinctive forms have been given names and in some cases were originally described as distinct species. It should be noted, however, that according to the *International Code of Zoological Nomenclature*, infrasubspecific forms have no status. Therefore, in the present work, the term "form" is used in a neutral sense and has been applied to ecotypes and seasonal forms as well as to genetical aberrations.

The highly polymorphic species of the Tortricidae offer wide scope for genetical studies. Notable contributions to the study of the genetics of this group are those on *Acleris comariana* (Lienig & Zeller) by Fryer (1926; 1931) and Turner (1968).

Melanism is evident in comparatively few species, but is frequent in *Archips podana* (Scopoli), *Cnephasia interjectana* (Haworth) and *C. stephensiana* (Doubleday).

Seasonal variation is rarely found in the British Tortricoidea, but is pronounced in *Acleris boscana* (Fabricius). This species has two generations of moths in a year, one in the spring, the other in the autumn. Imagines of the spring generation have the head, thorax and forewing ground colour white, whilst those of the autumn generation are grey.

The incidence of minor individual variation is very high in many species and is usually most apparent in the development or coloration of the basic markings of the forewing, and in the intensity of the suffusion and strigulation on the ground colour. Such variation is usually limited and produces only minor forms and these tend to intergrade.

General characteristics of the Tortricoidea

Imago

Head (text-figs. 7–14) with crown clothed with loosely appressed or raised scales, often with rising side-tufts, roughened or spreading, more or less meeting over vertex and projecting between antennae; frons receding, scales appressed. Antennae less than two-thirds length of forewing; flagellum or shaft of antenna simple, dentate, serrate, pectinate, ciliate, hairy or bristly (ciliation longer in male). Compound eyes glabrous. Ocelli and chaetosemata present. Proboscis or tongue variably developed, naked (not scaled). Labial palpi 3-segmented, short to very long, porrect or ascending; second segment usually more or less rough-scaled, sometimes with a tuft beneath, or expanded distally with roughened scales above and below; terminal segment short, obtuse, with appressed scales. Maxillary palpi developed, basal segment with a setose patch on inner surface. Thorax sometimes with a posterior crest of raised scales.

Fore legs (text-fig. 15) with epiphysis present on tibiae. Middle legs (text-fig. 16) with apical spurs present on tibiae, median spurs absent. Hind legs (text-fig. 17) with median and apical spurs present on tibiae.

Fore- and hindwings usually developed in both sexes; brachyptery is known only in the female of a few species. Frenulum of hindwing simple in male (text-fig. 18),

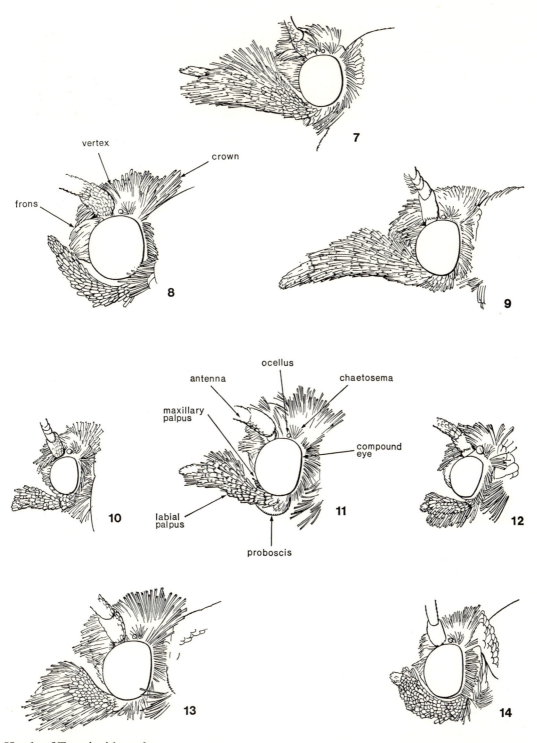

Heads of Tortricoid moths

Fig. 7, Cochylidae—*Agapeta hamana* (Linnaeus), ♂. Fig. 8, Tortricinae: Archipini—*Archips podana* (Scopoli), ♂. Fig. 9, Tortricinae: Sparganothini—*Sparganothis pilleriana* (Denis & Schiffermüller), ♂. Fig. 10, Tortricinae: Cnephasiini—*Cnephasia interjectana* (Haworth), ♂. Fig. 11, Tortricinae: Tortricini—*Tortrix viridana* (Linnaeus), ♂. Fig. 12, Olethreutinae: Olethreutini—*Olethreutes arcuella* (Clerck), ♂. Fig. 13, Olethreutinae: Eucosmini—*Eucosma cana* (Haworth), ♂. Fig. 14, Olethreutinae: Laspeyresiini—*Cydia pomonella* (Linnaeus), ♂

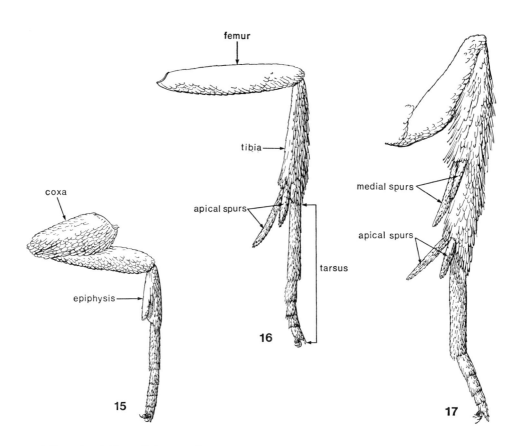

Legs of *Tortrix viridana* (Linnaeus)
Fig. 15, fore leg. Fig. 16, middle leg. Fig. 17, hind leg

multiple (usually 2–4 bristles) in female (text-fig. 19). Forewing (text-fig. 20) with twelve veins; anal veins 1a (1A) and 1b (2A) coalescent beyond base; anal vein 1c (CuP) absent in Cochylidae (text-figs. 20–22), present in Tortricidae (text-figs. 23–30); median vein (media, M) and chorda (stem of $R_4 + R_5$) sometimes present (text-fig. 28), the latter forming an accessory cell. Forewing of male sometimes with basal portion of costa enlarged and folded back to form a flap known as the costal fold (text-figs. 20, 23, 29). Hindwing (text-fig. 20) with eight veins; usually with three anal veins; cubital vein (cubitus, Cu) with a ridge of hairs (cubital pecten) present in the British species of Olethreutinae (text-figs. 28–30) but not in the Cochylidae nor Tortricinae except the solitary representative of the Sparganothini (text-fig. 24).

The system of numbering the veins to indicate their origin and the simplified numerical system used by Meyrick (1928) are correlated in the text-figures of the wing venation.

Genitalia

The genitalia of both sexes of nearly all the known British species of Tortricoidea are figured by Pierce & Metcalfe (1922; 1960 [facsimile]), and the appropriate references to these illustrations have been included under each species in the present work. A simplified technique for the preparation and dissection of genitalia is described on page 17.

8

Pierce (1909) gives schematic diagrams of male and female genitalia, indicating the various structures, and later (1914) provides a glossary of the terms used in his publications.

Students intending to study genitalia of the Tortricoidea will find it helpful to consult the descriptions and figures of the basic structures given by Razowski (1966; 1970), Bentinck & Diakonoff (1968) and Common (1970), and in particular the glossary by Klots (1970).

Ovum

Egg flattened, scale-like or lenticular, chorion usually smooth, sometimes reticulate; laid singly or in pairs, or in batches which are sometimes in the form of a large imbricate mass (Archipini), the micropyle of each egg being exposed. The eggs are usually covered with a secretion from the female accessory glands; in some species scales from the anal tuft or wings adhere to the egg mass.

Larva (text-figs. 31–36)

Varying from slender (external feeders) to rather stout (internal feeders); three pairs of thoracic legs, four pairs of prolegs on abdominal segments 3–6, and a pair of prolegs on abdominal segment 10.

Head (text-figs. 31–33) hypognathous or almost prognathous, often black in early instars, lighter in final instar; adfrontals reaching or nearly reaching vertical triangle; coronal suture usually short; six ocelli on each side, ocellar area rounded (internal feeders) or angular (internal and external feeders). Prothoracic plate variably sclerotized, often with distinct medial sulcus. Abdomen (text-figs. 34–36) with integument shagreened (spinulose (text-figs. 34, 35) or granular), usually without markings, without secondary setae; pinacula bearing primary setae often strongly sclerotized; spiracles round or broadly oval, with well-defined peritreme or peripheral sclerite; abdominal segment 10 often with sclerotized dorsal tergite (anal plate or shield) and a multi-pronged fork or comb below. Crotchets of prolegs arranged in a complete circle or ellipse, uni-, bi- or triordinal (uniordinal in the most primitive Tortricinae). Prothorax with three prespiracular setae in the lateral (L) group (text-fig. 35). Abdomen with the long subdorsal seta (SD1) on segment 8 almost invariably directly anterior to spiracle (text-fig. 35). Setae L1 and L2 always adjacent on segments 1–8 (text-fig. 35).

Hinton (1943: 203) states that the subventral (SV) setal group (text-fig. 35) on segment 8 is always bisetose except in the Cochylidae in which it is unisetose. However, examination of a number of larvae has shown that this character is unstable and that the SV group may be unisetose on one side of a larva and bisetose on the other, in both the Tortricidae and Cochylidae.

Swatschek (1958), in his studies of larvae of the German Tortricoidea, describes the chaetotaxy of many species which occur also in the British Isles. MacKay (1959; 1962), in her major papers on larvae of North American Tortricidae, also describes some species found in the British Isles. It is of interest that her extensive studies revealed no character which would sharply differentiate the Tortricinae and Olethreutinae, and showed that the Cochylidae could be separated only with difficulty. MacKay (1962: 7) gives a key to the Tortricinae tribes based on larval characters, and (1959: 11) a general key to late instar larvae of the Olethreutinae and certain species of Cochylidae.

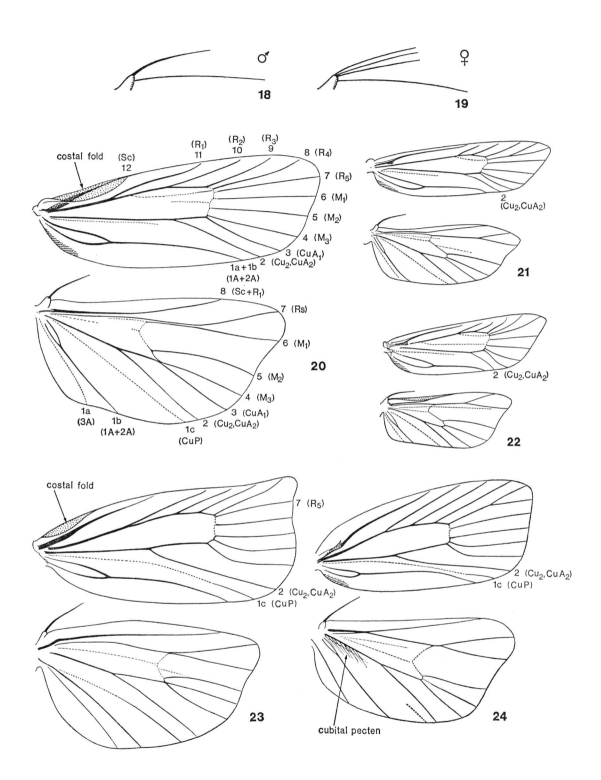

Frenula and venation of Tortricoid moths

Figs. 18, 19, frenula. Fig. 18, ♂. Fig. 19, ♀. Figs. 20–24, wing venation. Fig. 20, Cochylidae—
Hysterosia inopiana (Haworth), ♂. Fig. 21, Cochylidae—*Aethes francillana* (Fabricius), ♂. Fig. 22,
Cochylidae—*Falseuncaria ruficiliana* (Haworth), ♂. Fig. 23, Tortricinae: Archipini—*Archips podana*
(Scopoli), ♂. Fig. 24, Tortricinae: Sparganothini—*Sparganothis pilleriana* (Denis & Schiffermüller), ♂

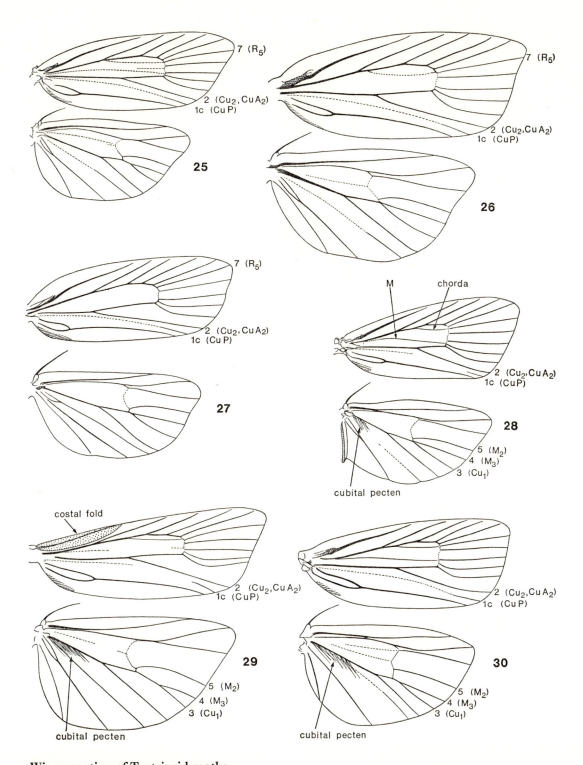

Wing venation of Tortricoid moths

Fig. 25, Tortricinae: Cnephasiini—*Cnephasia interjectana* (Haworth), ♂. Fig. 26, Tortricinae: Tortricini—*Tortrix viridana* (Linnaeus), ♂. Fig. 27, Tortricinae: Tortricini—*Acleris variegana* (Denis & Schiffermüller), ♂. Fig. 28, Olethreutinae: Olethreutini—*Olethreutes arcuella* (Clerck), ♂. Fig. 29, Olethreutinae: Eucosmini—*Eucosma cana* (Haworth), ♂. Fig. 30, Olethreutinae: Laspeyresiini—*Cydia pomonella* (Linnaeus), ♂

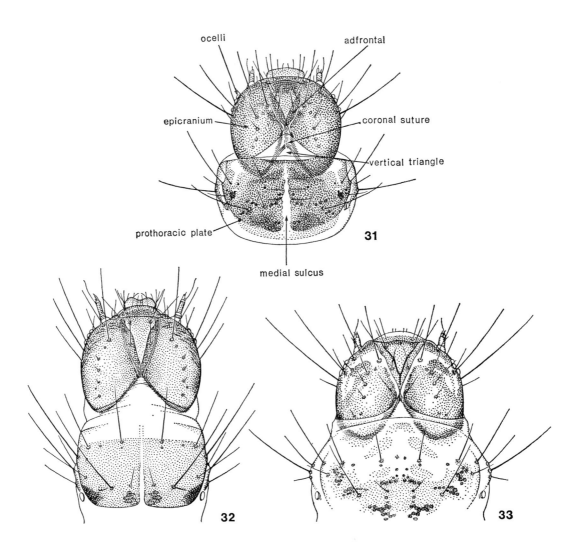

Heads and prothoracic segments of Tortricoid larvae
Fig. 31, Cochylidae—*Stenodes alternana* (Stephens). Fig. 32, Tortricidae: Tortricinae—*Pandemis heparana* (Denis & Schiffermüller). Fig. 33, Tortricidae: Olethreutinae—*Cydia pomonella* (Linnaeus)

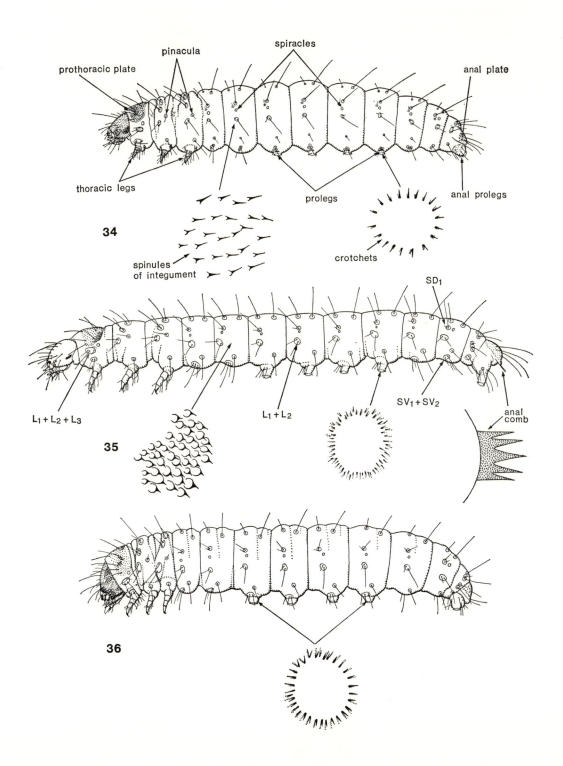

Tortricoid larvae

Fig. 34, Cochylidae—*Stenodes alternana* (Stephens). Fig. 35, Tortricidae: Tortricinae—*Pandemis heparana* (Denis & Schiffermüller). Fig. 36, Tortricidae: Olethreutinae—*Cydia pomonella* (Linnaeus)

There are a number of systems of setal nomenclature, one of the earliest being that of Fracker (1915). Swatschek (1958) follows that of Gerasimov (1952), whose key was translated into English by Martin (1962). Currently, the most widely used system is that introduced by Hinton (1946; 1948), and this has been adopted in the present work.

Pupa (text-figs. 37–39)

Abdominal segments 1 and 2 fused in both sexes, 8–10 fused in male, 7–10 in female; transverse rows of spines dorsally on segments 2–8 or 9; cremaster variably developed, often with fine hooked bristles, sometimes produced caudally, or in the form of a coronet of stout spines, or as lateral hooks. Wings broad, not sharply tapered, antennae reaching nearly to apices.

The structure of the cremaster and the arrangement of the tergal spines of the abdomen appear to have been little studied taxonomically and may offer good generic and specific characters.

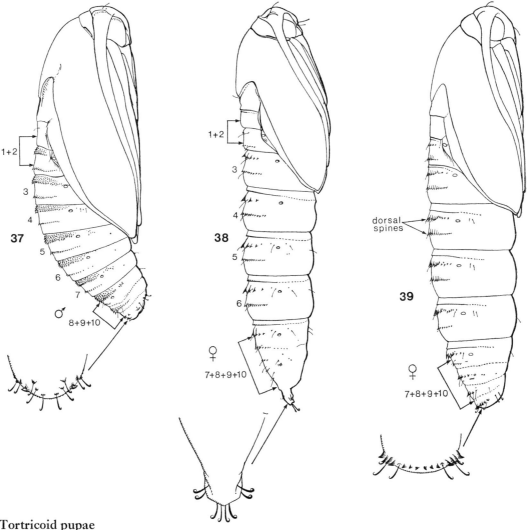

Tortricoid pupae

Fig. 37, Cochylidae—*Stenodes alternana* (Stephens), ♂. Fig. 38, Tortricidae: Tortricinae—*Pandemis heparana* (Denis & Schiffermüller), ♀. Fig. 39, Tortricidae: Olethreutinae—*Cydia pomonella* (Linnaeus), ♀

Biology

In the British Isles the majority of Tortricoid species have a single generation in a year (univoltine), but a number have two generations (bivoltine) and, in favourable seasons, some may have three or more generations (multivoltine). Differences in local and regional environmental conditions often have an influence on the life cycle, and due allowance should be made for the time of appearance of a species and the number of generations according to locality.

In most species imagines of the second generation emerge in summer or early autumn, remain on the wing until late autumn, then hibernate before pairing in the early spring. The eggs laid by the females of this second generation produce larvae which develop into the moths constituting the first generation in spring or early summer. Larvae of those species which are univoltine and hatch during the summer are usually full-fed by late autumn and overwinter in a cocoon, which also serves as a hibernaculum, before pupating in the spring. Larvae of univoltine species which hatch in late summer or autumn usually hibernate in an early instar in a silken hibernaculum and recommence feeding in the spring.

The larvae of Cochylids are mostly monophagous or oligophagous and almost invariably are internal feeders, living within flowers or seedheads, or boring in stems and roots. Those of the Tortricinae are much more polyphagous and usually live concealed within rolled or spun leaves or flowers, constructing shelters often characteristic of the particular species; in early instars a number are miners, boring into a bud or the base of a leaf, twig or petiole. Many species such as *Cacoecimorpha pronubana* (Hübner) and *Epiphyas postvittana* (Walker), are highly polyphagous, whilst others are more restricted in foodplant selection. *Clepsis consimilana* (Hübner) is an exception and unlike other British Tortricines it does not feed on living plant tissue but on the dead or shrivelled leaves of its foodplant. The larvae of the Olethreutinae feed in a greater variety of ways; besides rolling or spinning leaves or shoots, they may bore in stems, roots, berries, fruits, seeds and seed-pods, also occasionally in bark and galls. Many are monophagous or oligophagous and none is as polyphagous as some of the species in the Tortricinae.

In the Tortricoidea, pupation usually occurs in the larval habitation which is modified to form a cocoon. Species which hibernate as fully grown larvae, other than stem or root borers, normally construct a cocoon, which serves as a hibernaculum, amongst leaf litter or debris on the ground and pupate *in situ* after hibernation. On eclosion the pupa protrudes from the cocoon.

Collecting and preparing specimens

The advice of Ford (1945) that anyone commencing the study of Microlepidoptera should begin by obtaining specimens in the larval stage cannot be bettered. As Ford states, some of the advantages are: (a) by rearing imagines from larvae one obtains a more thorough knowledge of the habits of each species; (b) specimens for the cabinet are obtained in perfect condition; (c) if specimens are in perfect condition and have full data of foodplant and time of appearance of the larva and imago, they can be more easily identified; (d) reared specimens are usually easier to set than those that have been taken on the wing; (e) many species are more easily obtained in the larval stage; and (f) the larvae of many species can be collected during the winter months.

Killing and setting the imago

Tortricoid moths can be killed and set by the same methods as those used for butterflies and larger moths or Macrolepidoptera. The use of potassium cyanide (KCN) as a killing agent is not generally to be recommended for Microlepidoptera since it tends to stiffen the wing musculature, but specimens killed in this way may be relaxed by placing them in a laurel box for 12–24 hours. Ammonia (0·880) is frequently used since it is rapid in action and leaves the specimen perfectly relaxed for immediate setting. Ethyl acetate is a useful alternative.

To use these liquid killing agents a wide-mouthed vial or small jar with a cork bung is required. A small wad of cotton wool should first be moistened with the agent, then dropped to the bottom of the jar and covered with a layer of non-absorbent cotton wool. There are a number of variations of this method; some collectors prefer to pack cellulose wool at the bottom of the jar, others do not put the moistened wad at the bottom of the jar but insert it into a small glass tube or piece of tubing let into the cork, so that the wad is virtually suspended in the middle of the container. With any of these methods, it is important to keep the sides of the container free from condensation or moisture.

Small killing jars are usually more convenient for the Micros, and it is better to have two or three in operation simultaneously, and put only one or two specimens in a jar at a time. Too long exposure to ammonia fumes may affect the coloration of certain species; very bright green or pale pink insects should not be killed with ammonia for this reason.

The length and type of pin is a matter of personal preference; for the Tortricoidea minuten-pins 10 mm long and 0·0076 or 0·010 gauge are suitable.

Specimens to be set should be handled as little as possible. It is best to tip the specimen from the killing jar, or remove it by picking it up by the legs with fine forceps, onto a sheet of blotting paper or similar material. The specimen can be held steady for pinning through the thorax by placing the shaft of a needle or long pin at right angles across the thoracic region. On the setting board the wings may be spread into position with the tip of the needle. With most Micros the wings tend to slip back, but this can be prevented by dabbing the spot where the costa should rest with the tip of a moistened sable hair brush. Alternatively a setting bristle may be used to hold the wings in position while applying the strips of setting paper.

On removal from the board, the specimens should be staged on a strip of polyporous, Plastazote, or similar material. A data label giving locality, foodplant if reared, date, and name of the collector must be attached to the staging pin. This is most important since specimens without data have little scientific value.

Examination of wing venation

Venational characters used in the key below for the separation of families, subfamilies and tribes may be examined by applying a drop of toluene or alcohol to the wings with a sable hair brush and examining them with a hand lens from the underside, where the veins are more prominent. For more critical examination the wings should be denuded of scales and a stereoscopic microscope used. This requires the wings to be severed from the thorax, immersed briefly in a watch-glass or petri dish of alcohol, then transferred to a weak chlorine solution (domestic bleach will suffice) for 5–10 minutes, which is usually sufficient time for the scales to become loosened; if numerous small bubbles begin to form on the surface of the wings during this time the wings should be removed immediately. They are then placed in 80% alcohol and manually descaled by abrading with a fine brush or pointed feather. After descaling, the wings are washed in clean alcohol and floated onto a microscope slide, one end of the slide being dipped in the dish and the wings eased on with the brush. When manoeuvred into position on the slide, the surplus alcohol is allowed to evaporate and the wings are mounted dry under the coverslip. If a circular coverslip is used this may be held in position by a square slide label in which a hole of appropriate dimensions has been cut or punched; a square coverslip may be held in position by strips of slide label.

Dissection and preparation of genitalia

A low power stereoscopic microscope is essential for dissection and examination of the genitalia, and a high power monocular microscope may be necessary for critical examination of certain structures, such as the cornuti in the aedeagus.

The abdomen is first detached by applying slight upward pressure to the underside. It is then placed in a 10% solution of potassium hydroxide (KOH) at room temperature for 24 hours to soften the tissues. The process of maceration may be quickened by heating the solution, but this must be done with caution and it is essential that the tube containing the solution and abdomen is heated in a water bath. Simmering the water bath for about 10 minutes will suffice. The softened abdomen is then removed to a shallow dish or palette of water and the viscera are teased out by slight pressure with a needle or sable hair brush. After initial cleaning the abdomen is transferred to a dish of 70% alcohol, where the remainder of the body contents is removed. The sclerotized genital structure of the male can at this stage be disengaged from the body wall and examined. The female genitalia should be dissected out by cutting the membranous tissue between the 7th and 8th segments, care being taken not to sever the ductus bursae, and the whole pulled free so that the bursa copulatrix and signum are exposed.

After examination, the genitalia and abdomen should be preserved. A simple method is to glue them to a rectangular piece of card which is affixed to the staging pin of the specimen. A water-soluble glue should be used so that the genitalia can be easily floated off for re-examination if necessary. If, however, a permanent preparation is required the genitalia should be thoroughly cleaned of deciduous hairs and scales and extraneous tissue, likewise the abdomen, though where coremata or specialized hair-tufts are present in the male these should be kept intact so far as possible. The genitalia should then be dehydrated in 90–95% alcohol before mounting. A proprietary brand of

mounting medium which requires minimal dehydration, such as Euparal, may be conveniently used. It is sometimes an advantage to immerse the preparation in a few drops of the solvent or essence of the particular medium, as this helps to prevent air bubbles forming.

The genitalia of both sexes are usually mounted to show the ventral aspect. The valvae of the male are spread apart and pressed flat to the surface of the slide. In the Cochylidae and Tortricinae the aedeagus is usually removed, but in the Olethreutinae it is kept *in situ*. The abdomen often shows important characters and should always be preserved and mounted with the genitalia. It may be mounted laterally, ventrally or dorsally, according to the position of the structures it is desired to show. Where coremata are present either the dorsal or ventral aspect is preferable.

It is often an advantage to stain the preparation, especially when the genitalia are weakly sclerotized, as this enables the membranous portions to be more clearly seen. Staining can be conveniently carried out after the initial cleansing in water. A 1% aqueous solution of mercurochrome is suitable, a drop or two being added to a little water when required. Various other simple stains, such as chlorazol black, or even coloured inks, may be used, but in all cases care should be taken to avoid overstaining.

Permanent slide preparations must be labelled with the name of the species and given individual slide numbers, and a label bearing the slide number should be fixed to the staging pin of the specimen.

Preservation of larvae and pupae

The larvae of many Tortricoids are large enough in the final instar to be preserved by the "blowing" method often used for Macrolepidoptera. Larvae of all instars and also pupae may be satisfactorily preserved by freeze-drying.

However, if the morphology and chaetotaxy are to be studied then larvae and pupae must be preserved in fluid. A simple but satisfactory method is to drop the larva or pupa into a vial containing a solution of about 9 parts 80% alcohol and 1 part glacial acetic acid. The addition of acetic acid helps to prevent discoloration. After a period of at least 48 hours the specimens are transferred to 80% alcohol to which a little glycerine has been added; the glycerine helps to keep the cuticle pliable and prevents excessive shrivelling and hardening should the alcohol evaporate.

A stereoscopic microscope together with a good spotlight and substage illumination are necessary for examination of the chaetotaxy. The chaetotaxy of the major setal groups used for family diagnoses, such as the lateral (L) and subventral (SV) groups, may usually be examined without dissection of the larva, but transparent and stained preparations of the skin are necessary to distinguish all the setae and pinacula. These can be prepared by immersing the larva in a 10% solution of potassium hydroxide for 24 hours, or longer if necessary, until it is sufficiently softened for manipulation. The skin is then slit lengthwise laterally and the viscera removed. To do this an incision is made at the anal extremity with a blade-like dissecting needle, which is then forced beneath the integument and a brass or similar soft metal needle or pin is rubbed along the edge of the blade so as to cut the skin neatly. The head may be kept attached to the prothorax or removed before or after dissection of the abdomen, as desired.

The setae and pinacula of some species may be more clearly seen by using a staining technique as with genitalia preparations; permanent mounts are similarly prepared. A weak solution of carbol fuchsin stain gives better results with larvae, but this is a slow-acting stain and takes two or three days to penetrate. Hinton (1956: 252) describes the technique in detail.

Phylogeny

The Tortricoidea form a large, homogeneous group whose phylogenetic origins, and the evolution and interrelationships of the generic and suprageneric taxa, are problematic. In common with the Cossoidea they show a number of primordial characters, such as the retention in the imago of the media (M) vein in the forewing in some groups of the more primitive Tortricinae. Although probably originating from the same ancestral stock as the Cossoidea, the Tortricoidea are in many respects more specialized. They therefore seem to warrant a higher position in a linear arrangement, though their placing must necessarily be somewhat arbitrary.

A satisfactory linear arrangement reflecting the interrelationship of the generic taxa is difficult to formulate. It is now generally accepted that the two principal groups, the Tortricinae and Olethreutinae, are of subfamily and not family status, and that the Cochlidae are a sufficiently divergent and homogeneous group to warrant family rank.

The subfamily Tortricinae may be divided into a number of well demarcated tribes, and in the British fauna is represented by the Archipini, Sparganothini, Cnephasiini and Tortricini. The subfamily Olethreutinae can likewise be divided into tribal groups and, although the divisions are not always so clearly defined, is represented by the Olethreutini, Eucosmini and Laspeyresiini in the British fauna.

Meyrick (1895: 452) considered that the Olethreutinae may not improbably be derived from the Tortricinae. Later (1913: 2) he asserted dogmatically that the external relationships of the Tortricinae appeared to be clear and that the subfamily is a development of the Olethreutinae, being derived from the *Argyroploce* [*Olethreutes*=] generic group, and that the *Laspeyresia* [*Cydia*=] group is the more primitive. In his *Revised Handbook* (1928) he maintained this sequence and placed the Cochlidae after the Tortricidae as the most advanced group.

Pierce & Metcalfe (1922) treated the whole of the British representatives of the Tortricoidea as belonging to the single family Tortricidae. These authors recognized the following tribal categories: "Archipsidii, Cnephasidii, Peroneidii, Phaloniadii, Olethreutidii, Epiblemidii and Ephippiphoridii", in that sequence.

Heinrich (1923) interpreted the genitalic characters as showing that the Olethreutinae are derived from the Tortricinae, and that the Laspeyresiini are the most advanced of the Olethreutinae.

Diakonoff (1952: 28; 1953: 87) followed a similar interpretation that the Olethreutinae developed from the Tortricinae, but that the Laspeyresiini is the most primitive tribe of the Olethreutinae and forms the link between it and the ancestral Tortricinae. The systematic arrangement adopted by Bentinck & Diakonoff (1968) reflects this view.

MacKay (1959: 9), on the basis of larval characters, considers the *Laspeyresia* [*Cydia*=] group as apparently the most primitive of the Olethreutinae and places it first in her arrangement. However, she subsequently revised this opinion (MacKay, 1962: 6) and states that the *Laspeyresia* [*Cydia*=] group is not primitive as previously supposed but is more likely to be among the most highly evolved in the Tortricidae, and concludes that the Tortricinae is evolutionally older. In the Tortricinae she considers that the tribal sequence Tortricini, Cnephasiini, Archipini, Sparganothini reflects increasing specialization and perhaps a true phylogenetic order in some cases, as judged by a study of larval material. However, MacKay (1963: 1343) has since concluded that the Archipini and Sparganothini appear to have evolved directly from the primordial tortricid ancestor, and that the presence of an anal fork or comb, and the angular ocellar area of the head of the first instar of many species, suggests some form of miner as their immediate ancestor.

To a great extent the systematic arrangement in the present work reflects the phylogenetic interpretations of Pierce & Metcalfe (1922) and Heinrich (1923). However, although the family Cochylidae is treated as a separate family and is considered to be more advanced than the Tortricidae, and in a linear arrangement should succeed it, it has been placed first for convenience.

Nomenclature and synonymy

In this work the nomenclature of the Tortricoidea is based on the revised check-list of Kloet & Hincks (1972), and that of the larval foodplants on the illustrated British flora by W. Keble Martin (1965), since this is now widely used by lepidopterists, and the botanical publications of Dollimore & Jackson (1966) and the Royal Horticultural Society (1951–69).

A systematic list of the Cochylidae and Tortricidae: Tortricinae is given on page 22. In this will be found the principal synonyms relating to the families and subordinate taxa which have appeared in the British literature. Subgenera are included in the list but, as they are seldom used in general practice, have been excluded from the text.

In the text, the identity of each species has been correlated with the names used by Barrett (1904–07), Pierce & Metcalfe (1922), Meyrick (1928), and Ford (1949). Original references and full synonymy of the genera and British species may be found in the generic revision of the Palaearctic Tortricidae by Obraztsov (1954–57).

KEY TO FAMILIES, SUBFAMILIES AND TRIBES OF BRITISH TORTRICOIDEA

1   Forewing with vein 2 (Cu2, CuA2) arising beyond three-fourths of discal cell; vein 1c (CuP) obsolescent or obsolete (text-figs. 20–22) ....COCHYLIDAE

 –   Forewing with vein 2 (Cu2, CuA2) arising before three-fourths of discal cell; vein 1c (CuP) present at margin (text-figs. 23–30) 2

2   Hindwing with cubital pecten on upper surface (text-figs. 24, 28–30) 3

 –   Hindwing without cubital pecten 6

3   Labial palpus very long and prominent (approximately three times width of eye), porrect (text-fig. 9) .... TORTRICINAE: Sparganothini

 –   Labial palpus shorter, upturned or porrect (text-figs. 12–14) 4

4   Hindwing with vein 5 (M2) straight, separate and almost parallel with vein 4 (M3) (text-fig. 30) ....OLETHREUTINAE: Laspeyresiini

 –   Hindwing with vein 5 (M2) curved towards vein 4 (M3) basally 5

5   Hindwing with veins 3 (Cu1) and 4 (M3) stalked or connate (text-fig. 29); forewing of male often with costal fold ....OLETHREUTINAE: Eucosmini

 –   Hindwing with veins 3 (Cu1) and 4 (M3) separate (closely approximate) or connate (text-fig. 28); forewing of male without costal fold ..............................OLETHREUTINAE: Olethreutini

6   Forewing usually with roughened scaling or scale-tufts; vein 7 (R5) to costa (*Acleris*) (text-fig. 26), apex or termen (text-fig. 27) ....TORTRICINAE: Tortricini

 –   Forewing usually smooth-scaled, without scale-tufts; vein 7 (R5) to apex or termen 7

7   Forewing usually relatively broad, apex often strongly produced; vein 7 (R5) to termen; male often with costal fold (text-fig. 23) ....TORTRICINAE: Archipini

 –   Forewing relatively narrow; vein 7 (R5) to apex or termen; male without costal fold (text-fig. 25) ....TORTRICINAE: Cnephasiini

LIST OF BRITISH SPECIES OF COCHYLIDAE & TORTRICIDAE: TORTRICINAE

(The systematic arrangement in this list is followed throughout the volume. Synonyms are indented and in a chronological sequence, and are restricted mainly to those which have appeared in the British literature. An asterisk * denotes that a species is believed to be a casual introduction and non-resident in the British Isles.)

TORTRICOIDEA

COCHYLIDAE
CONCHYLIDAE
LOZOPERIDAE
PHALONIIDAE
COMMOPHILIDAE
AGAPETIDAE

Hysterosia Stephens
  Idiographis Lederer
  Propira Durrant
inopiana (Haworth)
schreibersiana (Frölich)
sodaliana (Haworth)
  amandana (Herrich-Schäffer)

Hysterophora Obraztsov
maculosana (Haworth)

Phtheochroa Stephens
rugosana (Hübner)
  v-albana (Donovan)

Phalonidia Le Marchand
Phalonidia (*Phalonidia* Le Marchand)
manniana (Fischer von Röslerstamm)
  udana (Guenée)
  notulana (Zeller)
minimana (Caradja)
  walsinghamana (Meyrick)
permixtana (Denis & Schiffermüller)
  mussehliana (Treitschke)

vectisana (Humphreys & Westwood)
  griseana (Haworth)
  geyeriana (Herrich-Schäffer)
  grisescens (Durrant & de Joannis)
alismana (Ragonot)
luridana (Gregson)
affinitana (Douglas)
  cancellana (Zeller)

Phalonidia (*Brevisociaria* Obraztsov)
gilvicomana (Zeller)
curvistrigana (Stainton)

Stenodes Guenée
  Euxanthoides Razowski
  Bleszynskiella Razowski
alternana (Stephens)
  gigantana (Doubleday)
straminea (Haworth)

Agapeta Hübner
  Agapete Hübner
  Euxanthis Hübner
  Xanthosetia Stephens
hamana (Linnaeus)
  diversana (Hübner)
zoegana (Linnaeus)
  ferrugana (Haworth)

Aethes Billberg
Aethes (*Aethes* Billberg)
  Phalonia Hübner
  Chlidonia Hübner
  Dapsilia Hübner
  Chrosis Guenée
  Argyridia Stephens

tesserana (Denis & Schiffermüller)
   aleella (Schultze)
   tesselana (Hübner)
rutilana (Hübner)
   sanguinella (Haworth)
piercei Obraztsov
hartmanniana (Clerck)
   baumanniana (Denis & Schiffermüller)
   subbaumanniana (Stainton)
margarotana (Duponchel)
   maritimana (Guenée)
williana (Brahm)
   zephyrana (Treitschke)
   dubrisana (Curtis)
   marmoratana (Curtis)
   luteolana (Stephens)
cnicana (Westwood)
rubigana (Treitschke)
smeathmanniana (Fabricius)
margaritana (Haworth)
   dipoltella (Hübner)

Aethes (*Lozopera* Stephens)
dilucidana (Stephens)
francillana (Fabricius)
   francillonana (Humphreys & Westwood)
beatricella (Walsingham)

Commophila Hübner
aeneana (Hübner)

Eugnosta Hübner
   Argyrolepia Stephens
*lathoniana (Hübner)

Eupoecilia Stephens
   Clysia Hübner
   Clysiana Fletcher
angustana angustana (Hübner)
   fasciella (Donovan)
angustana thuleana Vaughan
ambiguella (Hübner)

Cochylidia Obraztsov
implicitana (Wocke)
   anthemidana (Stainton)
heydeniana (Herrich-Schäffer)
   pudorana (Staudinger)
   erigerana (Walsingham)
   sabulicola (Walsingham)
subroseana (Haworth)
   flammeolana (Tengström)
rupicola (Curtis)
   marginana (Stephens)

Falseuncaria Obraztsov & Swatschek
ruficiliana (Haworth)
   ciliella (Hübner)
degreyana (McLachlan)

Cochylis Treitschke
Cochylis (*Cochylis* Treitschke)
   Conchylis Sodoffsky
roseana (Haworth)
   rubellana (Hübner)
flaviciliana (Westwood)
dubitana (Hübner)
hybridella (Hübner)
   carduana Zeller

Cochylis (*Cochylichroa* Obraztsov &
   Swatschek)
atricapitana (Stephens)

Cochylis (*Brevicornutia* Razowski)
pallidana Zeller
   albicapitana (Cooke)

Cochylis (*Acornutia* Obraztsov)
nana (Haworth)

TORTRICIDAE

TORTRICINAE
APHELIINAE
SCIAPHILINAE
SPARGANOTHINAE

ARCHIPINI

Pandemis Hübner
  Pandemia Stephens
corylana (Fabricius)
cerasana (Hübner)
  ribeana (Hübner)
  grossulariana (Stephens)
cinnamomeana (Treitschke)
heparana (Denis & Schiffermüller)
dumetana (Treitschke)

Argyrotaenia Stephens
pulchellana (Haworth)
  politana (Haworth)
  cognatana Stephens
  fuscociliana Stephens

Homona Walker
*menciana (Walker)

Archips Hübner
Archips (*Archippus* Freeman)
oporana (Linnaeus)
  piceana (Linnaeus)
  dissimilana (Bentley)
podana (Scopoli)
  fulvana (Denis & Schiffermüller)
betulana (Hübner)
  decretana (Treitschke)

Archips (*Archips* Hübner)
  Cacoecia Hübner
crataegana (Hübner)
  roborana (Hübner)
xylosteana (Linnaeus)
  westriniana (Thunberg)

rosana (Linnaeus)
  avellana (Linnaeus)
  ameriana (Linnaeus)
  laevigana (Denis & Schiffermüller)
  oxyacanthana (Hübner)
  acerana (Hübner)
  nebulana (Stephens)

Choristoneura Lederer
  Cornicacoecia Obraztsov
diversana (Hübner)
  transitana (Guenée)
hebenstreitella (Müller)
  sorbiana (Hübner)
lafauryana (Ragonot)

Cacoecimorpha Obraztsov
pronubana (Hübner)

Syndemis Hübner
musculana musculana (Hübner)
musculana musculinana (Kennel)

Ptycholomoides Obraztsov
aeriferanus (Herrich-Schäffer)

Aphelia Hübner
Aphelia (*Aphelia* Hübner)
viburnana (Denis & Schiffermüller)
  galiana (Stephens)
  teucriana (Tutt)
  donelana (Carpenter)

Aphelia (*Zelotherses* Lederer)
paleana (Hübner)
  flavana (Hübner)
  icterana (Frölich)
  intermediana (Herrich-Schäffer)
unitana (Hübner)

Clepsis Guenée
Clepsis (*Clepsis* Guenée)
senecionana (Hübner)
  helvolana (Frölich)

Clepsis (*Siclobola* Diakonoff)
rurinana (Linnaeus)
   modeeriana (Linnaeus)
   semialbana (Guenée)
spectrana (Treitschke)
   costana sensu auct.
   latiorana (Stainton)
consimilana (Hübner)
   unifasciana (Duponchel)
   ? *peregrinana* (Stephens)
trileucana (Doubleday)
   persicana (Fitch)
melaleucanus (Walker)
   ? *biustulana* (Stephens)

Epichoristodes Diakonoff
Epichoristodes (*Tubula* Diakonoff)
acerbella (Walker)
   iocoma (Meyrick)
   ionephela (Meyrick)

Epiphyas Turner
   Austrotortrix Bradley
postvittana (Walker)

Adoxophyes Meyrick
orana (Fischer von Röslerstamm)
   reticulana (Hübner)

Ptycholoma Stephens
lecheana (Linnaeus)
   obsoletana (Stephens)

Lozotaeniodes Obraztsov
formosanus (Geyer)

Lozotaenia Stephens
forsterana (Fabricius)
subocellana (Stephens)

Paramesia Stephens
gnomana (Clerck)
   costana (Denis & Schiffermüller)

Paraclepsis Obraztsov
cinctana (Denis & Schiffermüller)

Epagoge Hübner
grotiana (Fabricius)
   ochreana (Hübner)

Capua Stephens
vulgana (Frölich)
   rusticana (Hübner)
   favillaceana (Hübner)
   ochraceana Stephens

Philedone Hübner
   Amphisa Curtis
gerningana (Denis & Schiffermüller)
   pectinana (Hübner)

Philedonides Obraztsov
lunana (Thunberg)
   prodromana (Hübner)
   walkerana (Curtis)

Ditula Stephens
   Batodes Guenée
angustiorana (Haworth)
   rotundana (Haworth)

Pseudargyrotoza Obraztsov
conwagana (Fabricius)
   conwayana (Gmelin)
   hofmannseggana (Hübner)
   subaurantiana (Stephens)

SPARGANOTHINI

Sparganothis Hübner
   Oenectra Guenée
   Aenectra Doubleday
pilleriana (Denis & Schiffermüller)
   luteolana (Hübner)

CNEPHASIINI

Olindia Guenée
  Anisotaenia Stephens
schumacherana (Fabricius)
  ulmana (Hübner)

Isotrias Meyrick
rectifasciana (Haworth)
  curvifasciana (Stephens)

Eulia Hübner
  Lophoderus Stephens
ministrana (Linnaeus)
  subfascianus (Stephens)

Cnephasia Curtis
Cnephasia (*Cnephasia* Curtis)
  Sciaphila Treitschke
  Sphaleroptera Guenée
longana (Haworth)
  ictericana (Haworth)
  eganana (Haworth)
  expallidana (Haworth)
*gueneana (Duponchel)
communana (Herrich-Schäffer)
conspersana Douglas
  perterana (Doubleday)
  ? *cretaceana* Curtis
  decolorana Stephens
stephensiana (Doubleday)
  chrysantheana sensu auct.
  chrysanthemana sensu auct.
  octomaculana Curtis
  perplexana Stephens
interjectana (Haworth)
  virgaureana (Treitschke)
pasiuana (Hübner)
  pascuana (Hübner)
  obsoletana Stephens
genitalana Pierce & Metcalfe

Cnephasia (*Cnephasiella* Adamczewski)
incertana (Treitschke)
  subjectana (Guenée)

Tortricodes Guenée
  Oporinia Hübner
alternella (Denis & Schiffermüller)
  tortricella (Hübner)
  nubilea (Haworth)
  hyemana (Hübner)

Exapate Hübner
  Cheimaphasia Curtis
  Oxypate Stephens
  Oxapate Stephens
congelatella (Clerck)
  gelatella (Linnaeus)

Neosphaleroptera Réal
nubilana (Hübner)
  perfuscana (Haworth)

Eana Billberg
Eana (*Ablabia* Hübner)
argentana (Clerck)
osseana (Scopoli)
  quadripunctana (Haworth)
  cantiana (Curtis)

Eana (*Eana* Billberg)
  Nephodesme Hübner
  Nephodesma Stephens
incanana (Stephens)
  ? *cinerana* (Humphreys & Westwood)
penziana penziana (Thunberg & Becklin)
  bellana (Curtis)
penziana colquhounana (Barrett)

TORTRICINI

Aleimma Hübner
  Dictyopteryx Stephens
loeflingiana (Linnaeus)
  plumbeolana (Haworth)

Tortrix Linnaeus
  Heterognomon Lederer
viridana (Linnaeus)

Spatalistis Meyrick
bifasciana (Hübner)
   audouinana (Duponchel)
   apicalis (Humphreys & Westwood)

Croesia Hübner
   Argyrotoza Stephens
   Argyrotosa Curtis
   Argyrotoxa Agassiz
bergmanniana (Linnaeus)
forsskaleana (Linnaeus)
   forskaliana (Haworth)
holmiana (Linnaeus)

Acleris Hübner
   Peronea Curtis
   Lopas Hübner
   Rhacodia Hübner
   Eclectis Hübner
   Teleia Hübner
   Oxigrapha Hübner
   Amelia Hübner
   Peronaea Stephens
   Leptogramma Stephens
   Glyphisia Stephens
   Cheimatophila Stephens
   Teras Treitschke
latifasciana (Haworth)
   comparana (Hübner)
   labeculana (Freyer)
   costimaculana (Stephens)
   perplexana (Barrett)
comariana (Lienig & Zeller)
   proteana (Herrich-Schäffer)
   potentillana (Gregson)
caledoniana (Stephens)
sparsana (Denis & Schiffermüller)
   sponsana (Fabricius)
   favillaceana (Hübner)
   reticulana (Haworth)
   fagana (Curtis)
rhombana (Denis & Schiffermüller)
   reticulata (Ström)
   contaminana (Hübner)
   ciliana (Hübner)
   obscurana (Donovan)
   dimidiana (Frölich)

aspersana (Hübner)
   subtripunctulana Stephens
tripunctana (Hübner)
   notana (Donovan)
   bifidana (Haworth)
   proteana (Guenée)
   obliterana (Herrich-Schäffer)
ferrugana (Denis & Schiffermüller)
   rufana (Hübner)
   tripunctulana (Haworth)
   lythargyrana (Treitschke)
   brachiana (Freyer)
   selasana (Herrich-Schäffer)
   lithargyrana (Herrich-Schäffer)
   fissurana (Pierce & Metcalfe)
shepherdana (Stephens)
schalleriana (Linnaeus)
   tristana (Hübner)
   castaneana (Haworth)
   plumbosana (Haworth)
   falsana (Hübner)
   trigonana (Stephens)
   semirhombana (Curtis)
variegana (Denis & Schiffermüller)
   asperana (Fabricius)
   cirrana (Curtis)
   albana (Humphreys & Westwood)
permutana (Duponchel)
boscana (Fabricius)
   cerusana (Hübner)
   ulmana (Duponchel)
   parisiana (Guenée)
logiana (Clerck)
   niveana (Fabricius)
   treueriana (Hübner)
   scotana (Stephens)
umbrana (Hübner)
hastiana (Linnaeus)
   coronana (Thunberg)
   byringerana (Hübner)
   radiana (Hübner)
   albistriana (Haworth)
   eximiana (Haworth)
   aquilana (Hübner)
   divisana (Hübner)
   combustana (Hübner)
   autumnana (Hübner)

bollingerana (Hübner)
veterana (Hübner)
mayrana (Hübner)
ramostriana (Stephens)
centrovittana (Stephens)
subcristana (Stephens)
obsoletana (Stephens)
leacheana (Curtis)
arcticana (Guenée)
leucophaeana (Humphreys & Westwood)

cristana (Denis & Schiffermüller)
desfontainana (Fabricius)
profanana (Fabricius)
rossiana (Fabricius)
cristalana (Donovan)
ephippana (Fabricius)
sericana (Hübner)
spadiceana (Haworth)
striana (Haworth)
ruficostana (Curtis)
substriana (Stephens)
brunneana (Stephens)
vittana (Stephens)
consimilana (Stephens)
albovittana (Stephens)
fulvocristana (Stephens)
fulvovittana (Stephens)
albipunctana (Stephens)
lichenana (Curtis)
semiustana (Curtis)
sequana (Curtis)
subcristalana (Curtis)
insulana (Curtis)
bentleyana (Curtis)
alboflammana (Curtis)
chantana (Curtis)
ruficristana (Johnson)
capucina (Johnson)
gumpinana (Johnson)
unicolorana (Desvignes)
curtisana (Desvignes)

hyemana (Haworth)
mixtana (Hübner)

lipsiana (Denis & Schiffermüller)
sudoriana (Hübner)

rufana (Denis & Schiffermüller)
apiciana (Hübner)
bistriana (Haworth)
rubellana (Müller-Rutz)

similana (Curtis)
albicostana (Stephens)

lorquiniana (Duponchel)
uliginosana (Humphreys & Westwood)

abietana (Hübner)

maccana (Treitschke)
? *marmorana* (Curtis)

literana (Linnaeus)
squamana (Fabricius)
asperana (Denis & Schiffermüller)
romanana (Fabricius)
squamulana (Hübner)
irrorana (Hübner)
tricolorana (Haworth)
abjectana (Hübner)
aerugana (Hübner)
fulvomixtana (Stephens)
suavana (Herrich-Schäffer)

emargana (Fabricius)
caudana (Fabricius)
scabrana (Fabricius)
excavana (Donovan)
fasciana (Müller-Rutz)
ochracea (Stephens)

COCHYLIDAE (PHALONIIDAE)

Hysterosia inopiana (Haworth)
Pl. 22, figs. 1, 2

[*Idiographis inopiana*; Barrett, 1904: 190. *Hysterosia inopiana*; Pierce & Metcalfe, 1922: 25, pl. 10 (♂, ♀ genitalia); Meyrick, 1928: 498; Ford, 1949: 49]

DESCRIPTION

♂ ♀ 17–22 mm. Sexual dimorphism pronounced; antenna of male weakly dentate-ciliate, length of cilia approximately equal to width of flagellum, forewing with narrow costal fold from base to about three-fifths, coloration paler and markings stronger; female usually larger, antenna shortly ciliate, forewing almost unicolorous, without costal fold but margin sometimes slightly upturned and thickened near middle.

Male (fig. 1). Forewing ground colour ochreous-white, weakly suffused with pale grey and diffusely strigulate with brownish ochreous, suffusion and strigulation most pronounced distally; costal fold dark brown, several dark costal striae beyond; dorsum sparsely edged with dark brown dots; a usually prominent dark brown discocellular spot; markings ochreous-brown, obsolescent; basal and sub-basal fasciae represented by a pale suffusion from costal fold to basal tuft on dorsum; median fascia with inner edge very oblique, diffuse on costa; cilia ochreous-white or greyish white, with a greyish brown sub-basal line. Hindwing light grey, infuscate distally; cilia ochreous-white, with a greyish sub-basal line.

Female (fig. 2). Forewing almost unicolorous brown or reddish brown, obscurely strigulate; discocellular spot indistinct; costa narrowly thickened and darker to near middle; markings obsolete. Hindwing as in male.

Variation. The strigulation in the forewing of the male varies in strength; in some specimens it may be very weak or absent, the wing then having a more uniform ochreous-brown appearance, often with relatively clear ochreous-white ground colour in the sub-basal and discal areas. In the female the coloration of the forewing varies from brown or reddish brown to dark brown.

COMMENTS
The relatively large size of this species and the obscure markings and drab coloration of the forewing are characteristic.

BIOLOGY
Larva. Head and prothoracic plate brown; abdomen yellowish white, shagreened. The larva may be found from September to April, living in a silken tube in the roots of *Pulicaria dysenterica* and *Artemisia campestris*.

Pupa. May and June.

Imago. June to early August; frequenting especially situations where the foodplants grow in damp places and along the verges of woods. During the day the moths usually

29

remain well concealed amongst the vegetation, but occasionally may be found at rest on a leaf of the foodplant. The male flies about the fleabane and field southernwood in the early evening, but the female is more often found sitting on the leaves, where her presence is sometimes betrayed by the assembling males.

DISTRIBUTION

Generally distributed and locally common in the south of England to Cambridgeshire, Norfolk (King's Lynn) and Herefordshire, occurring abundantly in favoured localities such as Instow (Devon), Corfe, Swanage and Charmouth (Dorset), Yarmouth (Isle of Wight), Haslemere and Redhill (Surrey), and Chattenden and Folkestone (Kent). Scarce in Cheshire, Lancashire and Yorkshire (Spurn Head, Scarborough). Michaelis (1966: 109) records it from Wales as locally common on the coast of south Cardiganshire (Gwbert, Aberporth); also known from Anglesey. Although there are a few records from south-west Scotland (Clyde district), it has not been recorded in recent years. Likewise records from Ireland are few and none is recent. Beirne (1941: 79) cites records from Co. Mayo and Co. Armagh, and comments that the species is evidently rare.

North and central Europe to Siberia and Japan; North America.

Hysterosia schreibersiana (Frölich)

Pl. 22, fig. 3

[*Argyrolepia schreibersiana*; Barrett, 1905: 333. *Hysterosia schreibersiana*; Pierce & Metcalfe, 1922: 25, pl. 10 (♂, ♀ genitalia). *Phtheochroa schreibersiana*; Meyrick, 1928: 496; Ford, 1949: 47]

DESCRIPTION

♂♀ 11–15 mm. Sexual dimorphism not pronounced; antenna of male weakly dentate-ciliate, length of cilia equal to or slightly exceeding width of flagellum, forewing without costal fold; antenna of female minutely ciliate.

Male (fig. 3). Head and labial palpus ochreous-yellow. Forewing ground colour pale ochreous-white; markings dark fuscous-brown mixed with black and ferruginous, ill-defined; basal and sub-basal fasciae narrowly indicated on costa, extending to near dorsum as diffuse yellow-ochreous or orange-ochreous suffusion; median fascia narrow on costa, inner edge well defined, inwardly oblique, indented below costa; pre-apical spot indeterminate, preceded by a quadrate patch of ground colour suffused with ochreous; numerous slightly raised metallic bluish or plumbeous spots or striae in distal area; a small pale ochreous-white pre-tornal spot extending into cilia; cilia yellow mixed with fuscous, a blackish sub-basal line. Hindwing fuscous-brown; cilia light grey, a dark fuscous-brown sub-basal line.

Female. Similar to male.

Variation. The forewing pattern shows only minor variation in coloration.

COMMENTS

Readily distinguished by the bright ochreous basal area and contrasting dark coloration of the distal half of the forewing with the distinctive metallic spots, and the area of yellowish coloration forming a conspicuous quadrate spot on the costa.

BIOLOGY

Larva. Head and prothoracic plate light brown, plate speckled with black posteriorly. The foodplants are *Ulmus*, *Populus nigra* and *Prunus padus*. After hatching towards the end of June the larva first feeds on the leaves, spinning two together, but in later instars it eats into the shoots and leaf stalks and may be searched for in this situation in July and August. By the end of September it is full-fed and constructs a cocoon under the bark, hibernating until the following spring before pupating.

Pupa. April and May, in the hibernaculum. Empty pupal exuviae can sometimes be found protruding from the bark (Raynor, 1882: 44).

Imago. May and June; most frequent in fenland districts. The moth may sometimes be found sitting on the tree trunks, but apparently only immediately after emergence; usually it rests high up amongst the foliage of the foodplants but has been beaten from hawthorn bushes. Occasionally it may be seen on the wing on warm sunny days.

DISTRIBUTION

Very local and apparently much scarcer than formerly, having been recorded from Cambridgeshire (Ely, Downham, Wicken Fen), Huntingdonshire, Suffolk, Kent (Sittingbourne), Surrey (Haslemere) and Gloucestershire (Cheltenham, Forest of Dean).

Central and southern Europe to Asia Minor; southern Russia.

Hysterosia sodaliana (Haworth)
Pl. 22, fig. 4
[*Phtheochroa sodaliana*; Barrett, 1905: 344. *Hysterosia amandana*; Pierce & Metcalfe, 1922: 26, pl. 10 (♂, ♀ genitalia). *Phtheochroa sodaliana*; Meyrick, 1928: 495; Ford, 1949: 47]

DESCRIPTION

♂♀ 14–16 mm. Sexual dimorphism not pronounced; antenna of male weakly dentate-ciliate, length of cilia almost equal to width of flagellum, forewing with very narrow costal fold from base to about one-fifth; antenna of female minutely ciliate.

Male (fig. 4). Forewing ground colour white, partially suffused with grey in distal half, with scattered black striae along costal and dorsal margins; markings fuscous or greyish fuscous mixed with black, poorly defined; basal and sub-basal fasciae indicated on costa; median fascia slightly inward-oblique, narrow on costa, interrupted immediately below, dilated towards dorsum, a dense black admixture at middle and in median fold, with two greyish ferruginous semi-erect tufts near each margin immediately above median fold, two similar metallic plumbeous tufts a little below fold; pre-apical spot weakly represented on costa, extended to tornal area as an irregular fuscous-black stria; two small black dashes beyond cell, a conspicuous ferruginous-red apical spot contiguous with a small black dot on termen closely followed by two smaller black terminal dots; cilia grey with yellowish admixture, with a dark grey sub-basal line. Hindwing whitish grey, dappled or strigulate with grey, suffused with grey apically; cilia whitish, with a grey sub-basal line and pale grey sub-apical line.

Female. Similar to male.

Variation. Except for slight individual variation in the extent of the black admixture in the forewing markings the general coloration and pattern are very constant.

COMMENTS

In some respects reminiscent of a small *Phtheochroa rugosana* but at once distinguished by the conspicuous ferruginous-red apical spot and the generally white basal area of the forewing. The dappled hindwing distinguishes this species from *Hysterophora maculosana*.

BIOLOGY

Ovum. Deposited on the unripe green berries of the foodplant, *Rhamnus catharticus*, at the end of June; hatching in about 14 days.

Larva. Short and stout; head light brown; prothoracic plate shining black, divided medially; abdomen light shining green, mottled with purplish or reddish purple dorsally; pinacula whitish green, moderately prominent; anal plate inconspicuous, light green, spotted with black; thoracic legs black; prolegs green. July and August, in the berries, spinning two or more together (pl. 1, fig. 1) and passing from one to another, covering the entrance hole with silk, feeding on the hard seeds. A purple berry amongst a cluster of green berries usually indicates the presence of a larva. At the end of August the full-fed larva leaves the berries and descends the tree to spin a strong, silken cocoon in a crevice of the bark near the ground (in captivity it will spin up in cork), hibernating within until the spring, pupating in early May. On the Continent *Frangula alnus* has also been recorded as a foodplant (Swatschek, 1958: 221).

Pupa. Yellow, wing cases lighter; May and early June, in the hibernaculum.

Imago. June and early July; frequenting chalk downs and similar districts where buckthorn grows. During the day the moth is exceedingly lethargic hiding amongst dense vegetation, dropping to the ground when disturbed. In the early evening it may sometimes be found sitting on leaves of the foodplant or on grass-stems beneath. According to Sheldon (1891: 301) the flight period is short, commencing towards late evening, when specimens may be netted flying around the foodplant.

DISTRIBUTION

A local species, almost entirely restricted to chalk localities in the southern counties of England. Noted localities include the Croydon and Sanderstead districts (Surrey), Gravesend (Kent) and Wicken Fen (Cambridgeshire). Barrett (1905: 345) also records Oxfordshire, Middlesex, Sussex, Somerset, Gloucestershire and Worcestershire. The most northerly locality appears to be Witherslack (Westmorland) and it is apparently unknown in Wales and Scotland. In Ireland several specimens were taken at Rinnamona Lough, in the Burren, Co. Clare, in June 1970.

Central and southern Europe to Asia Minor.

Hysterophora maculosana (Haworth)

Pl. 22, figs. 5, 6

[*Conchylis maculosana*; Barrett, 1905: 278. — *maculosana*; Pierce & Metcalfe, 1922: 26, pl. 10 (♂, ♀ genitalia). *Phtheochroa maculosana*; Meyrick, 1928: 496; Ford, 1949: 48]

DESCRIPTION

♂ ♀ 11–14 mm. Sexual dimorphism moderately pronounced; antenna of male weakly dentate-ciliate, length of cilia approximately equal to width of flagellum, forewing with narrow costal fold from base to middle or slightly beyond, hindwing white basally; antenna of female minutely ciliate, hindwing wholly fuscous.

Male (fig. 5). Forewing ground colour white, sparsely suffused with greyish; markings dark grey mixed with black, ill-defined; basal and sub-basal fasciae forming a diffuse basal patch; median fascia narrow, slightly outward-oblique, outer margin indeterminate, rather diffuse, fascia weak or constricted below costa, black admixture mostly concentrated near middle, forming spots edged with ferruginous and shining plumbeous above and below median fold; pre-apical spot small, produced nearly to termen; apex and termen with diffuse ferruginous spots; cilia white mixed with grey and pale ochreous except at tornus, with a blackish grey sub-basal line. Hindwing white, strongly infuscate in apical area and along termen; cilia white, with a faint fuscous sub-basal line along termen.

Female (fig. 6). Forewing pattern similar to that of male, but grey suffusion usually heavier; hindwing wholly greyish fuscous; cilia dark grey, with darker sub-basal line.

Variation. In both sexes only minor individual variation is found in the forewing pattern, but in the male the white ground colour is seldom strongly suffused as in the female. The greyish fuscous suffusion on the hindwing of the male is sometimes more extensive; this form appears to be most frequent in Scotland and in the west of Ireland (Co. Galway) (Emmet, 1971: 17).

COMMENTS

Distinguished from *Hysterosia sodaliana* by its smaller size, the generally darker coloration of the forewing, and the relatively whiter hindwing of the male and dark greyish fuscous hindwing of the female.

BIOLOGY

Ovum. Deposited singly on the seed capsules of the foodplant in May and June (Wood, 1878: 149).

Larva. Head light brown; prothoracic and anal plates black; abdomen white, with ferruginous dorsal, spiracular and two subspiracular lines; pinacula pale ferruginous. July and August, feeding on the seeds of *Endymion non-scriptus*, living within the seed capsules. On moving to a fresh capsule the larva seals the entrance hole with silk; the exit holes in two evacuated capsules can be seen in the illustration (pl. 2, fig. 2). When full-grown the larva leaves the foodplant and bores into nearby dead stems of umbellifers, brambles and other plants, where it overwinters until the following spring, pupating in April. On the Continent *Chondrilla juncea* has also been recorded as a foodplant (Hannemann, 1964: 16), the larva feeding in the young flowers.

Pupa. April and May; in the hibernaculum.

Imago. May and June; frequenting open woodland and other situations where the bluebell is plentiful. The male flies in sunshine and is conspicuous by its white hindwings and underside.

DISTRIBUTION

Locally common and very often abundant in suitable woodland localities throughout the southern and eastern counties of England, extending westwards to the Isles of Scilly (Richardson & Mere, 1958: 141), becoming less common northwards; in Wales it is known locally from Denbighshire, Flintshire, Caernarvonshire, Merionethshire, Monmouthshire and Pembrokeshire. It is locally common in the west of Scotland and occurs as far north as Inverness-shire and Sutherland (Invershin). In Ireland it is apparently widespread but very local; Beirne (1941: 78) records it from Co. Cork, Co. Dublin (Howth), Co. Sligo (Sligo), Co. Donegal (Donegal, Portnoo); Emmet (1971: 17) found an isolated colony in Co. Galway (Ballyconneely).

Europe to Asia Minor and the Ukraine.

Phtheochroa rugosana (Hübner)
Pl. 22, fig. 7
[*Phtheochroa rugosana*; Barrett, 1905: 342; Pierce & Metcalfe, 1922: 30, pl. 11 (♂, ♀ genitalia); Meyrick, 1928: 495; Ford, 1949: 47]

DESCRIPTION

♂ ♀ 16–20 mm. Sexual dimorphism not pronounced; antenna of male weakly dentate-ciliate, cilia approximately equal in length or slightly longer than width of flagellum; antenna of female minutely ciliate.

Male (fig. 7). Forewing rough-scaled, with conspicuous tufts and groups of raised scales; ground colour white extensively suffused with grey, the greyish suffusion overlaid with black strigulation often forming irregular black spots; markings fuscous-brown, strigulate with black, poorly defined; basal and sub-basal fasciae indeterminate, represented by black costal striae and roughened groups of plumbeous or black-tipped scales medially; median fascia inward-oblique, obliquely incised with white from costa, two closely approximate, elongate black dashes above middle, a small ferruginous-red spot in the angle formed by the upper dash and white incision, two pale ferruginous tufts above median fold, two larger plumbeous tufts edged with white a little below fold; pre-apical spot developed as an irregular brown mixed with black stria reaching to tornus, often interrupted below costa; apex broadly edged with ferruginous-ochreous; cilia pale ochreous irregularly mixed with grey, with a dark grey sub-basal line. Hindwing white suffused with fuscous, with indistinct darker strigulation, apical area infuscate; cilia whitish, with dark grey sub-basal and sub-apical lines.

Female. Similar to male.

Variation. The obscure but very distinctive and characteristic forewing pattern of this species is remarkably constant, showing only minor variation, notably in the extent of the black strigulation and in the irregular blackish brown stria from the pre-apical spot.

COMMENTS

A distinctive species with a characteristic roughened texture and variegated pattern of the forewing.

BIOLOGY

Ovum. Deposited on the flowers of female plants of *Bryonia dioica* in June.

Larva. Head chestnut-brown, sometimes edged with black posteriorly; prothoracic and anal plates green; abdomen pale yellowish green, shagreened, with wrinkled and somewhat swollen segments; pinacula inconspicuous; legs green. June to September, feeding on *Bryonia dioica*. The larva may be found from the latter part of June to August, living in the spun flowers (pl. 2, fig. 3) and immature berries (pl. 2, fig. 4), eating out the pulp and leaving only the skins of the berries spun together. In August it leaves the berries and gnaws the stem, to which it spins a leaf with white paper-like silk. By the end of September it is full-grown and spins a tough silken cocoon in the earth, overwintering until the following spring before pupating.

Pupa. Light brown. April and early May, in a cocoon in the earth.

Imago. May and June, frequenting hedgerows, copses, open woodland and similar situations, especially on well-drained soils, where white bryony grows. Inactive by day; it often sits on a leaf of the foodplant during the afternoon, resembling a bird-dropping. It may sometimes be seen on warm evenings flying about its habitat but is usually more active towards dusk; occasionally taken at light.

DISTRIBUTION

Widespread in southern England from Kent to Dorset and Somerset, northwards to Herefordshire, Gloucestershire, Worcestershire and Norfolk, occurring in suitable localities but seldom commonly; also recorded from Yorkshire and Durham.

Central and southern Europe to Asia Minor; North Africa; Canary Islands.

Phalonidia manniana (Fischer von Röslerstamm)
Pl. 22, figs. 8, 9

[*Eupoecilia notulana*; Barrett, 1905: 293.—*manniana*; Pierce & Metcalfe, 1922: 27, pl. 10 (♂, ♀ genitalia). *Phalonia manniana*; Meyrick, 1928: 489; Ford, 1949: 44]

DESCRIPTION

♂♀ 10–13 mm. Sexual dimorphism not pronounced; antenna of male shortly ciliate; female usually larger.

Male (fig. 8). Forewing ground colour cream to whitish cream, suffused with patches of pale ochreous, delicately irrorate and strigulate with fuscous; markings fuscous; basal and sub-basal fasciae indeterminate, forming a diffuse greyish fuscous basal patch often weakly suffused with ochreous, usually with blackish costal striae; median fascia moderately broad, angulate below costa, costal part narrow or partially obliterated, sometimes narrowly interrupted by ochreous suffusion, dorsal part of fascia blackish with variable brownish ochreous admixture; pre-apical spot small, produced as a thick stria reaching to tornal area; a weak, triangular or subquadrate ochreous postmedian marking from dorsum; a small pre-tornal spot; cilia concolorous with ground colour, tinged with ochreous. Hindwing grey; cilia paler, with a dark sub-basal line.

Female (fig. 9). Similar to male.

Variation. In both sexes the ground colour of the forewing varies from whitish cream to brownish cream; in darker specimens the strigulation is usually more conspicuous and sometimes is connected with darkened venation, forming a reticulation; the colour pattern varies from pale ochreous-brown to blackish fuscous and in heavily marked specimens the dorsal part of the median fascia appears black.

COMMENTS

Distinguished from other species of the genus by the comparatively light ground colour of the forewing, and contrasting blackish admixture in the dorsal part of the median fascia.

BIOLOGY

Larva. Head and prothoracic plate shining black, plate with medial sulcus; abdomen pale yellow or yellowish green; pinacula grey, indistinct; anal plate light brown (Barrett, 1880: 37). August to October, feeding in the stems of *Mentha aquatica* and *Lycopus europaeus*; overwintering in the stem and pupating in the spring. On the Continent it is recorded also on *Alisma plantago-aquatica* (Hannemann, 1964: 28).

Pupa. Light brown. May and early June, in a stem of the foodplant, spun up in a cocoon formed of white silk and frass. The emergence hole is concealed by a thin layer of bark (Richardson, 1890: 299).

Imago. June and July; frequenting marshy areas along the banks of streams, canals and ditches where water mint grows. The moth flies freely about the habitat in the evening and at night, when it comes to light.

DISTRIBUTION

Locally common in the southern counties of England from Essex and Kent to Devon and Gloucestershire, East Anglia northwards to Lancashire and Yorkshire. Known localities include King's Lynn (Norfolk), Thorpeness (Suffolk), Wicken Fen (Cambridgeshire), Folkestone (Kent), Byfleet (Surrey), Ditchling, Eastbourne, Lewes and Tilgate Forest (Sussex), New Forest (Hampshire), Bloxworth and Portland (Dorset), Brislington, Clevedon, Ascott and Shapwick (Somerset) (Turner, 1955: 126), Wilmslow, Clough and Marple (Cheshire), Oldham (Lancashire), and Spurn (Yorkshire). Elsewhere in the British Isles it is only known from North Wales (Ddôl near Bodfari, Flintshire).

Europe to central and eastern Asia.

Phalonidia minimana (Caradja)

Pl. 22, figs. 10, 11

[*Eupoecilia geyeriana*; sensu Barrett, 1905: 297.—*walsinghamana*; Pierce & Metcalfe, 1922: 38, pl. 13 (♂, ♀ genitalia). *Phalonia walsinghamana*; Meyrick, 1928: 489; Ford, 1949: 44]

DESCRIPTION

♂ ♀ 11–14 mm. Sexual dimorphism not pronounced; male antenna shortly ciliate; female usually larger.

Male (fig. 10). Forewing ground colour whitish ochreous, sparsely suffused with ochreous, sprinkled with fuscous-black distally, radial veins usually dark-lined, dorsal margin dotted with fuscous-black, costa suffused with grey, weakly strigulate; markings ochreous-brown; basal fascia indeterminate, confluent with sub-basal fascia on costa; sub-basal fascia well developed, strongly angulate below costa; median fascia well defined, angulate below costa, outer margin edged with plumbeous, a rather diffuse inconspicuous blackish patch on outer margin at angle below costa mixed with plumbeous suffusion extending to tornal area, a rather diffuse blackish patch in median fold; pre-apical spot produced as a broad stria to tornal area, often confluent with a similar small patch near tornus; postmedian patch reaching to blackish patch in fold, immediately followed by a small blackish pre-tornal marking; cilia pale ochreous, grey basally. Hindwing grey; cilia paler, with a grey sub-basal line.

Female (fig. 11). Forewing pattern similar to that of male, coloration often lighter.

Variation. A similar range of minor variation in the general coloration of the forewing, which varies from whitish ochreous to yellowish or greyish ochreous, is found in both sexes.

COMMENTS

Similar to *P. alismana* but distinguished by the more angulate median fascia.

BIOLOGY

Larva. Head shining, dark greyish brown or wholly black; prothoracic plate similar, mottled with brown, divided medially; abdomen light grey-brown or purplish grey; pinacula large and conspicuous, blackish brown, those on posterior segments lighter; peritreme of spiracles black; thoracic legs dark greyish brown; prolegs light grey-brown (Richardson, 1891: 239). July, and another generation in September; feeding on *Pedicularis palustris*, living in the seed capsules. Richardson (1892: 173), in a note on the life history, depicts in colour the foodplant and larva. A series of specimens in the E. R. Bankes collection in the British Museum (Natural History) has the data "from *Menyanthes trifoliata* seedpods."

Pupa. May and, for the second generation, July; in a strong, white silken cocoon in the earth or amongst surface debris.

Imago. Bivoltine, the first generation emerging in May and June, the second at the end of July and August. Frequenting boggy heathland and fens, hiding during the day amongst patches of lousewort and difficult to disturb; most often seen on warm still evenings when it takes short flights around the foodplant.

DISTRIBUTION

In England, apparently restricted to a few localities in the south and south-east, occurring commonly in suitable spots in the fens of Norfolk (Horning) and Cambridgeshire; also in Suffolk and on boggy heathland in Dorset (Corfe Castle, Purbeck, Bloxworth). It is recorded by Barrett (1905: 298) "on the bank of the river Leith, in Scotland" but this record has never been confirmed. E. C. Pelham-Clinton suggests that this record may in fact refer to the River Teith, in Perthshire, which is a more suitable area than the Water of Leith. The only authentic record from Scotland is of four specimens taken by E. C. Pelham-Clinton in 1953, 1954, 1955 and 1962—one in each year—at Port Appin, Argyll, flying in the late afternoon.

Europe to the Ukraine and Siberia; Japan.

Phalonidia permixtana (Denis & Schiffermüller)

Pl. 22, fig. 12

[*Eupoecilia mussehliana*; Barrett, 1905: 295. — *permixtana*; Pierce & Metcalfe, 1922: 38, pl. 13 (♂, ♀ genitalia). *Phalonia mussehliana*; Meyrick, 1928: 489; Ford, 1949: 44]

DESCRIPTION

♂ ♀ 10–12 mm. Sexual dimorphism not pronounced; antenna of male ciliate; female usually larger.

Male (fig. 12). Forewing relatively narrow, ground colour pale ochreous-white, suffused with grey along costa and in distal half, sparsely strigulate with fuscous, most evident in terminal area and as strigulae on costa and along dorsum; markings ochreous-brown; basal and sub-basal fasciae diffuse, weakly indicated; median fascia well defined, narrow, angulate below costa, slightly dilated medially, often darkened with grey at costa and on dorsum, a small patch or spot of grey or blackish grey suffusion at middle of outer margin, a lighter grey suffusion extending obliquely to tornal area; pre-apical spot well developed, subquadrate, emitting a weak stria often reaching tornal area; a slender, triangular postmedian marking from dorsum moderately well defined, its apex reaching to median fold and contiguous with median fascia; an indistinct blackish pre-tornal spot; cilia pale ochreous, with a darker sub-basal line. Hindwing grey; cilia paler, with a grey sub-basal line.

Female. Similar to male.

Variation. The extent and intensity of the greyish suffusion and strigulation on the forewing varies; in some specimens it may be very weak, the pale ochreous-white ground colour then being more evident.

COMMENTS

The blackish grey suffusion at the middle of the outer margin of the median fascia and the grey suffusion in the distal half of the forewing are characteristic.

BIOLOGY

Larva. The immature stages and life history of this species do not appear to be known in the British Isles; Meyrick (1928: 490) states that larvae on *Butomus* or other plants hitherto attributed to this species were based on wrong identifications. However, on the Continent, Swatschek (1958: 237) describes the larva as follows: head light brown; prothoracic plate yellowish; abdomen dirty yellowish white. June and July, another generation in September, overwintering until the following April; feeding in the flowerheads, seeds and stems of *Butomus umbellatus*, *Pedicularis*, *Alisma plantago-aquatica*, *Gentiana lutea*, *Euphrasia* and *Rhinanthus*.

Imago. Bivoltine, the first generation occurring in May and June, the second in late July and August; frequenting rough, damp ground where the vegetation is sparse. During the day the moth hides amongst low-growing herbage, flying readily if disturbed, becoming active in the evening and later coming to light.

DISTRIBUTION

A local species the distribution of which is little known; it has been recorded in England from Kent (Deal, Walmer, Sandwich, Folkestone), Sussex (Eastbourne), Somerset and Devon. In Wales it is known from Pembrokeshire and Carmarthenshire. There is one record from Scotland (King, 1887: xlix) but no locality data are given and the record needs confirmation. In Ireland, Beirne (1941: 77) cites Kane's record from Co. Sligo and confirms the identity of the specimens; Bradley (1953: 17) records this species from the Bantry-Glengarriff area in Co. Cork.

West and central Europe to Scandinavia and north-west Russia; Asia Minor; central Asia; China and Japan.

Phalonidia vectisana (Humphreys & Westwood)

Pl. 22, figs. 13–16

[*Eupoecilia vectisana*; Barrett, 1905: 300. —*vectisana*; Pierce & Metcalfe, 1922: 38, pl. 13 (♂, ♀ genitalia). *Phalonia vectisana*; Meyrick, 1928: 488; Ford, 1949: 44]

DESCRIPTION

♂ ♀ 9–11 mm. Sexual dimorphism not pronounced; antenna of male strongly ciliate; female usually larger.

Male (fig. 13). Forewing ground colour whitish ochreous, variably suffused with grey or plumbeous, sometimes heavily, strigulate distally with olive-brown, sometimes with blackish admixture; radial veins often finely lined with blackish, forming a faint reticulate pattern; markings olive-brown; basal and sub-basal fasciae diffuse, often indeterminate but sub-basal fascia usually developed on dorsum; median fascia rather narrow, angulate below costa, thence inward-oblique to dorsum, often darkened with blackish admixture, especially in median fold; a variable grey suffusion extending from outer edge of fascia near middle to tornus; a small, triangular postmedian marking on dorsum, usually poorly developed and diffuse; pre-apical spot variably developed, usually obsolescent and indicated only by fuscous striae on costa; an inconspicuous blackish pre-tornal spot; cilia whitish ochreous, darker basally. Hindwing grey; cilia paler, with a dark sub-basal line.

Female (fig. 14). Similar to male.

Variation. An extremely variable species showing considerable minor variation in the forewing pattern. In strongly marked specimens (fig. 15) the blackish admixture may be heavy and extensive, especially on the median fascia and in the distal area, the pre-apical spot may be developed, sometimes very strongly and produced as a bifurcate stria extending to the terminal and tornal areas, and the dorsal margin may be dotted with black; in some heavily marked specimens the markings may be obscured or obliterated with grey suffusion. A common weakly marked form occurs in which the forewing appears almost unicolorous olive-brown mixed with ochreous and blackish, the only conspicuous marking being the diffuse patch in the median fascia (fig. 16).

COMMENTS

Although extremely variable in wing pattern, the small size of this species and the olive-brown general coloration of the forewing and the reticulation in the distal area are characteristic.

BIOLOGY

Larva. Head brownish yellow; prothoracic and anal plates light greenish yellow, prothoracic plate with a weak medial division; abdomen light green, dorsal and subdorsal areas tinged with pink; pinacula very small, black; thoracic legs brownish yellow; prolegs light green. June, feeding on the flowerheads of *Triglochin maritima* (pl. 2, fig. 1) and *T. palustris*; another generation occurs in September, burrowing first in the shoots just above the crown and eating out the pith, then working downwards into the crown and roots, where they overwinter until the following spring. The presence of a larva is indicated by the yellow appearance of the affected shoot and traces of straw-coloured frass on the exterior (Bankes, 1899: 178).

Pupa. Short and stout; dark orange-brown, wing cases and appendages browner. April and May, in a silken cocoon in a dead flower stem of the foodplant, or amongst ground debris, and again in June and July amongst ground debris.

Imago. Bivoltine, the first generation on the wing from mid-May to the end of June, the second from July to September. Frequenting wet boggy heaths, fens and marshes where marsh arrow-grass grows, and coastal salt marshes and grassy places on rocky shores where sea arrow-grass grows. During the day the moth rests close to the ground among herbage and seldom flies even if disturbed; in the evening as sunset approaches it becomes active and during warm calm weather may often be seen in numbers flying until dusk.

DISTRIBUTION

Locally common and sometimes abundant, especially in coastal salt marshes, in England as far north as Cheshire, Lancashire, Westmorland and south Durham. Localities include the banks of the River Medina and Yarmouth (Isle of Wight), Shoreham, Lancing and Greatham (Sussex), Swale marshes and Gravesend (Kent), Benfleet, Colne and Shoeburyness (Essex), Hemley, Shingle Street, Orford, Southwold and Thorpeness (Suffolk), Wicken Fen (Cambridgeshire), Bidston Marsh, Burton and

Parkgate (Cheshire), Bolton-le-Sands (Lancashire), Doncaster (Yorkshire), and the Humber Estuary. In Wales it is known from Anglesey (Traeth Dulas), Caernarvonshire (Aber), Merionethshire and Cardiganshire. Apparently unknown from Scotland and only recently discovered in Ireland, where moths were collected in June and August at Ballyconneely in Co. Galway (Emmet, 1968: 48, 55; 1971: 9).

North and central Europe; China and Japan.

Phalonidia alismana (Ragonot)

Pl. 22, fig. 17

[*Eupoecilia alismana*; Barrett, 1905: 299.—*alismana*; Pierce & Metcalfe, 1922: 37, pl. 13 (♂, ♀ genitalia). *Phalonidia alismana*; Meyrick, 1928: 489; Ford 1949: 44]

DESCRIPTION

♂ ♀ 11–14 mm. Sexual dimorphism not pronounced; antenna of male strongly ciliate; female usually larger.

Male (fig. 17). Forewing ground colour whitish ochreous, sparsely strigulate with ochreous and fuscous, mostly distally; costal and dorsal margins variably dotted with blackish; markings olive-brown, variably suffused with grey; basal and sub-basal fascia diffuse, sub-basal fascia outward-oblique from dorsum to beyond middle, sometimes darkened with greyish fuscous, angulate below costa; median fascia moderately broad, well defined, darkened with greyish suffusion on costa and dorsum, angulate below costa, outer edge slightly indented, a sprinkling of black scales sometimes forming a blackish spot adjacent to indentation, a diffuse blackish patch below median fold; a triangular postmedian marking on dorsum, its apex contiguous with outer edge of median fascia in fold; pre-apical spot fasciate, reaching nearly to termen, continuing as a stria or striae to tornal area; cilia whitish ochreous, darker basally, irregularly irrorate with blackish. Hindwing grey, infuscate apically and narrowly edged around margins; cilia whitish, with a dark grey sub-basal line.

Female. Similar to male.

Variation. Considerable minor variation is found in the strength of the diffuse blackish patch or spot in the median fascia; in some specimens this patch may appear as a heavy elongate black streak along the median fold while in others it may be broader, sometimes extending as a heavy suffusion to the dorsum.

COMMENTS

Similar to *P. minimana* but distinguished by the less angulate median fascia of the forewing and the pre-apical spot which is usually strigulate or fractured in the tornal area.

BIOLOGY

Ovum. Deposited on the upper part of the flower stem of the foodplant in June and July.

Larva. Head and prothoracic plate blackish brown, plate divided medially; abdomen pale green, becoming bright yellowish pink or pinkish brown when the larva is full-grown, with a faint greyish brown dorsal line; pinacula inconspicuous; anal plate light brown. September and October, feeding on *Alisma plantago-aquatica* and probably other species of *Alisma*, burrowing down the flower stem (pl. 2, fig. 2), eating the pith and leaving frass scattered irregularly along the burrows. Several larvae may feed in a single stem. The larva hibernates in a grey silken cocoon spun up in the lower part of the stem, after gnawing the stem nearly through for emergence of the imago, pupating in early spring.

Pupa. Slender, dull brown, wing cases and appendages shining blackish brown, cremaster blunt and flattened, bristling with hooked hairs. May, in the stem.

Imago. June to August, with a prolonged emergence period. Frequenting the margins of lakes, ponds, streams, canals, ditches, fens and river estuaries where water plantain grows. Resting amongst the foodplant and other herbage during the day, but flying freely in the evening at sunset.

DISTRIBUTION

Widely distributed and locally common in the coastal and inland counties of southern England, ranging from Norfolk and Kent to Cornwall and Herefordshire; occurring as far north as Cheshire (Sealand) and Lancashire (Formby). This species was formerly abundant at Hackney Marsh in the Lea Valley (London). Known localities where the species may be found are Thorpeness (Suffolk), Bowers Gifford (Essex), Faversham and Folkestone (Kent), Byfleet and Effingham (Surrey), Camber (Sussex), Studland and Wareham (Dorset), Abbots Leigh and Shapwick (Somerset), and Madeley (Staffordshire). Elsewhere in the British Isles it is known only from Wales (Pembrokeshire).

North and central Europe.

Phalonidia luridana (Gregson)

Pl. 22, fig. 18

[*Eupoecilia manniana*; sensu Barrett, 1905: 294. — *luridana*; Pierce & Metcalfe, 1922: 37, pl. 13 (♂, ♀ genitalia). *Phalonia luridana*; Meyrick, 1928: 488; Ford, 1949: 44]

DESCRIPTION

♂ ♀ 11–13 mm. Sexual dimorphism not pronounced; antenna of male strongly ciliate; female usually larger, hindwing darker.

Male (fig. 18). Forewing ground colour whitish ochreous, sparsely strigulate with brown, mostly distally; dorsum dotted with blackish; markings ochreous-brown, variably suffused with grey; basal and sub-basal fasciae confluent on costa, sub-basal fascia outward-oblique from dorsum to beyond middle; median fascia narrow, well defined, angulate below costa, its outer edge slightly indented below angulation, darkened with fuscous suffusion at costa and dorsum and on outer margin below median fold; a diffuse greyish streak from angulation of fascia to tornal area; a well-defined, triangular post-median marking on dorsum, its apex approximate to median fascia in fold; pre-apical spot well developed, fasciate, often excised on costa, strigulate in tornal area; pre-tornal spot obsolescent; cilia whitish ochreous, darker basally. Hindwing pale grey; cilia whitish, with a dark sub-basal line.

Female. Similar to male; hindwing darker.

Variation. Forewing pattern and general coloration very constant, showing only minor variation.

COMMENTS

The comparatively uniform whitish ochreous ground colour of the forewing and the pale ochreous-brown markings are characteristic.

BIOLOGY

Larva. Undescribed. Apparently little is known of the early stages and life history of this species. However, a moth was bred on 7th June, 1939 by A. E. Wright (1947: 69) from flowerheads of *Matricaria recutita* gathered in September of the previous year.

Imago. May to August, possibly bivoltine. Frequenting dry grassy banks, woodland rides and waysides where wild chamomile grows. Resting by day amongst herbage and on flowerheads, flying at sunset and coming to light.

DISTRIBUTION

Rare and local, first described from Witherslack, in Westmorland, by C. S. Gregson (1870: 80) from specimens taken in May of the previous year by J. B. Hodgkinson. It has since been recorded from Lancashire (Grange-over-Sands), Durham, Norfolk, Kent, Sussex, Surrey, Hampshire (New Forest), Wiltshire (Ramsbury), Berkshire (Hungerford), Dorset (Corfe), Devon (Strete, near Dartmouth), and Gloucestershire, where the species was taken in abundance on the Cotswolds by J. W. Metcalfe in June, 1917 (Fletcher & Clutterbuck, 1945: 65).

North, central and southern Europe to Asia Minor; China and Japan.

Phalonidia affinitana (Douglas)

Pl. 22, figs. 19, 20

[*Eupoecilia affinitana*; Barrett, 1905: 302.—*affinitana*; Pierce & Metcalfe, 1922: 27, pl. 10 (♂, ♀ genitalia). *Phalonia affinitana*; Meyrick, 1928: 488; Ford, 1949: 44]

DESCRIPTION

♂ ♀ 11–14 mm. Sexual dimorphism moderately pronounced; antenna of male strongly ciliate; female with forewing coloration usually darker and more leaden.

Male (fig. 19). Forewing ground colour whitish ochreous or pale light buff, suffused with plumbeous-grey distally, rather sparsely strigulate with ochreous mixed with blackish brown; markings ochreous varying to ochreous-brown; basal and sub-basal fasciae obsolescent, represented by ochreous-brown suffusion; median fascia deep ochreous, slender, obsolescent from costa to near middle, well developed and edged with submetallic plumbeous dorsally; pre-apical spot obsolete; a small blackish brown pre-tornal spot; cilia concolorous with ground colour. Hindwing grey; cilia paler, with an indistinct grey sub-basal line.

Female (fig. 20). Forewing pattern as in male but general coloration often leaden.

Variation. A wide range of minor variation is found in both sexes. In the male the strigulation and median fascia of the forewing vary in strength, and in very lightly marked forms the wing may appear almost unicolorous. The coloration of the female is almost invariably greyer or more leaden than in the male; exceptionally specimens occur with pale ochreous coloration as in the male but usually with at least a trace of the leaden suffusion distally. In Essex (Leigh-on-Sea) a melanic form occurs not uncommonly (Huggins, 1958b: 108).

COMMENTS

Distinguished from other species of *Phalonidia* by the median fascia being strongly developed in the dorsal half of the wing and obsolescent costally.

BIOLOGY

Ovum. Deposited on the flowers of the foodplant in June and July.

Larva. Head light brown; prothoracic plate darker, with medial sulcus; abdomen white or greyish white, more greyish dorsally and with faint grey dorsal and spiracular lines; when full-grown it becomes slightly tinged with reddish prior to pupation; anal plate light brown; anal comb weakly developed (Wright, 1935: 264). Late July to October, overwintering fully grown until the following spring. Feeding on *Aster tripolium*, at first on the pappus, later boring into the upper part of the flower stalk. In early autumn, before the flower stalk breaks off, the larva bores down to the crown of the plant, and feeds in the rootstock until late autumn when it is full-grown (Huggins, 1958b: 107).

Pupa. Late March to May; in the larval habitation in the rootstock, or in the immediate vicinity amongst tidal debris held by broken stems of the foodplant.

Imago. Apparently univoltine, June to August; however, some authors consider that specimens occurring in August belong to a small second brood. Frequenting coastal and estuarine salt marshes where sea aster grows. The moth sits close to its foodplant during the day and usually is not readily disturbed though occasionally it may be found on the wing on sunny afternoons; the normal flight period is in the late evening and at sunset.

DISTRIBUTION

Locally common in southern England, occurring in salt marshes northwards to Lancashire, Yorkshire (Middlesbrough and Humber salt marshes) and Durham; in the south known localities include King's Lynn (Norfolk), Hollesley, Thorpeness and Southwold (Suffolk), Benfleet, Canvey Island, Great Wakering, Leigh-on-Sea, Pitsea and Wivenhoe (Essex), Dungeness and Graveney (Kent), Brighton (Sussex), and Yarmouth (Isle of Wight); also recorded from Kent, Dorset, Gloucestershire (Fletcher & Clutterbuck, 1939: 299) and Somerset. Locally common in Wales in Caernarvonshire, Cardiganshire, Merionethshire and Pembrokeshire. Apparently unknown in Scotland, and in Ireland known only from the Burren coast, Co. Clare, where several specimens were taken at mercury vapour light in June, 1970 by D. W. H. Ffennell.

North, central and southern Europe to the Ukraine.

Phalonidia gilvicomana (Zeller)

Pl. 22, fig. 21

[*Eupoecilia gilvicomana*; Barrett, 1905: 292. — *gilvicomana*; Pierce & Metcalfe, 1922: 28, pl. 11 (♂, ♀ genitalia). *Phalonia gilvicomana*; Meyrick, 1928: 493; Ford, 1949: 47]

DESCRIPTION

♂♀ 11–13 mm. Sexes similar; antenna of male shortly ciliate.

Male (fig. 21). Forewing ground colour yellowish white, strongly suffused distally with plumbeous mixed with fuscous; markings ochreous-yellow variably overlaid with fuscous and plumbeous mixture; basal and sub-basal fasciae indeterminate, forming a rather diffuse patch, angulate below costa, narrowly infuscate along costa; median fascia moderately broad, dilated on costa, heavily overlaid and suffused with plumbeous mixed with black; a diffuse triangular or subquadrate postmedian patch on dorsum contiguous with a small blackish pre-tornal spot; an extensive diffuse ill-defined ferruginous-ochreous mixed with blackish patch above tornus reaching to middle or beyond; pre-apical spot fasciate, heavily infuscate; cilia grey mixed with whitish, darker basally, whitish at tornus. Hindwing dark grey; cilia paler, with a dark sub-basal line.

Female. Similar to male.

Variation. Only minor variation is found in the forewing markings and in the distribution of the infuscation.

COMMENTS

P. gilvicomana and *P. curvistrigana* are readily distinguished from other *Phalonidia* species by their characteristic ochreous-yellow forewing markings and heavy infuscation. *P. gilvicomana* may be separated by the more extensive infuscation in the postmedian area.

BIOLOGY

Ovum. Yellowish green, later changing to grey. End of June and early July, deposited on the flower buds and stems of the foodplant.

Larva. Head and prothoracic plate light brown, the latter marked with dark brown; abdomen light yellowish or brownish green, translucent (Sheldon, 1937: 198). July and August, feeding on the seeds of *Mycelis muralis* and *Lapsana communis*. The larva is generally concealed within the pappus (pl. 3, figs. 1, 2) but when full-grown may often be seen extruding from the seedhead. *Chenopodium* is also recorded as a foodplant on the Continent (Hannemann, 1964: 30). When full-grown in August the larva spins a cocoon amongst ground litter, in which it hibernates before pupating in the spring. Wakely (1937: 225) gives an account of rearing this species in captivity.

Pupa. Yellowish brown. May, in a slight silken cocoon amongst ground litter or in the soil.

Imago. Normally univoltine, June to early July; Sheldon (1937: 199) bred a small number of moths in August, following a hot early season. Frequenting waysides, walls, wood margins, waste places and similar situations, usually in chalk localities, where wall lettuce and nipplewort are found. The moths fly over the herbage in the evenings, often quite high for a Cochylid, at heights of around five feet.

DISTRIBUTION

Very local and apparently restricted to southern England, occurring in Surrey (Box Hill, Mickleham, Headley Lane and Riddlesdown), Kent (High Rocks, near Tunbridge Wells), and Devon (Lynton and Lynmouth). The locality of the specimens taken by Standish (1879: 205), who first recorded the species from the British Isles, is believed to be Cheltenham (Gloucestershire), but this has never been confirmed; Fletcher (1945: 21) records a worn female taken at Rodborough on 22nd June, 1944.

North and central Europe.

Phalonidia curvistrigana (Stainton)

Pl. 22, fig. 22

[*Eupoecilia curvistrigana*; Barrett, 1905: 291. — *curvistrigana*; Pierce & Metcalfe, 1922: 28, pl. 10 (♂, ♀ genitalia). *Phalonia curvistrigana*; Meyrick, 1928: 493; Ford, 1949: 46]

DESCRIPTION

♂ ♀ 13–14 mm. Sexes similar; antenna of male strongly ciliate.

Male (fig. 22). Forewing ground colour whitish ochreous, dotted and strigulate with fuscous along costa and dorsum, apical area and terminal margin overlaid with plumbeous; markings ochreous-yellow; basal and sub-basal fasciae indeterminate, forming a rather diffuse patch angulate above median fold, infuscate along costa; median fascia moderately broad, dilated on costa, heavily overlaid with plumbeous mixed with black, the black forming irregular diffuse patches; a slender postmedian marking on dorsum reaching to fold; an irregular patch above tornus reaching beyond middle, weakly suffused with plumbeous or fuscous; pre-apical spot fasciate to middle of terminal margin, continuing along termen as a stria to tornus, weakly suffused with brown, variably overlaid with black, especially on costa; pre-tornal spot obsolescent; cilia plumbeous mixed with ochreous. Hindwing grey, infuscate distally; cilia paler, with a dark sub-basal line.

Female. Similar to male.

Variation. In the forewing the markings show only minor variation, and the ochreous-yellow coloration is often darker and tinged with brown.

COMMENTS

Distinguished from *P. gilvicomana* by the comparatively clear ground colour in the postmedian area of the forewing.

BIOLOGY

Ovum. July and early August; deposited on the buds, flowerheads and seedheads of the foodplant. The egg is laid on a closed seedhead or late green bud; these seed vessels do not open and the seeds are not dispersed.

46

Larva. Head and prothoracic plate brown, plate minutely spotted with dark brown or black, more densely posteriorly; abdomen light brown with a pinkish flush, or pale yellowish pink; pinacula slightly more pink than surrounding integument; anal plate light brown, minutely dotted with black. August and September, feeding in the flowers of *Solidago virgaurea*, eating the unripe seeds and passing from flower to flower, sometimes uniting them slightly with silk. There is otherwise little external evidence of feeding but unopened seedheads usually indicate the presence of a larva. In September the larva leaves the seedhead and spins a silken cocoon on the ground, hibernating until the early spring before pupating.

Pupa. Yellowish brown. April to June, in a cocoon on the ground.

Imago. July and August; frequenting glades, clearings, particularly in two-year-old and three-year-old coppiced woods, rides, and open spaces amongst scrub where golden rod grows. During the day the moth hides amongst dense vegetation, dropping to the ground or flying quickly to a fresh hiding place when disturbed. During the evening and at sunset it may be seen flying about its habitat.

DISTRIBUTION

Local, but often common where it occurs, in the southern counties of England, ranging northwards to Lancashire (Silverdale and Grange-over-Sands) and Westmorland (Fairclough, 1961: 16). Known localities include Thetford (Norfolk), Bexley, Blean Woods, Darenth and Folkestone (Kent), Gomshall (Surrey), Polegate (Sussex), Barnstaple (Devon), and Abbots Leigh (Somerset). Turner (1955: 127) considers the species to be probably extinct in Somerset. It has also been recorded from Dorset, Herefordshire and Cambridgeshire, and in Wales it is known from Flintshire and Pembrokeshire.

Europe to southern Russia and Siberia; Japan.

Stenodes alternana (Stephens)

Pl. 23, figs. 1, 2

[*Lozopera alternana*; Barrett, 1905: 325.—*alternana*; Pierce & Metcalfe, 1922: 31, pl. 11 (♂, ♀ genitalia). *Euxanthis alternana*; Meyrick, 1928: 497; Ford, 1949: 48]

DESCRIPTION

♂ ♀ 18–25 mm. Sexual dimorphism not pronounced; antenna of male shortly ciliate; female usually larger, hindwing slightly darker.

Male (fig. 1). Forewing ground colour whitish yellow, irregularly suffused and strigulate with pale ochreous, costa and dorsum strigulate with fuscous, termen edged with ochreous and fuscous mixture; markings ochreous-brown; basal and sub-basal fasciae indeterminate, forming a diffuse basal patch; median fascia comparatively well developed and conspicuous, fractured on costa, angulate below, a variable admixture of fuscous-brown from near middle to dorsum, two black or fuscous-black scale-tufts costad of median fold and two a little below; pre-apical spot poorly defined, extended as a diffuse fasciate stria to tornal area, strigulate on costa, its inner margin irrorate with black;

47

indication of small brown or blackish pre-tornal spot; cilia whitish yellow, darker basally. Hindwing whitish grey, infuscate apically, along terminal margin and on veins; cilia white, yellowish basally, with a grey sub-basal line.

Female (fig. 2). Similar to male; hindwing darker.

Variation. In both sexes minor variation is found in the extent of the ochreous suffusion over the ground colour of the forewing, and in the blackish admixture in the markings. In lightly marked specimens the suffusion may be weak, the wing then appearing distinctly whitish ochreous, the scale-tufts are vestigial and the fuscous-brown suffusion is reduced; in the other extreme the forewing is ochreous-grey in general appearance and the black scaling is more extensive and is usually heavy at the middle of the median fascia; in some specimens the raised scale-tufts in the median fascia may be greyish fuscous or plumbeous.

COMMENTS

A larger species than *S. straminea*, and further distinguished by the presence of scale-tufts in the comparatively broad median fascia.

BIOLOGY

Ovum. Deposited in the crown of the foodplant during late summer.

Larva. Head, prothoracic plate and thoracic legs varying from dark brown to black, plate divided medially; abdomen yellowish white, or pale yellow, tinged with greyish green; pinacula very small, conspicuous, dark greyish green; spiracular peritremes blackish; anal plate very small, dark brown ringed with greyish green. The larva hatches in September, overwintering in an early instar in the green crown of the foodplant, *Centaurea scabiosa*; in late spring it moves to a developing flower bud and feeds internally on the immature ovaries and florets, lining the cavity with silk. In July the presence of a larva in the last instar is indicated by frass which is ejected and adheres to the stem immediately below the base of the flower bud (pl. 3, fig. 3).

Pupa. Dark chestnut-brown. July, in the larval habitation in the flower bud. After the moth has emerged the pupal exuviae may be seen protruding from the exit hole.

Imago. July and August; frequenting rough ground on coastal and inland chalk downs where greater knapweed grows. The moth may be netted flying over the foodplant at dusk.

DISTRIBUTION

In the British Isles this species is apparently restricted to Kent and although very local it is often abundant where it occurs. Known localities include Deal, Dover, Folkestone, Kingsdown, St. Margaret's Bay, Walmer and Wye. Its occurrence in Essex (Southend) (Barrett, 1905: 326) and Yorkshire (Meyrick, 1928: 498) has not been confirmed and seems doubtful. The record from the Isle of Barra, in the Outer Hebrides (Thomas & Waterston, 1936: 281) has also not been confirmed and undoubtedly refers to the widely distributed and common *S. straminea*.

Europe to southern Russia; North Africa.

Stenodes straminea (Haworth)

Pl. 22, fig. 3

[*Lozopera straminea*; Barrett, 1905: 324. — *straminea*; Pierce & Metcalfe, 1922: 31, pl. 11 (♂, ♀ genitalia). *Euxanthis straminea*; Meyrick, 1928: 497; Ford, 1949: 48]

DESCRIPTION

♂ ♀ 15–20 mm. Sexual dimorphism not pronounced; antenna of male shortly ciliate; hindwing of female darker, cilia more contrasting.

Male (fig. 3). Forewing ground colour whitish yellow, weakly suffused and strigulate with pale ochreous, costa and dorsum strigulate with fuscous, termen edged with ochreous and fuscous mixture; markings ochreous-brown; basal and sub-basal fasciae obsolescent, forming a weak diffuse patch, strongest on costa; median fascia strongly developed in dorsal half, moderately broad on dorsum, tapering and reaching obliquely to beyond middle, obsolescent costally, present as a small costal patch darkened with fuscous, sometimes with dark brown or blackish admixture in vicinity of median fold and on dorsum; pre-apical spot obsolescent or weakly developed, represented by greyish and fuscous costal striae which may be confluent internally and form a diffuse fasciate stria extending to tornal area; indication of a small brown or blackish pre-tornal spot; cilia pale yellow, a faint greyish sub-basal line. Hindwing grey, paler basally; cilia whitish, with a grey sub-basal line.

Female. Similar to male; hindwing unicolorous grey, with contrasting cilia.

Variation. The extent of the ochreous suffusion over the ground colour of the forewing varies in intensity as also does the blackish admixture in the markings. When the ochreous suffusion is weak the wing may have a distinctly whitish appearance, particularly when the markings are correspondingly pale. Well-marked specimens usually have the blackish scaling in the median fascia more extensive and sometimes forming a small patch in the fold.

COMMENTS

Large specimens with the markings well developed may resemble small *S. alternana* but may be recognized by the comparatively unmarked basal and terminal areas of the forewing and the absence of scale-tufts in the median fascia.

BIOLOGY

Larva. Head dark brown or black; prothoracic plate brown, divided medially; abdomen very pale yellow or almost white; pinacula inconspicuous, concolorous with integument; anal plate light brown; thoracic legs black. Feeding on *Centaurea nigra*; the larvae from the spring generation of moths occur in July and early August, living in the upper part of the stem immediately below the flowerhead, eating the developing seeds (pl. 3, fig. 4). Larvae from the second generation of moths have been found in the shoots in May, feeding on the pith of the stems, causing the tips to turn brown; the growth of these shoots is retarded and they are usually hidden by the healthy stems (Wakely, 1936: 198). These larvae apparently hatch the previous autumn and overwinter in an early instar in the crown of the foodplant, but Ford (1949: 48) implies that the September larvae are full-grown by late autumn, since he states that they pupate in a cocoon on the ground, hibernating in the cocoon.

Pupa. Brown. April and May, in a silken cocoon in the earth or in surface debris (Ford, 1949: 48), or in the larval habitation in the stem (Wakely, 1936: 198); for the second generation, at the end of July and in early August, in the larval habitation in the seedhead (Ford, 1936b: 85).

Imago. Bivoltine, the first generation emerging from the end of May to early July, the second generation in late August and September. Frequenting meadows, grassland, waysides, hills and other situations where the lesser knapweed is found, sitting on the foodplant during the day, and when disturbed flying briskly to another plant. The moth flies actively about the flowerheads from dusk onwards and is frequently taken at light.

DISTRIBUTION

Widely distributed throughout the British Isles as far north as Sutherland, but in the more northerly areas apparently restricted to low elevations and especially common on the coast; absent from the higher parts of Derbyshire and mountainous areas to the north although the foodplant may be plentiful. In North Wales it is widespread, and also occurs in mid-July up to 1200 ft on mountains.

Europe and North Africa to Turkestan; Canary Islands.

Agapeta hamana (Linnaeus)
Pl. 23, figs. 4–7
[*Xanthosetia hamana*; Barrett, 1905: 341. *Pharmacis hamana*; Pierce & Metcalfe, 1922: 30, pl. 11 (♂, ♀ genitalia). *Euxanthis hamana*; Meyrick, 1928: 498; Ford, 1949: 48]

DESCRIPTION

♂ ♀ 15–24 mm. Sexual dimorphism not pronounced; antenna of male strongly ciliate; female usually larger, hindwing darker.

Male (fig. 4). Forewing ground colour pale yellow, costa narrowly suffused with ferruginous-brown, darker basally; markings ferruginous-brown, often with plumbeous admixture; basal, sub-basal and median fasciae obsolete; pre-apical spot usually reduced to diffuse striae on costa, usually with several variable striae dorsad; a conspicuous streak from tornus inward-oblique to middle, its margins slightly irregular, angulate apicad in discal area, usually terminating on upper margin of cell, extending into cilia at tornus; cilia otherwise concolorous with ground colour, with a weak pale ochreous sub-basal line. Hindwing dark grey; cilia whitish ochreous, with a dark grey sub-basal line.

Female. Similar to male; hindwing darker.

Variation. A variable species in size and general coloration, with strongly marked and intermediate forms. In the more common forms the forewing ground colour varies from whitish yellow to deep yellow. An increase in the strength of the markings often corresponds with the darker ground coloration (figs. 5, 6, f. *diversana* Hübner). In

50

strongly marked specimens the median fascia may be represented by an elongate ferruginous-fuscous spot on the costa and a similar orbiculate patch between the median and submedian folds; exceptionally these markings are extended and form the complete transverse fascia; in specimens with the pre-apical spot strongly developed, this is excised on the costa and its apex is often connected with the tornal streak.

In the Burren area, Co. Clare, specimens are found not infrequently with the ground colour deep ochreous-yellow, reminiscent of *A. zoegana* f. *ferrugana* Haworth; in the same locality a further form also occurs devoid of markings except for the tornal streak, which is usually narrower, and the apical area of the wing is finely irrorate with black (fig. 7).

COMMENTS

This species is readily distinguished by the pale yellow ground coloration and the reduced, ferruginous-brown markings of the forewing.

BIOLOGY

Larva. Head reddish brown; abdomen greyish white; apparently little known and otherwise undescribed. August to October, feeding in the roots of *Carduus* (*sensu lato*) and possibly various species of *Cirsium*, overwintering in a silken cocoon and pupating in the spring. In Cornwall it has been reported feeding in the roots of *Serratula tinctoria*, but this has not been confirmed.

Pupa. Reddish brown. May, in a whitish silken cocoon spun up in the larval habitation or in the earth nearby.

Imago. Occurring from the end of May to early September, with an apparently prolonged period of emergence, there being no clear evidence of more than a single generation. Most frequently found amongst rough herbage in fields, waysides and waste places. It is readily disturbed from rest among thistles, knapweeds and other herbage at any time of the day, when it will dart to a fresh resting place, its light coloration rendering it conspicuous whilst in flight. In the evening and at dusk it takes short flights about the habitat, and at night will come to light.

DISTRIBUTION

Generally distributed and common at lower elevations throughout the British Isles as far north as Sutherland; in Scotland it is uncommon except on the coast, and in the higher parts of Derbyshire and other mountainous areas it is scarce although the foodplants are plentiful.

Europe to Asia Minor and Siberia.

Agapeta zoegana (Linnaeus)

Pl. 22, figs. 8–11

[*Xanthosetia zoegana*; Barrett, 1905: 340. *Agapeta zoegana*; Pierce & Metcalfe, 1922: 31, pl. 11 (♂, ♀ genitalia). *Euxanthis zoegana*; Meyrick, 1928: 498; Ford, 1949: 48]

DESCRIPTION

♂ ♀ 15–23 mm. Sexual dimorphism not pronounced; antenna of male strongly ciliate. *Male* (fig. 8). Forewing ground colour bright yellow, costa narrowly suffused with ferruginous-brown with plumbeous admixture; basal and sub-basal fasciae obsolete except for costal suffusion; median fascia represented by a conspicuous orbiculate subdorsal patch; pre-apical spot indeterminate, its apex confluent with a broad inward-oblique streak from the tornus, both being confluent with a broad terminal band, these markings enclosing a semicircular patch of ground colour; cilia concolorous with wing basally, grey apically. Hindwing grey; cilia paler, with a dark sub-basal line.

Female. Similar to male.

Variation. The nominate form of this species varies little except in size. The bright yellow ground colour and the pattern of the forewing are remarkably constant, the most variable feature being the size and shape of the patch of ground colour in the distal area. A distinct form occurs (fig. 10, f. *ferrugana* Haworth) in which the forewing ground colour is orange-brown or ferruginous-brown. Intermediates between this and the nominate form occur (fig. 9). An almost unicolorous whitish or greyish buff form (fig. 11) is known but is extremely rare.

COMMENTS

A distinctive species readily recognized by the bright yellow coloration and characteristic pattern of the forewing.

BIOLOGY

Larva. Head light brownish yellow; prothoracic and anal plates brownish yellow; abdomen yellowish white. September and October, feeding in the roots of *Centaurea nigra* and *Scabiosa columbaria*, and probably also other species of *Centaurea* and also *Knautia arvensis*; living in a whitish silken spinning, overwintering, and pupating in the spring. On the Continent, *Serratula salicina* is recorded as a foodplant (Swatschek, 1958: 230).

Pupa. Rather slender, brown. April and May, spun up in the larval habitation.

Imago. Occurring from May to August, with an apparently prolonged period of emergence, there being no clear evidence of more than a single generation. Usually plentiful where lesser knapweed and field scabious are found in meadows, fields, hillsides, chalk-pits and rough ground generally. The moth is seldom seen during the day, resting amongst the low herbage, but from sunset to dusk it makes short flights about the habitat, and at night occasionally comes to light.

DISTRIBUTION

Generally distributed and common in England and Wales, but less frequent in the northern counties; uncommon in the south of Scotland, occurring as far north as Perthshire, and along the coast of Kincardineshire. According to Beirne (1941: 79) common and generally distributed in Ireland.

Europe to Asia Minor and southern Russia.

Aethes tesserana (Denis & Schiffermüller)

Pl. 23, figs. 12–16

[*Argyrolepia tesserana*; Barrett, 1905: 330. *Phalonia aleella*; Pierce & Metcalfe, 1922: 29, pl. 11 (♂, ♀ genitalia). *Phalonia tesserana*; Meyrick, 1928: 488; Ford, 1949: 44]

DESCRIPTION

♂ ♀ 11–17 mm. Sexual dimorphism not pronounced; antenna of male shortly ciliate; female usually larger.

Male (fig. 12). Forewing ground colour yellow, variably overlaid with ochreous-yellow or dull orange-red; markings ferruginous-ochreous varying to ochreous-yellow and greenish yellow, dull orange, and ferruginous-brown, variably edged with metallic plumbeous, sometimes with dark grey or blackish admixture; basal and sub-basal fasciae indeterminate, forming a diffuse patch on costa; median fascia displaced medially, formed of a quadrate patch on dorsum and a similar patch on costa; pre-apical spot developed as a subquadrate or triangular marking often extending to apex; a variable, often large, subquadrate tornal patch; cilia pale yellow, grey at tornus, a ferruginous or ochreous sub-basal line. Hindwing dark grey; cilia whitish, with a grey sub-basal line.

Female. Similar to male.

Variation. A variable species with several strongly divergent forms. The markings may be reduced (fig. 13) or enlarged and confluent in the dorsal half of the wing (fig. 14). An unusual form has the markings greenish yellow, conspicuously edged with metallic plumbeous (fig. 15); lighter forms lacking the plumbeous edging are not infrequent (fig. 16).

COMMENTS

The bold chequered pattern formed by the markings and contrasting ground colour of the forewing are characteristic.

BIOLOGY

Larva. Head brownish yellow or black; prothoracic plate yellow; abdomen varying from yellowish white to brownish white. September to April, feeding in the rootstock and roots of *Picris hieracioides*, *P. echioides*, *Hieracium*, *Crepis* and *Inula conyza*. The larva is difficult to find as there is no external evidence of feeding; it is best to dig up rootstocks in the late autumn and pot them.

Pupa. Yellowish brown. May and June.

Imago. Apparently univoltine, with a prolonged period of emergence, occurring from the end of May to August but most plentiful in late June and July. Frequenting rough hillsides, especially calcareous slopes, railway embankments, quarries, field margins, coastal cliffs and waste places where the vegetation is sparse. The moth flies actively in the sunshine; when at rest it is easily disturbed.

DISTRIBUTION

Widely distributed and locally common in suitable localities throughout the southern counties of England, scarce in the Midlands and recorded only as far north as Lancashire and Yorkshire. There are old records from Renfrewshire, Arran and Argyll, but otherwise the species is unknown in Scotland. It has probably been overlooked in Wales, since the only known records are from Anglesey, Cardiganshire and Pembrokeshire. Apparently unknown from Ireland.

Europe to southern Russia and northern Iran.

Aethes rutilana (Hübner)

Pl. 24, fig. 1

[*Dapsilia rutilana*; Barrett, 1905: 338. *Phalonia rutilana*; Pierce & Metcalfe, 1922: 29, pl. 11 (♂, ♀ genitalia); Meyrick, 1928: 486; Ford, 1949: 42]

DESCRIPTION

♂ ♀ 9–11 mm. Sexual dimorphism not pronounced; antenna of male shortly ciliate; female larger, hindwing sometimes weakly tinged with ferruginous distally along veins and margins.

Male (fig. 1). Forewing ground colour deep yellow; markings ferruginous-brown mixed and extensively overlaid with violaceous; basal and sub-basal fasciae forming a more or less solid patch; median fascia well defined, varying from narrow to moderately broad, extended along dorsum to basal patch; pre-apical spot strongly developed, its apex produced to dorsum and confluent with a thick streak from tornus; a fasciate terminal stria from apex to tornal area; cilia ochreous-yellow, grey at tornus, with a dark sub-basal line. Hindwing grey; cilia whitish, with a dark sub-basal line.

Female. Similar to male; hindwing sometimes weakly tinged with ferruginous distally, usually along veins and margins.

Variation. Only minor variation is found in the strength and development of the forewing markings.

COMMENTS

The conspicuous fasciate purplish markings, which contrast with the deep yellow ground colour of the forewing, are characteristic.

BIOLOGY

Ovum. July and August, deposited singly in the leaf axil.

Larva. Head light brown; prothoracic plate brown, irrorate with dark brown, a group of heavier dots each side of the medial sulcus; abdomen varying from yellow to brownish white; anal plate brownish grey. September to June, feeding on *Juniperus communis.* The newly hatched larva eats its way into the leaf near the base, moving from leaf to leaf and ejecting frass which adheres to silken strands and webbing spun between neighbouring leaves. The larva overwinters in a short, frass-covered, silken tube spun against the twig, recommencing to feed in the spring and hollowing out the concave surface of the leaves (Freeman, 1967: 434, figs. 23, 24). The foodplant is illustrated (pl. 4, fig. 1).

Pupa. Light yellow. July, in a slight silken cocoon in the larval habitation.

Imago. July and August; occurring amongst juniper on chalk hills and downs, flying in bright sunshine from late morning to early evening, and again at dusk. It is a difficult species to net, flying close to the ground, but may be more easily taken by beating juniper over a beating-tray in the early morning.

DISTRIBUTION

Local and known only from the chalk downs in southern England. It has been recorded from various localities in Surrey (Betchworth, Box Hill, Caterham, Croydon, Mickleham, Purley, Riddlesdown, Sanderstead and Westcott) and Berkshire (Moulsford and Thurle Down); also known from Buckinghamshire and Hampshire. Following the decline of juniper in recent years the species has become rare and may now be extinct in some of these localities.

Northern, central and south-east Europe to central Russia; North America.

Aethes piercei (Obraztsov)

Pl. 24, figs. 2–4

[*Argyrolepia baumanniana*; sensu Barrett, 1905: 331. *Aethes baumanniana*; sensu Pierce & Metcalfe, 1922: 32, pl. 12 (♂, ♀ genitalia). *Chlidonia baumanniana*; sensu Meyrick, 1928: 494; sensu Ford, 1949: 47]

DESCRIPTION

♂ ♀ 15–24 mm. Sexual dimorphism not pronounced; antenna of male weakly dentate-ciliate.

Male (fig. 2). Forewing ground colour pale ochreous-yellow, variably suffused with grey or fuscous, sometimes mixed with ferruginous, termen edged or narrowly suffused with ferruginous or ferruginous-brown, costa strigulate and dorsum dotted with ferruginous or fuscous; markings ferruginous varying to ferruginous-brown, often, suffused with grey or fuscous, bordered with slightly raised white or silver-white spots sometimes coalescing and forming striae; basal and sub-basal fasciae indeterminate, often obsolescent but usually forming a heavy diffuse patch strongest on costa and dorsum; median fascia broad on costa, narrowing dorsally, interrupted below costa, angulate near middle and strongly produced or dilated distally, often with grey admixture especially on costa and in median fold, dorsal half sometimes thinly edged with blackish; pre-apical spot triangular, reaching apex of wing, apex of spot often extended obliquely inward towards median fascia; a series of white or silver-white spots forming broken striae in distal area, interrupting pre-apical spot and tornal patch; cilia ochreous, suffused with grey, with a dark grey sub-basal line. Hindwing grey or dark grey; cilia whitish, with a dark grey sub-basal line.

Female. Similar to male.

Variation. Very variable like *A. hartmanniana* but showing a stronger tendency for the yellowish ground colour of the forewing to become obliterated with grey or ferruginous suffusion. The markings are frequently stronger and blackish brown in colour (fig. 3), and in some heavily marked specimens they are diffuse (fig. 4).

COMMENTS

This species is closely related to *A. hartmanniana* and some authors have suggested that it may be no more than an ecological form. Obraztsov (1952: 157) discusses the synonymy and distinguishes two taxa, separating them on genitalic characters, notably on the structure of the aedeagus in the male, and established the present application of the names.

A. piercei differs from *hartmanniana* by its larger size, the generally darker coloration of the forewing, the more strongly angulate median fascia, and the silver-white spots which usually do not form continuous striae.

BIOLOGY

Larva. Head light brown; prothoracic plate yellowish brown; abdomen yellow, strongly shagreened; pinacula large, conspicuous, dark-coloured (Swatschek, 1958: 231). August to March and April, feeding on *Succisa pratensis*, living in the rootstock.

Pupa. April and May; in the larval habitation.

Imago. Late May to early July. Frequenting damp habitats such as marshes, fens, rough pastures and open or thinly wooded boggy country and similar situations where the foodplant grows; easily disturbed from rest during the day and flying freely at dusk.

DISTRIBUTION

Widespread and locally common in suitable habitats in Britain north to the Orkneys. In England it has been recorded in the southern counties from Kent and Dorset northwards to Durham and Northumberland; in Wales known from Glamorgan, Pembrokeshire, Denbighshire, Anglesey and Montgomeryshire; and in Scotland from inland localities in the south, but absent from the Highlands and restricted to coastal districts in the north. In Ireland it is widely distributed and locally common, occurring also on the predominantly limestone area of the Burren, Co. Clare, where *Succisa* and various calcifuges are found in association with calcicoles.

Central Europe.

Aethes hartmanniana (Clerck)

Pl. 24, figs. 5, 6

[*Argyrolepia subbaumanniana*; Barrett, 1905: 332. *Aethes subbaumanniana*; Pierce & Metcalfe, 1922: 32, pl. 12 (♂, ♀ genitalia). *Chlidonia subbaumanniana*; Meyrick, 1928: 494; Ford, 1949: 47]

DESCRIPTION

♂ ♀ 11–17 mm. Sexual dimorphism not pronounced; antenna of male weakly dentate-ciliate.

Male (fig. 5). Forewing ground colour pale ochreous-yellow, usually weakly suffused with grey, termen edged or narrowly suffused with ferruginous, costa sparsely strigulate and dorsum minutely dotted with ferruginous or fuscous; markings ferruginous-yellow varying to ferruginous-brown, bordered with slightly raised white or silver-white spots, often coalescing and forming nearly continuous striae; basal and sub-basal fasciae obsolescent, most pronounced on costa; median fascia moderately broad on costa, narrowing towards dorsum, interrupted below costa, obtusely angulate near middle and with outer margin dilated or produced distad, sometimes with grey or fuscous admixture on costa, along margins and in median fold; pre-apical spot well developed, triangular, reaching to apex; a large ferruginous patch or band in tornal area; a series of white or silver-white spots forming slightly sinuate striae in distal area, interrupting pre-apical spot and tornal patch; cilia pale yellow suffused with grey, sprinkled with ferruginous at base, with a dark grey sub-basal line. Hindwing grey or dark grey; cilia white mixed with grey, with a grey sub-basal line.

Female (fig. 6). Similar to male.

Variation. Although extremely variable in minor details of colour and pattern the forewing in this species usually retains its characteristic ochreous-yellow general coloration; the grey suffusion is sometimes sufficiently strong to subdue the ochreous-yellow ground colour, the fasciae then appearing very dark ferruginous or ferruginous-brown; in such dark-coloured specimens the whitish spots or striae may appear submetallic.

COMMENTS

This species differs from *A. piercei* by its smaller size, the lighter and often distinctly yellow general coloration of the forewing, the more obtusely angled median fascia, and the silver-white spots which coalesce and form continuous striae.

BIOLOGY

Little is known about the biology of this species in the British Isles, and on the Continent there appears to be some confusion with the closely related *A. piercei*. In Britain *hartmanniana* is usually found amongst *Knautia arvensis* and *Scabiosa columbaria*, but neither has been confirmed as a foodplant. The larva is assumed to live in the rootstock of the foodplant like that of *piercei*.

Pupa. April and May; probably in the larval habitation.

Imago. June to early August. Frequenting chalk downland and similar well-drained localities where the plants mentioned above grow; easily disturbed from rest during the day and flying freely at dusk.

DISTRIBUTION

Because of confusion with *A. piercei* the range and distribution of *hartmanniana* in the British Isles are not fully known. Available data indicate that the species is generally distributed and locally common in chalk and limestone districts in the southern counties of England from Norfolk and Kent to Somerset, Gloucestershire and Herefordshire; it also occurs in Lancashire and Westmorland, but more northerly records in England and Scotland are at present unconfirmed. In Wales it is widely distributed in the north, occurring from Flintshire to Caernarvonshire. Records from Ireland are unconfirmed.

North and central Europe to Asia Minor.

Aethes margarotana (Duponchel)

Pl. 24, fig. 7

[*Argyrolepia maritimana*; Barrett, 1905: 337. *Aethes maritimana*; Pierce & Metcalfe, 1922: 33, pl. 12 (♂, ♀ genitalia). *Phalonia maritimana*; Meyrick, 1928: 487; Ford, 1949: 43]

DESCRIPTION

♂ ♀ 14–18 mm. Sexual dimorphism moderately pronounced; antenna of male weakly dentate, strongly ciliate; female usually larger, hindwing less white.

Male (fig. 7). Forewing ground colour pale yellow with faint olive tinge, thickly strigulate with yellow, often weakly suffused with greyish, with scattered somewhat shining whitish spots or striae, margins obscurely dotted or strigulate with dull reddish brown, those along terminal and dorsal margins sometimes mixed with black; markings reddish brown with plumbeous admixture, poorly developed; basal and sub-basal fasciae obsolete; median fascia narrow, weak on costa and dorsum, angulate at middle, margins irregular and diffuse, usually roughened plumbeous scales on inner margin above and below median fold; a similar tuft below fold on inner margin of pre-tornal patch; pre-apical spot obsolescent, represented by striae, sometimes connected to a suffused pre-tornal patch on dorsum; cilia pale yellow, usually a dark grey, often interrupted, subapical line around apex to tornus. Hindwing white, heavily suffused and irrorate with dark brown, the irroration becoming weaker and mottled in the basal half; cilia white, with a dark brown sub-basal line.

Female. Similar to male; hindwing usually more strongly irrorate and less white basally and with cilia wholly grey around apex.

Variation. Minor variation is found in the development of the markings; weakly marked specimens are generally yellow in appearance and lack the olive tinge; strongly marked specimens are often suffused with grey and have a ferruginous admixture in the markings, and the plumbeous tufts may be more conspicuous. The brown irroration on the hindwing varies in both sexes; in the female the whole wing is sometimes heavily irrorate and appears almost unicolorous brown.

COMMENTS

Similar to *A. williana* but distinguished by its larger size and the yellower general appearance of the forewing, and in the male the presence of mottling on the hindwing.

BIOLOGY

Ovum. Usually deposited in a leaf axil in the lower part of the stem, but sometimes just below the flower, in July.

Larva. Head brown; prothoracic and anal plates concolorous with abdomen, anal plate dotted with brown or black; abdomen translucent, yellowish white or yellow, pinacula concolorous; peritreme of spiracles black. July to April, feeding on *Eryngium maritimum*; when the egg is deposited in the leaf axil, the larva at first burrows downwards into the stem and eats the pith; on reaching a node it may eat its way out and crawl down the outside of the stem to re-enter a little above ground level, then working its way down into the root; when the egg is deposited below the flower the larva first eats into the flower bracts, causing a slight discoloration which gradually extends as the larva works downwards in the stem, till the whole of the immature flowerhead becomes withered and brown. By the beginning of September the larvae are all in the roots, where they continue to feed until they are full-grown in October. The interior of the root is eaten out leaving only a thin shell or skin, and the larva overwinters in this situation until the end of March, when it works its way to the upper part of the root, where it hollows out a pupal chamber, pupating by early May. The life history of this species has been fully described and illustrated by Elisha (1891), who notes that the best time to collect larvae is from October onwards, and that it is important to get up the roots without injuring the mined portion. The roots are very brittle and of a light brown colour, the mined portion being black and discoloured, very soft and easily broken. Roots gathered—about 15 cm of root and 8 cm of stem—should be potted upright and left out of doors until the following spring.

Pupa. Yellowish brown. April and May, in a pupal chamber constructed in the rootstock of the foodplant. Following eclosion the pupal exuviae remain extruded from the crown of the rootstock (Elisha, 1891: 278, pl. 5, fig. 4).

Imago. June and early July. Frequenting coastal sandhills and shingle where sea holly is found. The moth is very sluggish and is seldom to be seen on the wing but is more often found sitting on the foodplant or close by on the sand or shingle.

DISTRIBUTION

In the British Isles this species is apparently restricted to coastal areas in Kent (Deal to Sandwich), Essex (Clacton, St. Osyth) and Suffolk (where it has been found in recent years (Aston, 1966: 160) at Aldeburgh, Felixstowe and Thorpeness). It has become rarer and is probably extinct in some of its old haunts such as Deal, Sandwich and Clacton.

Europe; North Africa; Asia Minor.

Aethes williana (Brahm)

Pl. 24, fig. 8

[*Argyrolepia zephyrana*; Barrett, 1905: 336. *Aethes dubrisana*; Pierce & Metcalfe, 1922: 33, pl. 12 (♂, ♀ genitalia). *Phalonia zephyrana*; Meyrick, 1928: 487; Ford, 1949: 43]

DESCRIPTION

♂ ♀ 10–16 mm. Sexual dimorphism moderately pronounced; antenna of male weakly dentate, strongly ciliate; female usually larger, hindwing darker.

Male (fig. 8). Forewing ground colour pale yellow strongly tinged with olive, or in dark specimens suffused with grey, with transverse series of shining white spots or striae; costa strongly strigulate or dotted with blackish brown, termen and dorsum lightly sprinkled or suffused with grey or black; markings fuscous mixed with plumbeous, with a ferruginous admixture; basal and sub-basal fasciae obsolete; median fascia narrow, usually interrupted below costa, angulate at middle, margins poorly defined, roughened plumbeous scale-tufts on inner and outer margins above and below median fold, a variable sprinkling of black scales in fold; pre-apical spot small, extending as a thick stria and usually confluent with a diffuse pre-tornal patch; cilia yellow, slightly darker basally. Hindwing white, suffused with brownish grey, weaker basally; cilia white, with a grey sub-basal line.

Female. Similar to male; hindwing usually wholly suffused with brownish grey and with cilia wholly grey around apex.

Variation. In the forewing the usually rather distinctive greenish tinge may be obscured by greyish suffusion. The strength of the brownish grey suffusion in the hindwing of the male varies; in specimens where it is weak the wing is distinctly white, the suffusion being restricted to the apical area.

COMMENTS

Smaller than *A. margarotana* and further distinguished by the more pronounced olive tinge of the forewing and the more conspicuous shining white transverse spots or striae.

BIOLOGY

Larva. Head varying from light to dark brown; prothoracic and anal plates pale yellow, anal plate dotted with minute black spots; abdomen yellow, segments deeply divided and ridged, integument shagreened; pinacula inconspicuous. August to April, living in the lower part of the stem of *Daucus carota*, not more than 10 to 12 cm from the ground. The larva enters the stem soon after hatching and bores downwards into the rootstock, feeding on the pith and filling the vacant space with frass. In the autumn exudation of frass from the lower part of the stem indicates the presence of a larva. On the Continent *Eryngium campestre*, *E. maritimum*, *Helichrysum stoechas* and *Gnaphalium*, and in Morocco *Ferula communis*, have been recorded as foodplants; in Italy the larva is injurious to cultivated carrots, boring into the roots from the rootstock (Balachowsky, 1966: 488).

Pupa. Pale yellow, wings and appendages thick, brilliantly glossy; abdomen dull pale yellow, with a frosted appearance (Barrett, 1905: 336). Late April to June; in a brownish cocoon in the rootstock. Prior to eclosion of the moth the pupa protrudes through a circular emergence hole previously prepared by the larva.

Imago. May to July and early August. Frequenting rough fields near woods, railway embankments and waste ground, especially on chalk downs and sandy areas. Flying actively in the evening and at sunset and during the day if the weather is warm and dry.

DISTRIBUTION

In the British Isles this species is apparently restricted to the southern counties of England, and is generally distributed and common in suitable localities from Suffolk and Kent westward to Devon, Monmouthshire (Chepstow) and Worcestershire. The record by Meyrick (1928: 487) from Cumberland is unconfirmed.

Central and southern Europe to Siberia; North Africa.

Aethes cnicana (Westwood)

Pl. 24, fig. 9

[*Argyrolepia cnicana*; Barrett, 1905: 328. *Phalonia cnicana*; Pierce & Metcalfe, 1922: 29, pl. 11 (♂, ♀ genitalia); Meyrick, 1928: 488; Ford, 1949: 43]

DESCRIPTION

♂ ♀ 14–17 mm. Sexual dimorphism not pronounced; antenna of male shortly ciliate; female with hindwing slightly darker.

Male (fig. 9). Forewing ground colour shining white, suffused with patches of ochreous-yellow or pale yellow, often mixed with grey; markings ferruginous-brown; basal and sub-basal fasciae represented by an elongate patch on costa; median fascia slightly inward-oblique, moderately well defined, angulate and constricted or narrowly interrupted above middle, margins of dorsal part of fascia variably edged with dark brown or fuscous; pre-apical spot well developed, subquadrate, emitting a thick greyish ochreous stria to tornus; a small triangular pre-tornal spot, often extended to middle or beyond; cilia pale yellow, darker basally. Hindwing grey; cilia paler, with a dark sub-basal line.

Female. Similar to male.

Variation. The normally lightly suffused ground colour varies considerably, and in some specimens may be almost clear shining whitish yellow with only a trace of ochreous suffusion, while in others it may be heavily suffused and appear distinctly pale yellow or ochreous-yellow, sometimes tinged with olive.

COMMENTS

Very similar to *A. rubigana*, but differing in the slightly smaller size and the comparatively narrow, less oblique and usually uninterrupted median fascia of the forewing.

BIOLOGY

Ovum. Deposited on the flower of the foodplant in July.

Larva. Head light brown; prothoracic plate pale yellow varying to light brown, divided medially, a few blackish dots near the middle of the mesal and lateral margins, posterior margin broadly edged with black; abdomen pale yellow or brownish white, with a weak greenish tinge; pinacula greyish brown or brown; anal plate pale yellow or light brown, unmarked; thoracic legs dark brown; prolegs concolorous with abdomen. September to April; feeding on various species of thistle, including *Cirsium vulgare*, *C. palustre* and *C. oleraceum*, eating first the seeds and later boring into the stem and consuming the pith; overwintering in the stem of the foodplant and pupating in the spring.

Pupa. Pale brown. May and June, in the dead stems of the foodplant.

Imago. End of May to July. Frequenting fields, waysides and waste places, especially marshy spots near woods. The moth is often disturbed from rest amongst the herbage during the day and with care can be boxed on a thistle; in the evening it may be netted flying sluggishly about its habitat and at night will come to light.

DISTRIBUTION

Widely distributed and generally common throughout the British Isles to the Orkneys and Shetlands, though more local in the north and less common in its Scottish haunts than in the south. In Ireland it is widely distributed but local.

Europe, occurring as far north as Lapland, to Siberia and Japan.

Aethes rubigana (Treitschke)

Pl. 24, fig. 10

[*Argyrolepia badiana*; sensu Barrett, 1905: 327. *Phalonia badiana*; sensu Pierce & Metcalfe, 1922: 29, pl. 11 (♂, ♀ genitalia); sensu Meyrick, 1928: 487; sensu Ford, 1949: 43]

DESCRIPTION

♂ ♀ 15–19 mm. Sexual dimorphism not pronounced; antenna of male shortly ciliate; female usually larger, hindwing darker.

Male (fig. 10). Forewing ground colour shining white, with patches of ochreous suffusion, often mixed with grey; markings ferruginous-brown; basal and sub-basal fasciae forming an elongate patch on costa; median fascia inward-oblique, well defined, angulate and interrupted above middle, dorsal part edged and partially suffused with dark brown or fuscous; pre-apical spot well developed, subquadrate, emitting a thick greyish ochreous stria to tornus; a usually strong triangular pre-tornal spot, sometimes extended to middle; cilia pale yellow, darker basally. Hindwing grey; cilia paler, with a dark sub-basal line.

Female. Similar to male.

Variation. The forewing ground colour varies from whitish yellow to ochreous-yellow; in darker specimens the areas suffused with ochreous often have a weak olive tinge.

COMMENTS

Distinguished from *A. cnicana* by its usually larger size and the broader, more oblique and almost invariably clearly interrupted, median fascia of the forewing.

BIOLOGY

Ovum. Deposited on the young burrs of the foodplant in early July.

Larva. Head light brown varying to dark brown, with irregular blackish markings posteriorly which are sometimes very heavy; prothoracic plate light brown varying to dark brown, with darker pigmented spots posteriorly; abdomen pale yellow faintly tinged with greenish, developing a reddish flush when the larva is full-grown; pinacula concolorous with abdomen, becoming light greyish brown in final instar; anal plate brown, sometimes speckled or mottled with dark brown or black; anal comb usually with three acutely pointed blackish prongs. September and October, living in seedheads of *Arctium*. The foodplant is generally considered to be greater burdock, *Arctium lappa*, but in the British Isles this plant has a more restricted range than the moth, and the larva must also feed on some other plant, probably burdock, *Arctium minus* agg., which occurs throughout the British Isles (Perring & Waters, 1962: 287). The larva feeds on the seeds (pl. 4, fig. 2) until full-grown at about the end of October, when it usually leaves the seedhead and spins a cocoon amongst ground debris, but occasionally it spins up in the seedhead, hibernating until the following spring before pupating. On the Continent the species is said to be bivoltine, and the larvae may be found from September to April in roots and stems of *Arctium lappa* and *Cirsium oleraceum*. The larvae producing the second generation may be found in the flowers and seedheads in June and July (Swatschek, 1958: 248). Hannemann (1964: 50) records the larva on *Cirsium vulgare*.

Pupa. Yellowish brown. May and early June, in a silken cocoon spun amongst ground debris near the foodplant. When reared in captivity the larva will sometimes spin up and pupate in the seedhead.

Imago. End of June to August; frequenting waysides and open places where burdock is found, but rarely in woods. Seldom seen during the daytime, being rather sluggish and retiring in habits, but on the wing at dusk and into the night, occasionally coming to light.

DISTRIBUTION

Widely distributed but somewhat local throughout the British Isles as far north as Perthshire and Aberdeenshire. Locally common in East Anglia and the southern counties of England to the Isles of Scilly, and in the west to Cumberland; uncommon in the Midlands, but more plentiful in Yorkshire and Durham. Local throughout North Wales and local but widely distributed in Ireland.

North and central Europe to Siberia and Japan.

Aethes smeathmanniana (Fabricius)
Pl. 24, fig. 11

[*Lozopera smeathmanniana*; Barrett, 1905: 322. *Aethes smeathmanniana*; Pierce & Metcalfe, 1922: 33, pl. 12 (♂, ♀ genitalia). *Phalonia smeathmanniana*; Meyrick, 1928: 487; Ford, 1949: 43]

DESCRIPTION

♂♀ 12–18 mm. Sexes similar; antenna of male ciliate.

Male (fig. 11). Forewing ground colour yellowish white, irregularly suffused with pale ochreous-yellow, sometimes extensively; terminal margin faintly edged with grey; markings ochreous varying to ochreous-brown and ferruginous-brown, often partially suffused with grey; basal and sub-basal fasciae obsolescent, sometimes indicated by a diffuse pale ochreous or greyish patch, narrowly infuscate on costa; median fascia weakly developed and strongly suffused with grey on costa, interrupted below, angulate before middle, thence strongly developed and inward-oblique to dorsum, often with dark brown admixture in median fold; a slender fasciate pre-tornal marking parallel with median fascia, reaching to near middle of wing below pre-apical spot, usually constricted or interrupted above tornus; pre-apical spot semi-ovate or subquadrate, usually greyish and comparatively weak; cilia whitish yellow, with an indistinct sub-basal line. Hindwing grey, paler basally; cilia paler, with a dark sub-basal line.

Female. Similar to male; hindwing often darker and more uniformly grey.

Variation. Considerable variation is found in the general coloration of the forewing, the ground colour sometimes being weakly suffused with ochreous, the wing then having a pronounced whitish appearance, or the suffusion may be strong, giving the wing a distinctly yellow general appearance.

COMMENTS

Readily distinguished from *A. cnicana* and *A. rubigana* by the narrower and more elongate forewing and the narrower median fascia which is more outward-oblique from the dorsum. In general appearance it somewhat resembles *Stenodes straminea* but is distinguished by the presence of the well-developed fasciate pre-tornal marking.

BIOLOGY

Larva. Head blackish brown or black; prothoracic plate divided medially, dark brown, posterior margin broadly blackish brown; abdomen dull green, with a slight pinkish tinge when full-grown; pinacula inconspicuous, concolorous with integument; anal plate blackish brown, dotted with black; anal comb with four or five somewhat flattened prongs; thoracic legs black; prolegs dark green. September to April and early May, feeding on *Achillea millefolium*, *Centaurea nigra* and probably other knapweeds, also *Anthemis arvensis*. The larva lives in the flower heads or seedheads of the foodplants, eating the seeds; on *Achillea* it lives in a tube of silk and seed-refuse spun among the flowers and seed capsules (pl. 4, fig. 3), where it overwinters, pupating in the spring. The presence of the species in a locality can be ascertained by the examination of seedheads during the winter when the larval spinnings are clearly visible.

Pupa. Abdomen dull reddish brown, wings and appendages black. May, in a silken cocoon amongst debris.

Imago. Apparently univoltine, with a prolonged period of emergence, occurring from the end of May to early August. Although this species is bivoltine on the Continent there is no definite evidence that two generations occur in the British Isles. Frequenting rough open ground, hill sides, chalk downs, railway banks and similar situations where yarrow and knapweeds abound. On warm sunny days the moth is readily disturbed from rest on the footplants and other herbage, flying freely in the early evening.

DISTRIBUTION

Local but widely distributed in Britain as far north as south-west Scotland to Perthshire and Fife; recorded from most of the southern and eastern counties of England, including the Isles of Scilly, to Oxfordshire and Norfolk, though apparently less common in recent years. Turner (1955) does not record this species from Somerset, but it may have been overlooked as it has recently been discovered in Gloucestershire at Woodchester Park (Newton, 1961: 87). Scarce in Cheshire, Lancashire and Yorkshire. In Wales it is known from Denbighshire, Caernarvonshire and Cardiganshire (Michaelis, 1966: 109). Apparently unknown from Ireland.

Europe to Asia Minor and central Asia; North America.

Aethes margaritana (Haworth)

Pl. 24, fig. 12

[*Lozopera dipoltella*; Barrett, 1905: 317. *Phalonia dipoltella*; Pierce & Metcalfe, 1922: 29, pl. 11 (♂, ♀ genitalia); Meyrick, 1928: 486; Ford, 1949: 43]

DESCRIPTION

♂ ♀ 12–16 mm. Sexual dimorphism not pronounced; antenna of male shortly ciliate; hindwing of female more uniformly grey.

Male (fig. 12). Forewing ground colour clear silver-white; markings well defined, deep ochreous, narrowly infuscate along costa; basal and sub-basal fasciae usually confluent on dorsum; median fascia bifurcate on costa, often darker in median fold; pre-apical spot isolated, or confluent with a slender postmedian streak and forming with it a complete transverse fascia parallel with median fascia; a variable tornal marking usually reaching postmedian fascia in middle but often fractured and incomplete; a well-developed subterminal streak, often bifurcate or fractured apically; cilia pale ochreous, with a darker sub-basal line. Hindwing grey, paler basally; cilia whitish, with a dark grey sub-basal line.

Female. Similar to male; hindwing more uniformly grey.

Variation. The general coloration of this species is very constant and although considerable minor variation is found in the development of the markings of the forewing, particularly in the distal area, the distinctive and characteristic multifasciate pattern is retained.

COMMENTS

A distinctive species, readily recognized by the characteristic silver-white ground colour and sharply contrasting deep ochreous markings of the forewing.

BIOLOGY

Ovum. Deposited on the flowerheads of the foodplant in July and August.

Larva. Head and prothoracic plate yellowish grey, the latter with an oblique, elongate blackish patch laterally, and with the posterior margin broadly edged with blackish; abdomen greenish grey, pinacula concolorous; anal plate brownish grey. September to April, feeding on the flowers and seeds of *Achillea millefolium*, living in a silken spinning, overwintering in the feeding place and pupating in the spring. Other foodplants are *Matricaria recutita* and *Chrysanthemum vulgare*.

Pupa. Reddish brown. May and June, spun up in debris on the ground; it has also been recorded in the seedheads of the foodplant but this probably only occurs when reared in captivity.

Imago. July and August. Frequenting open waste ground on chalk downs and shingle beaches; it has also been found rarely in clay districts. The moth is readily disturbed from rest among low-growing plants during the day and flies freely in the evening and at dusk.

DISTRIBUTION

Apparently restricted to the southern and south-eastern counties of England, occurring from Suffolk and Kent to Dorset. Known localities include Thorpeness (Suffolk), Deal, Dover, Folkestone, Margate, Sandwich, Sittingbourne and Walmer (Kent), Falmer, Moulsecombe and Pevensey (Sussex), Southampton and Sway (Hampshire), Freshwater (Isle of Wight), and Arne (Dorset). Formerly this species occurred in numbers near Southend (Essex) and at Guildford, Croydon and Box Hill (Surrey) but became scarce before the turn of the century (Barrett, 1905: 318).

Europe, including Norway and Sweden, to Asia Minor and central Asia.

Aethes dilucidana (Stephens)

Pl. 24, fig. 13

[*Lozopera dilucidana*; Barrett, 1905: 318. *Aethes dilucidana*; Pierce & Metcalfe, 1922: 33, pl. 12 (♂, ♀ genitalia). *Lozopera dilucidana*; Meyrick, 1928: 484; Ford, 1949: 42]

DESCRIPTION

♂ ♀ 12–15 mm. Sexual dimorphism not pronounced; antenna of male shortly ciliate; forewing of female with median fascia sometimes developed as a small isolated dot or bar on costa.

Male (fig. 13). Forewing ground colour pale yellow, edge of costa dotted with minute ferruginous-brown striae in basal half, termen narrowly edged with grey; markings ferruginous-brown delicately sprinkled with dark brown and plumbeous; basal and sub-basal fasciae obsolete; median fascia incomplete, very oblique and narrow, well developed from dorsum to beyond middle, obsolete in costal half except for indistinct bar on edge of costa; a well-developed, slender postmedian fascia parallel with median fascia, sometimes weakened below costa; cilia pale yellow, with a faint sub-basal line. Hindwing pale grey, whitish basally; cilia whitish.

Female. Similar to male; median fascia of forewing usually more strongly developed on costa.

Variation. This species shows little variation except in the forewing ground colour which varies from pale yellow to yellowish white.

COMMENTS

Distinguished from *A. francillana* and *A. beatricella* by having the median fascia of the forewing obsolescent in the costal half.

BIOLOGY

Ovum. Deposited singly on the flowers and seeds of the foodplant in July and August.
Larva. Head shining black; prothoracic plate divided medially, light brown anteriorly, black posteriorly; abdomen varying from yellowish white to yellowish brown, pinacula concolorous; peritreme of spiracles only slightly darker than integument; anal plate light brown, marked with black; thoracic legs dark brown or black. August and September, feeding on the seeds of *Pastinaca sativa*, spinning several seeds together and

66

eating out their interiors. When full-fed, usually before the end of September, the larva leaves the seedhead and descends part of the way down the stem, then bores into the interior and tunnels upwards through the pith until a node is reached, where it hibernates through the winter. In late spring the larva gnaws a passage towards the outside of the stem, leaving only a very thin skin entire, before pupating. During late autumn the presence of a larva in a stem is indicated by a small accumulation of whitish powder in a leaf axil (pl. 15, fig. 1). Sheldon (1888: 103) describes the life history of this species and Walsingham (1898: 76) records *Heracleum sphondylium* as an additional foodplant.

Pupa. Pale yellowish brown. May and June, in the stem of the foodplant.

Imago. July and August. Frequenting grassland in chalk and limestone areas where wild parsnip grows; it has also been taken in gravel pits. Towards sunset the moth may be seen flying gently about the foodplant; during the day it may be disturbed from rest by shaking or tapping the plants.

DISTRIBUTION

Locally common in England northwards to Lancashire and Yorkshire; in the southern counties it is apparently restricted to chalk and limestone localities. The exact distribution of this species is uncertain because of confusion with *A. francillana* but on the mainland it has been recorded as far west as Gloucestershire (Chapman's Cross and Tetbury), Dorset and Somerset (Clevedon), although Turner (1955: 126) does not confirm the last record. Records from the Isles of Scilly (Barrett, 1905: 319; Blair, 1925: 9) have not been confirmed. The species is apparently unknown from Wales and Scotland. Beirne (1941: 77) considers that the records from Howth and Sutton (Co. Dublin) are probably erroneous and referable to *francillana*.

Sweden; Sicily; North Africa; southern Russia; doubtfully recorded from Germany.

Aethes francillana (Fabricius)

Pl. 24, fig. 14

[*Lozopera francillana*; Barrett, 1905: 319. *Aethes francillonana*; Pierce & Metcalfe, 1922: 32, pl. 12 (♂, ♀ genitalia). *Lozopera francillana*; Meyrick, 1928: 484; Ford, 1949: 42]

DESCRIPTION

♂ ♀ 13–18 mm. Sexes similar; antenna of male shortly ciliate.

Male (fig. 14). Forewing ground colour pale yellow, basal half of costa edged with ferruginous-brown and dotted with minute fuscous-brown striae, termen sometimes very weakly suffused or dotted with ochreous or grey, dorsum sometimes similarly dotted; markings ferruginous-brown suffused with plumbeous, sometimes lightly, sprinkled with dark brown or blackish; basal and sub-basal fasciae obsolete; median fascia complete, narrow and very oblique, somewhat dilated on costa and dorsum, margins irregular; a well-developed postmedian fascia more or less parallel with median fascia; cilia pale yellow, a faint sub-basal line. Hindwing grey, paler basally; cilia whitish, a faint sub-basal line.

Female. Similar to male.

Variation. Little variation is found in this species except in the forewing ground colour which varies from whitish yellow to pale ochreous-yellow.

COMMENTS

Differing from *A. dilucidana* in having the median fascia of the forewing complete. Compared with *A. beatricella* the median and postmedian fasciae are narrower and more oblique, the costal end of the median fascia extending beyond the dorsal end of the postmedian.

BIOLOGY

Larva. Head dark brown or black; prothoracic plate yellowish brown, posterior margin narrowly dark brown, with mid-dorsal sulcus; abdomen yellowish white tinged with green; pinacula light grey; anal plate small, yellowish brown, darker posteriorly. September to April and May, on *Daucus carota*, living at first in spun flowers and seeds (pl. 5, fig. 2), then eating into the seeds; in the late autumn the larva leaves the seedhead and bores into the lower part of the stem, burrowing downwards through the pith, returning to the upper part of the stem in the spring before pupation a little below the entrance hole (pl. 5, fig. 3). Often several larvae may be found in a single stem; their presence is usually indicated by slight exudations of whitish frass from the entrance holes. The larva is also known to feed on *Peucedanum officinale*, which is found locally in Kent and Essex.

Pupa. Pale brown. June and July, in the stem of the foodplant.

Imago. Late June to early September, the peak period of emergence occurring in July and August. Frequenting rough grassland, quarries and hillsides, particularly on chalk downs and on the coast. During the day the moth flies up actively when disturbed from rest on wild carrot or other herbage, and from sunset onwards flies freely about the habitat.

DISTRIBUTION

Generally distributed and locally common in the coastal and inland counties of the south of England from Suffolk and Kent to Somerset, Devon, the Isles of Scilly (Mere, 1959: 109) and Worcestershire, becoming scarcer northwards, ranging to Lancashire, where it has apparently not been recorded since about 1890, Yorkshire (Spurn Head), and Northumberland (Bolam, 1929: 125). The only records from Scotland are Hawick (Guthrie, 1897: 344) and Jedburgh (Bolam, 1929: 125), both in Roxburghshire. In Wales it is known from Cardiganshire (Michaelis, 1966: 109) and Merionethshire. According to Beirne (1941: 77) the species is very common on the coast at Howth (Co. Dublin).

Central and south-east Europe to Asia Minor; North Africa; Canary Islands.

Aethes beatricella (Walsingham)

Pl. 24, fig. 15

[*Lozopera beatricella*; Barrett, 1905: 321. *Aethes beatricella*; Pierce & Metcalfe, 1922: 32, pl. 12 (♂, ♀ genitalia). *Lozopera beatricella*; Meyrick, 1928: 484; Ford, 1949: 42]

DESCRIPTION

♂ ♀ 14–17 mm. Sexual dimorphism not pronounced; antenna of male shortly ciliate; hindwing of female darker.

Male (fig. 15). Forewing ground colour pale ochreous-yellow, basal half of costa dotted with minute ferruginous-brown striae; markings ferruginous-brown darkened with blackish brown and sprinkled with plumbeous; basal and sub-basal fasciae obsolete except for a narrow band on costa; median fascia complete, moderately oblique, constricted or interrupted below costa; postmedian fascia well developed, parallel with median fascia, often slightly dilated in tornal and apical areas; cilia pale yellow, with an indistinct sub-basal line. Hindwing grey, slightly paler basally; cilia whitish, with a grey sub-basal line.

Female. Similar to male; hindwing darker.

Variation. Forewing ground colour and pattern very constant; a rare aberration occurs in which the ground colour is brownish ochreous.

COMMENTS

Distinguished from *A. francillana* by the broader and less oblique median and postmedian fasciae of the forewing.

BIOLOGY

Larva. Head black; prothoracic plate brown; abdomen varying from pale yellow to pinkish buff. August to April, on *Conium maculatum*, feeding first on the seeds until the autumn, then boring into the stem, preferring the upper part (pl. 6, fig. 1), and hibernating before pupating the following spring. The larvae appear to be somewhat gregarious and as many as twenty have been found in a single stem. It has also been found in the stems of *Smyrnium olusatrum* near Margate where *Conium* does not grow. This species was originally described from specimens believed to have been bred from larvae on *Pastinaca sativa* (Walsingham, 1898: 75), which was also erroneously recorded as the foodplant by Purdey (1899a; 1899b) and Barrett (1905: 321). Apparently Thurnall (1911: 260) was the first to show that the foodplant is *Conium maculatum*, and Fryer (1926: 114) and subsequent authors, with the exception of Razowski (1970: 363), have confirmed this.

Pupa. Light brown. May, in the stem of the foodplant.

Imago. June and July. Frequenting margins of fields and woods, hedgerows and waste ground where hemlock is found, the moth flying freely from late evening to sunset.

DISTRIBUTION

In the British Isles this species is locally common in the southern and south-eastern counties of England, and apparently is not found elsewhere although the foodplant is widespread. Known localities include Ipswich and Leiston (Suffolk), Chatteris (Cambridgeshire), Burnham-on-Crouch, Mucking, Stanford-le-Hope, Wanstead and Woodford (Essex), Dartford, Folkestone and Margate (Kent), Misterton, Shapwick,

Road and Steepholme Island (Somerset), and Claypits, Rodborough and Woodchester Park (Gloucestershire) (Fletcher & Clutterbuck, 1939: 298; 1945: 65).
Europe; North Africa.

Commophila aeneana (Hübner)
Pl. 24, fig. 16

[*Argyrolepia aeneana*; Barrett, 1905: 334. *Commophila aeneana*; Pierce & Metcalfe, 1922: 30, pl. 11 (♂, ♀ genitalia). *Euxanthis aeneana*; Meyrick, 1928: 498; Ford, 1949: 48]

DESCRIPTION

♂ ♀ 13–17 mm. Sexual dimorphism not pronounced; antenna of male weakly dentate, strongly ciliate; female usually larger.

Male (fig. 16). Forewing ground colour bright orange-yellow, a sprinkling of shining whitish scales in discal area, sometimes forming a small patch, and near middle of inner edge of median fascia, costa and to a lesser extent dorsum strigulate with black; markings black embossed with submetallic plumbeous spots; basal fascia developed medially as a small patch; sub-basal fascia obsolete; median fascia broad, obsolescent on costa, irregularly interrupted above and below middle, usually forming three more or less distinct longitudinal dashes terminating in plumbeous spots; a broad diffuse subterminal fascia, irregularly spotted with plumbeous, some of the spots coalescing to form thick striae; cilia concolorous with ground colour of wing basally, paler apically, dark grey at tornus. Hindwing dark brown; cilia yellow, suffused with grey along inner margin, with a dark brown sub-basal line.

Female. Similar to male.

Variation. Minor variation is found in the depth of the ground colour and strength of the markings of the forewings. In heavily marked specimens with dark general coloration the median fascia may be more or less solid though usually rather diffuse.

COMMENTS

The bright orange-yellow ground colour and submetallic markings of the forewing are characteristic.

BIOLOGY

Larva. Head varying from light to dark brown; prothoracic plate pale yellow; abdomen slender, yellowish or greyish white; pinacula shining greyish white; peritreme of spiracles blackish brown; anal plate greyish white marked with yellow. September to April, on *Senecio jacobaea*, living in the rootstock (pl. 6, fig. 2). When full-fed in the autumn the larva burrows up the stem to about 8 cm above ground level and, after partially gnawing through the walls of the stem, seals the top of the burrow with silk. Here it overwinters before pupating the following spring, the weakened upper part of the stem having in the meantime broken off. Meyrick (1928: 498) records the larva on *Senecio paludosus*, a plant which is now extinct in the British Isles (Perring & Waters, 1962: 272).

Pupa. Pale brown. May, in the larval habitation.

Imago. Late May to early July. Frequenting waste ground, waysides and railway embankments. During the day the moth rests low down amongst the ragwort and surrounding herbage and is rarely disturbed except when the sun is shining. At sunset it sits on the tops of the plants, and takes short flights, usually keeping close to the ground.

DISTRIBUTION

Local and scarce in the British Isles, apparently occurring only in the southern half of England, preferring localities on heavy clay. Known localities include Lincoln (Lincolnshire), Lowsonford and Offchurch (Warwickshire), Aston Wold (Northamptonshire), Warboys (Huntingdonshire), Benfleet, Eastwood, Laindon, Mucking, Romford and Thames Haven (Essex), Chattenden, Folkestone, Herne Bay, St. Mary Marsh and Strood (Kent), Lewes (Sussex), Box Hill, Caterham, Haslemere, Guildford, Limpsfield and Oxted (Surrey), and Yarmouth (Isle of Wight). There is no record of this species from Scotland and the origin of two specimens labelled "Aberdeen, 1910 Purdey Coll." in the British Museum (Natural History) is doubtful.

West and central Europe.

Eugnosta lathoniana (Hübner)

This south European species was recorded as British by Haworth (1811: 402), who states that he had seen one example. Stephens (1834: 176) refers to Haworth's record and records a pair of specimens believed to have been taken near Tunbridge Wells, Kent, in July, 1831. Wood (1839: 164, pl. 37, fig. 1120n) illustrated a specimen and refers to Stephens' record.

There are no further records of *lathoniana* from the British Isles and its occurrence here was probably adventitious.

Eupoecilia angustana (Hübner)

Pl. 25, figs. 1–3

[*Eupoecilia angustana*; Barrett, 1905: 289. *Clysia angustana*; Pierce & Metcalfe, 1922: 26, pl. 10 (♂, ♀ genitalia). *Euxanthis angustana*; Meyrick, 1928: 497; Ford, 1949: 48]

Throughout most of its range in the British Isles this species is represented by two ecotypes; in the Shetlands a geographically distinct subspecies occurs.

DESCRIPTION

♂ ♀ 10–15 mm. Sexual dimorphism moderately pronounced; antenna of male shortly ciliate, hindwing pale grey, whitish basally, or uniformly grey; hindwing of female uniformly dark grey.

E. angustana angustana (Hübner) (open woodland and downland form).

Male (fig. 1). Forewing ground colour cream-white, basal half of costa narrowly suffused with fuscous; markings ochreous; basal and sub-basal fasciae poorly defined, usually forming a diffuse patch, strongest on costa and outwardly; median fascia moderately broad, diffuse, extensively overlaid with plumbeous, thickly strigulate with black on dorsum, an admixture of ferruginous-brown above dorsum, a black dash or stigma in median fold; pre-apical spot represented by costal striae; a small usually isolated postmedian patch; a comparatively large, irregular patch above tornus; pre-tornal spot indicated by a black dot; a rather broad terminal streak from apex, tapering towards tornus; cilia pale ochreous mixed with grey, cream-white on dorsum before tornus, with a dark grey sub-basal line. Hindwing white, suffused and obscurely strigulate with fuscous; cilia whitish, suffused with grey along termen, with a dark grey sub-basal line.

Female. Similar to male; hindwing uniformly dark grey.

Variation. Minor variation is found in the markings of the forewing and in the shade of the ground colour, which in some specimens appears distinctly yellow.

E. angustana angustana f. *fasciella* Donovan (heathland and moorland form).

Male (fig. 2), *female*. Somewhat smaller than the nominate form and with the forewing ground colour distinctly white, the markings often lacking the ferruginous-brown admixture in the median fascia.

Variation. Showing minor variation in forewing markings and ground colour as in the nominate form, but retaining its characteristic lighter appearance.

E. angustana thuleana Vaughan (Shetland subspecies).

Male (fig. 3), *female*. Wings narrower and with apices more produced than in the mainland forms. The forewing markings are obsolescent, being reduced to brownish and ferruginous-brown suffusions; the ground colour is more or less evenly suffused with greyish ochreous except for a delicate transverse crescent-shaped area beyond the cell; cilia greyish ochreous, with a darker sub-basal line. Hindwing uniformly dark grey in both sexes.

Variation. This distinct subspecies is very constant in general coloration and forewing pattern, but intermediates approaching the nominate form occasionally occur.

COMMENTS

The two mainland forms of this species are readily distinguished from *E. ambiguella* by the narrower and relatively diffuse median fascia of the forewing and the presence of a terminal fascia. The attenuate wings and obsolescent markings of the forewing are characteristic of ssp. *thuleana*.

BIOLOGY

Larva. Head varying from light to dark brown; prothoracic plate yellowish brown varying to dark brown; abdomen whitish tinged with pink, or yellowish brown in larvae on *Calluna*. July to September and October, feeding in spun flowers and seeds of the foodplants, overwintering in the spinning and pupating *in situ* the following spring. The larva of the nominate form feeds on *Plantago*, *Achillea*, *Origanum*, *Solidago* and

Thymus, that of f. *fasciella* on *Calluna*. In Scotland the larva has been found on *Picea sitchensis* at Caithness, but this is apparently exceptional. The biology of ssp. *thuleana* in the Shetlands appears to be unknown.

Pupa. Shining dark brown. April and May, in the larval habitation.

Imago. Apparently univoltine, with a prolonged period of emergence lasting from June to September. The nominate form frequents margins of woods, open woodland, meadows and downland where plantain, yarrow, marjoram, golden rod and thyme grow, and in these biotopes flies at dusk; f. *fasciella*, considered a distinct ecotype associated with *Calluna*, occurs on heathland and moorland habitats, and flies in abundance on warm afternoons over the ling. In the Shetlands the subspecies *thuleana* inhabits rocky coasts.

DISTRIBUTION

Widely distributed and common throughout the British Isles, including the Outer Hebrides, Orkneys and Shetlands.

Europe and Asia Minor to China and Japan.

Eupoecilia ambiguella (Hübner)

Pl. 25, fig. 4

[*Eupoecilia ambiguella*; Barrett, 1905: 287. *Clysia ambiguella*; Pierce & Metcalfe, 1922: 26, pl. 10 (♂, ♀ genitalia); Meyrick, 1928: 495; Ford, 1949: 47]

DESCRIPTION

♂ ♀ 12–15 mm. Sexual dimorphism moderately pronounced; antenna of male shortly ciliate, hindwing paler basally; female usually larger, hindwing more uniformly grey.

Male (fig. 4). Forewing ground colour pale ochreous-white, extensively suffused and strigulate with yellow-ochreous, costa and dorsum strigulate or dotted with black, costa narrowly suffused with fuscous-black from base to beyond middle (outer margin of median fascia); markings dark grey; basal and sub-basal fasciae obsolescent, indicated only by costal striae; median fascia conspicuous, inward-oblique, dilated on costa, diffusely strigulate or mixed with blackish, a variable ferruginous admixture above and below median fold, sometimes extending to dorsum; pre-apical spot obsolescent, indicated by weak costal striae; apex variably suffused with blackish; a small diffuse blackish spot in terminal margin near middle; cilia pale ochreous, a dark sub-basal line. Hindwing grey, paler basally; cilia light grey, with a dark sub-basal line.

Female. Forewing pattern similar to that of male; hindwing grey.

Variation. The general coloration and pattern of the forewing is very constant, showing only slight variation in the intensity of the markings.

COMMENTS

A distinctive species, readily recognized by the pale ochreous general coloration of the forewing and the contrasting heavy median fascia.

BIOLOGY

Ovum. Opalescent, greyish brown when first laid, later becoming spotted with bright orange. Oviposition usually takes place on the flowers of the foodplant in June.

Larva. Head, prothoracic plate and thoracic legs dark brown or shining black; abdomen shagreened, varying from olive-green to reddish brown or brownish yellow; pinacula brown, large and moderately prominent; anal plate varying from brown to yellow. July and August, feeding on the berries of *Frangula alnus*, living within a berry and eating the pulp and seeds, usually attaching the berry to a leaf (pl. 6, fig. 3). In later instars the larva may move to a fresh berry or join two or three berries together with a silken tube. When full-grown in the autumn the larva leaves the feeding place and constructs a case from leaf fragments, either amongst debris on the ground or attached to a stem of the foodplant, overwintering in the case, but sometimes pupating in the autumn (Barrett, 1881: 153). Other foodplants of this species recorded by Ford (1949: 47) are *Hedera*, *Thelycrania sanguinea* and *Lonicera*. On the Continent, where this species has two or more generations, the larva is sometimes a pest on grape vine (*Vitis vinifera*). In the spring the larva spins the flowers and flowerheads of the vine and is known as "Heuwurm"; in late summer the larva lives in the fruits and is known as "Sauerwurm" (Swatschek, 1958: 227). The larva is also polyphagous on various trees and shrubs, including *Prunus* and *Ribes*. Balachowsky (1966: 461) describes the biology of this species in detail.

Pupa. Reddish brown. April and May, the larva generally overwintering before pupating, but sometimes pupation occurs in the autumn, in a leaf-case amongst ground debris or attached to a stem of the foodplants.

Imago. End of May and June. Frequenting moist heaths and scrubland, limestone scrub and open woods where alder buckthorn occurs. The moth may be beaten from rest amongst the foliage of the foodplant during the day; at dusk it may be netted flying round the blossom.

DISTRIBUTION

Local and scarce, occurring only in England from Essex and Kent to Dorset, and in Wales in Glamorgan (Llantrissant). Known localities include Tilgate Forest (Sussex), Durfold, Haslemere, Holmwood and Hindhead (Surrey), and the New Forest, Romsey and Southampton (Hampshire).

According to Beirne (1941: 78) records of this species from Ireland are erroneous and refer to *E. angustana*.

Widespread in the temperate zones of the Palaearctic and Indo–Oriental regions.

Cochylidia implicitana (Wocke)

Pl. 25, fig. 6

[*Eupoecilia implicitana*; Barrett, 1905: 310.—*implicitana*; Pierce & Metcalfe, 1922: 34, pl. 12 (♂, ♀ genitalia). *Phalonia implicitana*; Meyrick, 1928: 490; Ford, 1949: 45]

DESCRIPTION

♂ ♀ 10–13 mm. Sexual dimorphism not pronounced; antenna of male shortly ciliate; hindwing of female more uniformly darker grey.

Male (fig. 6). Forewing ground colour greyish white, tinged with cream-yellow in basal half, weakly flushed with pink distally, diffusedly strigulate with fuscous; markings olive-brown; basal and sub-basal fasciae obsolescent, forming a weak diffuse patch, strongest on costa; median fascia oblique and well developed from dorsum to beyond middle, obtusely angulate below costa and reduced to a greyish bar or spot; pre-apical spot obsolescent, indicated by greyish costal striae, confluent with a fasciate suffusion from tornal area; a black dot-like pre-tornal marking; cilia cream-yellow, mixed with grey at tornus, with a well-defined dark grey or blackish sub-basal line. Hindwing greyish white indistinctly strigulate with grey; cilia whitish, with a grey sub-basal line.

Female. Similar to male; hindwing more uniformly grey, usually darker.

Variation. The markings show only minor variation in strength; the extent and intensity of both the cream and the pink suffusions vary and in some specimens the latter may be absent.

COMMENTS

A slightly larger species than *C. heydeniana*, and usually readily distinguished by the presence of a weak pinkish flush to the forewing; differing further in the relatively slender median fascia and unmarked cilia. Similar to forms of *Falseuncaria degreyana* which lack the pink flush of the forewing but readily recognized by the conspicuous and well-defined sub-basal line in the cilia.

BIOLOGY

Larva. Head brown varying to light brown, darker posteriorly; prothoracic and anal plates yellowish, prothoracic plate with medial sulcus, each half marked with black on the postero-medial margin; abdomen whitish yellow, becoming pinkish white when full-grown, pinacula concolorous; thoracic legs grey. June to early September, over-wintering until the following April and early May; living in the stems and shoots, eating the pith, also in the flowers and seedheads, of *Matricaria recutita*, *Tripleurospermum maritimum* ssp. *inodorum*, *Anthemis cotula* and *Solidago virgaurea*. According to Barrett (1905: 311) and Ford (1949: 45) the larva may be found from May to October in three successive generations, but H. C. Huggins (*in litt.*) states that in his experience this species is univoltine and that the larvae occur in August and early September on *Solidago* growing in woods, overwinter, and pupate the following spring, the moths emerging in June. In drier situations such as railway embankments and open land the larva shows a preference for *Matricaria* and *Anthemis*. It is possible that such ecological and biological differences influence the times of appearance of the imago. On the Continent *Achillea*, *Chrysanthemum*, *Aster*, *Gnaphalium* and *Artemisia campestris* have been recorded as additional foodplants.

Pupa. April to June, in the larval habitation or spun up amongst ground debris.

Imago. Apparently univoltine, occurring from May to July and August, with a prolonged period of emergence. Frequenting wayside verges, banks and rough wasteland where scentless chamomile, wild chamomile and stinking chamomile grow; also in open woodland where golden rod flourishes. The moth rests amongst the foodplants and other herbage during the day, and is most easily disturbed in the afternoon, flying towards sunset and later coming to light.

DISTRIBUTION

In the British Isles this species appears to be almost entirely confined to the south of England, where it is local but widely distributed in the south-eastern and southern counties from Essex and Kent to Somerset and Herefordshire. The most northerly record known is from Walney Island (Lancashire), where it was taken in 1955 by N. L. Birkett (1955: 331). In Wales known only from Bangor (Caernarvonshire); apparently unknown from Scotland. Beirne (1941: 77) considers the single record from Belfast (Co. Antrim) to be unreliable.

Europe to central Asia; North Africa.

Cochylidia heydeniana (Herrich-Schäffer)

Pl. 25, fig. 6

[*Eupoecilia erigerana*; Barrett, 1905: 314.—*sabulicola*; Pierce & Metcalfe, 1922: 35, pl. 12 (♂, ♀ genitalia). *Phalonia sabulicola*; Meyrick, 1928: 491; Ford, 1949: 46]

DESCRIPTION

♂♀ 9–12 mm. Sexes similar; antenna of male shortly ciliate; hindwing of female darker.
Male. Forewing ground colour white, extensively suffused with grey, forming weak strigulae, costa marked with fuscous striae; markings ochreous-brown varying to olive-brown, weakly irrorate or suffused with fuscous; basal and sub-basal fasciae obsolescent, forming a diffuse patch, strongest on costa; median fascia moderately broad, oblique from dorsum to beyond middle, obtusely angulate below costa and reduced to a dark grey spot, often weakened or constricted below angle, an admixture of blackish suffusion in median fold and on dorsum; pre-apical spot obsolescent, indicated by fuscous striae on costa, emitting two diffuse fasciate striae to tornal area; a small black triangular pre-tornal marking; cilia pale ochreous, greyish at tornus and some indistinct greyish bars along termen, with a blackish sub-basal line. Hindwing grey; cilia paler, with an indistinct sub-basal line.
Female (fig. 6). Similar to male; hindwing darker.
Variation. Showing only minor variation in general coloration and markings.

COMMENTS

A usually smaller species than *C. implicitana*, lacking the pink flush to the forewing and further distinguished by the relatively broad median fascia and the barred cilia.

BIOLOGY

Larva. Head brown; abdomen ivory white. June and early July, and another generation in late August and September; on *Erigeron acer* and *Conyza canadensis*. Larvae from the first generation of moths live in the flowers and seedheads; a seedhead compressed at the tip with some seed-down protruding indicates the presence of a larva (pl. 7, fig. 1); larvae from the second generation of moths live in the central shoot, eating out the immature seeds from the seedhead and drawing together some of the florets above, or in the shoot, stunting the growth, hibernating fully grown in the seedhead and pupating *in situ* the following spring.

Pupa. Glossy light brown, very delicate. Late April and May; July; both generations in a slight cocoon in the larval feeding place.

Imago. Bivoltine, the first generation emerging in June, the second in late July and early August. Frequenting rough grassland, waysides, walls, banks and dunes where blue fleabane and the naturalized Canadian fleabane are found. During the day the moth is easily disturbed from rest amongst low vegetation; in the evening it flies actively about the foodplants.

DISTRIBUTION

In the British Isles this locally common species is known only from England, occurring from East Anglia to Oxfordshire and Sussex, and in Cheshire. Known localities include Merton (Norfolk), North Shoebury (Essex), Bexley and Shoreham (Kent), Limpsfield, Mickleham and Westcott (Surrey), Tilgate (Sussex), and Birkenhead and Hoylake (Cheshire) (Michaelis, 1954a: 57; 1954b: 53).

Europe to Asia Minor.

Cochylidia subroseana (Haworth)

Pl. 25, fig. 7

[*Eupoecilia subroseana*; Barrett, 1905: 309. — *subroseana*; Pierce & Metcalfe, 1922: 35, pl. 13 (♂, ♀ genitalia). *Phalonia subroseana*; Meyrick, 1928: 490; Ford, 1949: 45]

DESCRIPTION

♂♀ 11–15 mm. Sexes similar; antenna of male shortly ciliate.

Male (fig. 7). Forewing ground colour yellowish white or cream, suffused with pale ochreous along costa and distally, costa shortly strigulate and dorsum indistinctly spotted with fuscous or black; markings pale olive-brown variably suffused with grey; basal and sub-basal fasciae indeterminate; median fascia moderately broad, slightly outward-oblique from dorsum, diffuse and more greyish towards costa, thinly edged with blackish dorsally; an indistinct pale postmedian patch on dorsum or slightly above; pre-apical spot indicated by blackish costal striae, confluent with diffuse sinuous ferruginous-ochreous striae in tornal and terminal areas, these interspersed with plumbeous; a moderately distinct blackish pre-tornal marking; cilia pale yellow-ochreous, with a dark grey sub-basal line. Hindwing grey; cilia whitish, with a darker sub-basal line.

Female. Similar to male.

Variation. Minor variation is found in the general coloration and markings of the forewing; in some specimens the outer margin of the median fascia is more broadly bordered with clear ground colour and is better defined, especially towards the costa; the plumbeous coloration between the striae in the distal area may be stronger and combine with the striae to form fasciate markings.

COMMENTS

Compared with *C. rupicola* the forewing of this species is more angulate apically and the median fascia is narrower.

BIOLOGY

Ovum. Deposited singly on the flowerhead of the foodplant in July.

Larva. Head varying from light to dark brown; abdomen deep yellow. August to October, feeding on *Solidago virgaurea*, living in the flowers, retarding their development and eating the unripe seeds. Towards the end of September or in early October the full-grown larva leaves the flowerhead and spins a cocoon in debris on the ground, pupating the following spring.

Pupa. May and June; spun up amongst ground debris.

Imago. June to early August. The moth frequents woods, especially those which are coppiced, with grassy clearings where golden rod grows; on the wing in the evening and easily netted.

DISTRIBUTION

In the British Isles this species is known only from the south-eastern and southern counties of England, and is locally common in suitable woodland habitats from Essex and Kent to Hampshire, Somerset and Gloucestershire. Records from Aberdeenshire and Kincardineshire (Reid, 1893b: 29) are unconfirmed and probably erroneous, and most likely result from misidentification of rose-coloured forms or narrow-banded examples of *Falseuncaria ruficiliana*.

Northern Europe to Asia Minor; China and Japan.

Cochylidia rupicola (Curtis)

Pl. 25, figs. 8, 9

[*Eupoecilia rupicola*; Barrett, 1905: 303. — *rupicola*; Pierce & Metcalfe, 1922: 35, pl. 13 (♂, ♀ genitalia). *Phalonia rupicola*; Meyrick, 1928: 490; Ford, 1949: 45]

DESCRIPTION

♂ ♀ 11–15 mm. Sexes similar; antenna of male weakly dentate-ciliate.

Male (fig. 8). Forewing ground colour yellowish white extensively suffused with ochreous and grey, especially distally, costa obscurely strigulate; markings olive-brown variably suffused with grey; basal and sub-basal fasciae indeterminate, forming a diffuse patch, strongly suffused with grey on costa, outer margin angulate below costa; median fascia broad, slightly outward-oblique from dorsum, its outer edge shallowly concave, reduced and darkened with blackish grey suffusion towards costa, obtusely angulate below, margins often darkened with grey dorsally; a pale postmedian patch on dorsum or slightly above; pre-apical spot indicated by costal striae, produced as a narrow sinuous fasciate stria to tornus; a more or less parallel, slightly broader fascia from before tornus to costa, sometimes diffuse and dilated or produced medially to middle of median fascia; a moderately distinct triangular blackish pre-tornal marking; cilia pale yellow-ochreous, with a darker sub-basal line. Hindwing dark grey; cilia whitish, with a dark grey sub-basal line.

Female. Similar to male.

Variation. In some specimens the greyish suffusion of the forewing markings tends to be more extensive and the markings may appear black, the pale ground colour being extensively obscured (fig. 9).

COMMENTS

The comparatively broad forewing and the rounded apex are rather characteristic of this species. Distinguished from *C. subroseana* by the broader median fascia and the well-developed postmedian and subapical markings.

BIOLOGY

Larva. Head light brown; prothoracic plate yellowish; abdomen shagreened, whitish yellow, tinged with pink dorsally. August to October, on the flowers and seeds of *Eupatorium cannabinum*. When full-fed towards the end of September or in October the larva leaves the feeding place and constructs a cocoon in a short broken stump of dead stem of the foodplant, pupating *in situ* the following spring. Warren (1883: 17) found the larva in old broken and rotten pieces of stems which were lying prostrate on the ground and covered over with moss and debris. On the Continent *Crinitaria linosyris* and *Lycopus europaeus* are also recorded as foodplants (Swatschek, 1958: 235).

Pupa. May and early June; in a dead stem. The pupae may be collected by gathering the short broken stems or stumps of the previous year's growth of the foodplant.

Imago. June and July. Frequenting a variety of habitats, such as chalk cliffs, fens and marshes in East Anglia, and lanes, waysides and banks of streams in the south-west. On warm days the moth flies in the afternoon but its normal flight occurs at dusk. During the day a bee-smoker applied to clumps of hemp agrimony, especially in the drier coastal localities, will effectively disturb the moth from rest.

DISTRIBUTION

Widely distributed and locally common in southern England, occurring northwards to Westmorland and Durham. Known localities include Merton and Sturston (Norfolk), Chippenham and Wicken (Cambridgeshire), Shoebury (Essex), Deal, Ham Street, Folkestone and Kingsdown, near Deal (Kent), Hindhead (Surrey), Sea View (Isle of Wight), Bloxworth, Corfe, Portland, Studland and Swanage (Dorset), Abbots Leigh, Batcombe and Brislington (Somerset), Redruth (Cornwall), Bristol and Painswick

(Gloucestershire), Bidston Marsh and Wirral (Cheshire), and Hawswater, near Silverdale (Lancashire). In Wales the species is common in Pembrokeshire; it is unknown from Scotland. The only known locality from Ireland is the Burren, Co. Clare (Bradley & Pelham-Clinton, 1967: 136).

Europe to southern Russia; Asia Minor.

Falseuncaria ruficiliana (Haworth)

Pl. 25, figs. 10, 11

[*Eupoecilia ciliella*; Barrett, 1905: 312. — *ciliella*; Pierce & Metcalfe, 1922: 34, pl. 12 (♂, ♀ genitalia). *Phalonia ciliella*; Meyrick, 1928: 491; Ford, 1949: 45]

DESCRIPTION

♂♀ 11–15 mm. Sexual dimorphism not pronounced; antenna of male weakly dentate-ciliate; female with hindwing slightly darker.

Male (fig. 10). Forewing ground colour white suffused with grey, diffusedly strigulate with ferruginous-ochreous, especially in basal and distal areas, terminal margin broadly mixed with grey or blackish grey; basal and sub-basal fasciae represented by strigulation; median fascia varying from dark chestnut-brown to ochreous-brown, moderately broad, oblique from dorsum and edged with whitish ground colour, obsolescent on costa, mixed with grey especially on dorsum; pre-apical spot obsolescent, weakly indicated by several grey striae on costa; a small black-brown pre-tornal spot on dorsum; cilia yellow-ochreous, with a brownish sub-basal line. Hindwing light grey, darker apically and along margins; cilia whitish grey, with a grey sub-basal line.

Female. Similar to male; hindwing uniformly grey.

Variation. Considerable variation is found in the general coloration of the forewing which varies from whitish ochreous to reddish ochreous and in darker specimens may be tinged with purplish (fig. 11).

COMMENTS

The lack of rose coloration in the forewing and the comparatively broad whitish-edged median fascia distinguish this species from *F. degreyana*.

BIOLOGY

Larva. Head and prothoracic plate shining black, in early instars the head may be brown; abdomen white or yellowish white, becoming tinged with pinkish when full-grown; pinacula inconspicuous, concolorous with integument; anal plate small, light brown; anal comb absent. July and August, living in the seed capsules of *Primula veris* and *Pedicularis sylvatica*, feeding on the seeds. In northern and western Ireland, where the species occurs on bogs, it almost certainly feeds solely on *Pedicularis*, which is probably the foodplant on heaths. When full-grown at the end of August the larva leaves the seed capsule, spins up in ground debris, usually in a hollow stem or in rotten wood, and overwinters until the following spring before pupating. The larva has also been found on *Solidago*, and on the Continent, *Inula*, *Linum vulgare*, *Aster*, *Primula vulgaris*,

P. farinosa, *Bellis*, *Antirrhinum* and *Gentiana* are also recorded as foodplants.

Pupa. Light brown, with darker brown wings. April and May, spun up in debris on the ground.

Imago. May and June. The moth is found in various biotopes; in England it is most frequently found in chalk and limestone districts where cowslip occurs and, to a lesser extent, on damp heaths where lousewort grows; in Norfolk it flies commonly on heathland and may often be disturbed from rest on heather, while in Scotland and Ireland it frequents mainly bogs and moorland.

DISTRIBUTION

Widely distributed throughout the British Isles to the Orkneys and Shetlands, but very local though often common where it occurs. In the north of England it occurs most commonly in limestone districts; in Scotland it is common on moorlands but is most abundant in the west. Recorded from Breconshire (Chalmers-Hunt, 1969: 42) and Glamorgan in South Wales but it is more abundant in the north.

Europe to Asia Minor and central Asia.

Falseuncaria degreyana (McLachlan)

Pl. 25, figs. 12, 13

[*Eupoecilia degreyana*; Barrett, 1905: 306.—*degreyana*; Pierce & Metcalfe, 1922: 34, pl. 12 (♂, ♀ genitalia). *Phalonia degreyana*; Meyrick, 1928: 491; Ford, 1949: 45]

DESCRIPTION

♂ ♀ 12–14 mm. Sexual dimorphism not pronounced; antenna of male shortly ciliate; female with hindwing slightly darker.

Male (fig. 12). Forewing ground colour greyish white, variably suffused with pink along costa and in distal half; basal and sub-basal fasciae obsolete or indicated only on costa by weak strigulae; median fascia varying from ochreous-brown to chestnut-brown, narrow, oblique from dorsum, weak or obsolescent from middle to costa, usually darkened with grey admixture on dorsum; pre-apical spot obsolescent, weakly indicated by grey striae on costa; a small blackish pre-tornal spot on dorsum; cilia ochreous, with a brownish sub-basal line. Hindwing light grey; cilia paler, with a grey sub-basal line.

Female. Similar to male; hindwing uniformly grey.

Variation. In both sexes the strength of the characteristic pink flush of the forewing is variable; it is usually evident even in the palest specimens whilst in others it may be very pronounced (fig. 13).

COMMENTS

This species may be distinguished from *F. ruficiliana* by the delicate pink flush of the forewing and the distinctly whitish-edged median fascia. *F. degreyana* resembles *Cochylis roseana* and *C. flaviciliana* in having the forewing flushed with pink. It differs from *roseana* by the lack of a dark subapical line in the cilia of the forewing, and from *flaviciliana* by the ground colour of the forewing which is more subdued and greyish white compared with cream-white.

BIOLOGY

Larva. Head brown; prothoracic plate pale yellow, nearly concolorous with the abdomen, with a brown or black marking laterally on the posterior margin; abdomen pale yellow; pinacula small and inconspicuous; anal plate concolorous with abdomen (Warren, 1887b: 134). Larvae from the first generation of moths may be found in June feeding on the flowers of *Linaria vulgaris* and *Plantago lanceolata*, those from the second generation of moths in August and September in the ripening seed capsules of *Linaria* and the seedheads of *Plantago* (pl. 7, fig. 2). Larvae feeding on *Linaria* usually leave the flowerheads and seed capsules and pupate in the ground or amongst debris, but those on *Plantago* will spin up and pupate in the seedhead. Larvae occurring in August and September hibernate through the winter and pupate the following spring.

Pupa. May and July; in the larval habitation on *Plantago*, or in a silken cocoon spun amongst ground debris, or in the earth.

Imago. Bivoltine, May and early June; July and early August. Frequenting localities where the vegetation is sparse, especially roadside verges, margins of fields and heathland where the ground has been recently disturbed and colonized by toadflax and plantain. The moth is on the wing in the evening towards sunset.

DISTRIBUTION

In the British Isles this species is local and rare. It appears to be almost entirely restricted to East Anglia, where it occurs in sandy areas such as the Breckland and at Croxton, Denton, East Wretham, Merton and Thetford (Norfolk), near Wicken (Cambridgeshire), and Brandon (Suffolk). It has been recorded from Somerset (Portishead), but Turner (1955: 127) considers it to be very rare or possibly extinct.

North and central Europe to central Asia and Mongolia.

Cochylis roseana (Haworth)

Pl. 25, fig. 14

[*Eupoecilia roseana*; Barrett, 1905: 304. *Cochylis roseana*; Pierce & Metcalfe, 1922: 36, pl. 13 (♂, ♀ genitalia). *Phalonia roseana*; Meyrick, 1928: 490; Ford, 1949: 45]

DESCRIPTION

♂♀ 10–17 mm. Sexes similar; antenna of male shortly ciliate.

Male (fig. 14). Forewing ground colour whitish yellow, suffused with rose-pink, especially in costal and distal areas; costa obscurely strigulate; markings light ochreous-brown; basal and sub-basal fasciae indeterminate; median fascia slightly curved and outward-oblique from dorsum to beyond middle, obsolescent costally, outer margin indented above dorsum, heavily suffused with a mixture of plumbeous and black on dorsum and along inner margin; pre-apical spot obsolescent; a small blackish pre-tornal marking; cilia yellow-ochreous, marked with dark grey along termen, with a rather coarse dark grey or blackish subapical line and a paler sub-basal line. Hindwing grey; cilia paler, grey apically, with a darker sub-basal line.

Female. Similar to male.

Variation. The rather characteristic rose-pink suffusion of the forewing varies in strength. In some specimens the suffusion may be very weak and inconspicuous, the forewing being pale ochreous in general appearance; in others it may be very strong, the wing then having a very pronounced flush. The plumbeous and black mixture in the median fascia also shows slight variation in strength.

COMMENTS

Distinguished from *C. flaviciliana* by the darker median fascia of the forewing and the presence of a subapical line in the cilia; the presence of the subapical line also separates this species from *Falseuncaria degreyana*.

BIOLOGY

Ovum. Deposited singly on the seedhead of the foodplant in July and August.

Larva. Head brown; prothoracic plate green, anterior margin edged with black; abdomen light green mottled with white, especially between the segments; pinacula and anal plate concolorous with integument. August to May, in the flowers and seedheads of *Dipsacus fullonum*, burrowing transversely through the seeds, forming a tough silken tunnel from seed to seed through the scales of the seedhead (pl. 7, fig. 3), never entering the central cavity. On the Continent the larva is recorded also on the flowers of *Aster* and in the seed capsules of *Antirrhinum* (Swatschek, 1958: 240).

Pupa. Yellowish brown, wings and appendages shining. May to July, in a tough cocoon in the larval habitation; the moth emerges at the side, between the scales of the teasel head.

Imago. Univoltine, the emergence period lasting from the end of May to August. Occurring in a variety of biotopes on chalkland, clay soils and sandhills, especially rough grassland, waysides, copses and along banks of streams where teasels grow plentifully. The moth flies actively over the teasel heads in the afternoon sunshine, and at other times is easily disturbed from rest on the foodplant or nearby vegetation. When at rest on a teasel head it is difficult to detect because of the cryptic pink coloration of the forewings, which closely resembles that of the spinous receptacle of the foodplant.

DISTRIBUTION

Widely distributed in England as far north as Derbyshire, Lancashire and Yorkshire, occurring most commonly in the extreme southern counties. Its occurrence elsewhere in the British Isles seems uncertain, the record from Sligo (Co. Sligo), Ireland, having been found by Beirne (1941: 77) to be based on a misidentification of *Falseuncaria ruficiliana*; a record from near Stirling in Scotland (Stainton, 1845: 1090) is unconfirmed. Apparently unknown from Wales.

North and central Europe to Asia Minor.

Cochylis flaviciliana (Westwood)

Pl. 25, fig. 15

[*Eupoecilia flaviciliana*; Barrett, 1905: 308. *Cochylis flaviciliana*; Pierce & Metcalfe, 1922: 36, pl. 13 (♂, ♀ genitalia). *Phalonia flaviciliana*; Meyrick, 1928: 491; Ford, 1949: 45]

DESCRIPTION

♂ ♀ 11–16 mm. Sexes similar; antenna of male shortly ciliate.

Male (fig. 15). Forewing ground colour cream-white, suffused with bright rose-pink, especially in costal and distal areas, costa often tinged or faintly strigulate with greyish; markings ochreous-brown; basal and sub-basal fasciae obsolete; median fascia outward-oblique from dorsum to beyond middle, obsolescent costally, margins more or less straight and parallel dorsally, outer margin sometimes indented above dorsum; pre-apical spot and other fasciate markings in distal area diffuse and indeterminate; pre-tornal marking indicated by a brownish dot; cilia ochreous-yellow, suffused with greyish towards tornus, with a darker sub-basal line. Hindwing grey, paler basally; cilia whitish, with a dark sub-basal line.

Female. Similar to male; hindwing slightly darker.

Variation. In this species the pink flush of the forewing is invariably evident and does not show the same degree of variation as in *C. roseana*. When the flush is weak the cream-white ground colour is more apparent, when strong it may appear brilliant pink tinged with violet. In well-marked specimens the median fascia may be olive-brown and more strongly developed in the costal half of the wing.

COMMENTS

The contrasting bright rose-pink and cream-white forewing coloration is characteristic; differs from *C. roseana* by the lighter median fascia of the forewing and the absence of a subapical line in the cilia.

BIOLOGY

Larva. Head brown; prothoracic plate light brown; abdomen varying from pinkish brown to reddish brown, sometimes cream, or green with a slight pinkish tinge; pinacula concolorous with integument; anal plate and thoracic legs light brown. Late July to early October, in the flowers and seedheads of *Knautia arvensis*, feeding on the seeds and attaching them to the seedhead (pl. 7, fig. 4), sometimes spinning several seedheads together. When full-fed in October the larva leaves the seedhead and spins up amongst debris on the ground, pupating *in situ* the following spring.

Pupa. May to July, spun up amongst ground debris.

Imago. Late June to early August; Fassnidge (1936: 123) records two specimens emerging in captivity in September. Frequenting dry grassland, especially in chalk districts, where field scabious is plentiful. The moth is seldom seen by day but is more readily disturbed in the evening and at dusk flies about the scabious.

DISTRIBUTION

Very local in the British Isles, apparently occurring only in southern England from

Essex and Kent, Hampshire and the Isle of Wight to Oxfordshire, and in South Wales (Pembrokeshire). Known localities include St. Margaret's Bay (Kent), Addington, Caterham, Coulsdon and Sanderstead (Surrey), Basingstoke and Farley Mount, near Winchester (Hampshire), and Reading (Berkshire).

North and central Europe.

Cochylis dubitana (Hübner)
Pl. 25, fig. 16

[*Eupoecilia dubitana*; Barrett, 1905: 284. *Cochylis dubitana*; Pierce & Metcalfe, 1922: 36, pl. 13 (♂, ♀ genitalia). *Phalonia dubitana*; Meyrick, 1928: 493; Ford, 1949: 46]

DESCRIPTION

♂♀ 11–15 mm. Sexual dimorphism not pronounced; antenna of male shortly ciliate; hindwing of female darker.

Male (fig. 16). Forewing ground colour white, costa strigulate and dorsum dotted with blackish grey; markings pale greyish ochreous, sparsely irrorate with black, variably suffused with a heavy mixture of plumbeous and black; basal fascia well developed, heavily suffused; sub-basal fascia developed and confluent with basal fascia on costa; median fascia slightly oblique and well developed from dorsum to median fold, heavily suffused with black, often with an admixture of ferruginous in middle and more or less thickly bordered with shining plumbeous, fractured and strigulate beyond fold, particularly its outer margin, reappearing as a blackish bar on costa; a slender irregular postmedian streak from dorsum; pre-tornal marking inconspicuous; an irregular patch above tornus in discal area; pre-apical spot represented by blackish striae on costa, produced as a diffuse stria usually confluent with blackish and plumbeous suffusion along terminal margin, the whole forming a diffuse terminal band, broadest on costa, tapering to tornus; cilia pale ochreous, suffused and marked with grey and blackish grey, with a dark sub-basal line. Hindwing dark grey, paler basally; cilia grey, apices whitish, with a dark grey sub-basal line.

Female. Similar to male; hindwing more uniformly darker grey.

Variation. This species shows only slight variation in the strength and development of the forewing markings.

COMMENTS

Similar to *C. atricapitana* and *C. hybridella*. In *atricapitana* the head and labial palpus are blackish grey; in both *dubitana* and *hybridella* the head and labial palpi are white, but these two species may be separated by the blackish thorax of the former and the white thorax of the latter.

BIOLOGY

Larva. Head brown; prothoracic plate light brown, narrowly bordered with blackish brown or black posteriorly, with a narrow medial sulcus; abdomen white or brownish white, tinged with pink when full-grown; pinacula and thoracic legs concolorous with integument; anal plate light brown dotted with black. Larvae from the first generation of moths in July; those from the second generation in September, overwintering spun up amongst debris; feeding in the flowers and seedheads of Compositae, especially *Senecio*, *Crepis*, *Solidago* and *Hieracium*. On the Continent *Arctium lappa*, *Cirsium vulgare*, *Carduus acanthoides*, *Centaurea jacea* and *Picris* are recorded as foodplants (Swatschek, 1958: 240).

Pupa. Yellowish brown. April and July, spun up amongst debris, usually on the ground.

Imago. Bivoltine, the first generation emerging in June, the second in August. Frequenting rough grassy hillsides, railway embankments and waste ground generally. Hiding by day amongst the coarse herbage and most readily disturbed towards late afternoon, flying freely in the evening.

DISTRIBUTION

Widely distributed and locally common in the southern counties of England northwards to Lancashire, Yorkshire, Westmorland and the Isle of Man; in Wales known from Cardiganshire, Caernarvonshire and Denbighshire, occurring mainly in carboniferous limestone areas. Rare in Scotland and known only from the south, having been recorded from Glasgow (Lanarkshire), Edinburgh (Mid Lothian), Berwickshire and Perthshire. In Ireland it occurs locally on the coast at Howth (Co. Dublin) and Belfast (Co. Antrim), and has probably been overlooked elsewhere.

North and central Europe to Iceland and Siberia; North America.

Cochylis hybridella (Hübner)

Pl. 25, fig. 17

[*Eupoecilia hybridella*; Barrett, 1905: 285. *Cochylis hybridella*; Pierce & Metcalfe, 1922: 37, pl. 13 (♂, ♀ genitalia). *Phalonia hybridella*; Meyrick, 1928: 493; Ford, 1949: 46]

DESCRIPTION

♂♀ 12–15 mm. Sexual dimorphism not pronounced; antenna of male shortly ciliate; hindwing of female darker grey.

Male (fig. 17). Forewing ground colour white, sometimes weakly flushed with pink distally, costa and dorsum sparsely strigulate with fuscous; markings ferruginous-ochreous varying to olive-brown, variably mixed and edged with grey and black; basal and sub-basal fasciae forming a dark grey patch on costa, strigulate with blackish, becoming obsolete towards dorsum; median fascia irregular, narrow on costa, black-edged and suffused with grey, widely interrupted below, slightly inward-oblique to dorsum, a black dash in median fold and a heavier dash above sometimes contiguous, inner margin thickly mixed with plumbeous on dorsum and above; an irregular pale patch from tornus to middle of wing, preceded by a small black pre-tornal marking on dorsum; pre-apical spot represented by costal striae, produced as a fasciate stria to termen above tornus, mixed with black, inner margin bordered with grey suffusion, outer margin indistinct, more or less confluent with diffuse blackish mixed with ferruginous-ochreous spots along terminal margin; cilia pale ochreous irregularly marked with grey, white before tornus, with a dark grey sub-basal line. Hindwing pale grey; cilia concolorous, with a darker sub-basal line.

Female. Similar to male; hindwing darker grey.

Variation. In fresh specimens the pink flush in the distal area of the forewing may be more pronounced; minor variation is found in the dark markings.

COMMENTS

Distinguished from *C. dubitana* and *C. atricapitana* by the white thorax.

BIOLOGY

Larva. Head light brown; prothoracic plate yellow, with a medial sulcus, posterior margin of each half with two semicircular brown markings; abdomen yellow, paler ventrally, tinged with reddish dorsally; pinacula inconspicuous, nearly concolorous with integument; anal plate light brown; anal comb bearing four or five short prongs. August and September, living in the flowers and seedheads of *Picris hieracioides*, *P. echioides* (pl. 8, fig. 1) and *Crepis*, eating the unripe seeds, moving from seedhead to seedhead. When full-grown at the end of September or in October the larva leaves the seedheads and spins a cocoon on the ground or amongst debris, hibernating within and pupating the following spring.

Pupa. Yellowish brown. May, in a tough silken cocoon spun amongst debris or ground litter.

Imago. July and August. Frequenting rough grassland on downs and similar chalk and limestone districts, and also on sandhills. At sunset the moth flies freely about its habitat, but is otherwise seldom seen during the day unless disturbed from rest low down amongst the vegetation.

DISTRIBUTION

In the British Isles apparently favours chalk and limestone districts in the south of England, occurring locally but commonly from Cambridgeshire, Essex and Kent to Dorset, Somerset, Gloucestershire and Herefordshire; recorded as far north as Cheshire, Lancashire (Formby sandhills) and Yorkshire but apparently rare. In Somerset it is uncommon and has been recorded only in the north of the county (Abbots Leigh and Portishead). In Wales it is known from Cardiganshire, and Barrett (1905: 286) records it as common in quarries in Pembrokeshire. Beirne (1941: 78) cites one unconfirmed record from Ireland (Howth, Co. Dublin).

Central and south-east Europe to Asia Minor; China and Japan.

Cochylis atricapitana (Stephens)

Pl. 25, figs. 18, 19

[*Eupoecilia atricapitana*; Barrett, 1905: 281. —*atricapitana*; Pierce & Metcalfe, 1922: 27, pl. 10 (♂, ♀ genitalia). *Phalonia atricapitana*; Meyrick, 1928: 492; Ford, 1949: 46]

DESCRIPTION

♂♀ 12–16 mm. Sexual dimorphism moderately pronounced; antenna of male strongly ciliate, hindwing strigulate; forewing of female often more strongly tinged with pink, hindwing dark grey.

Male (fig. 18). Forewing ground colour white or ochreous-white, tinged with pink, mostly distally, costa and dorsum sparsely strigulate with blackish; markings ferruginous-ochreous varying to olive-brown, variably mixed and edged with grey and black; basal fascia complete; sub-basal fascia developed on costa, edged with blackish and often confluent with basal fascia, obsolescent below and towards dorsum; median fascia irregular, narrow on costa and thickly edged or mixed with blackish, interrupted below, slightly inward-oblique to dorsum, a black dash in median fold and a similar dash above, these variable and sometimes divided into two marginal spots, inner margin bordered with plumbeous dorsally; a slender postmedian streak from dorsum meeting median fascia in fold; a small patch above tornus; a small black pre-tornal spot; pre-apical spot represented by costal striae, produced as a fasciate stria to termen above tornus, outer edge indeterminate, confluent with plumbeous mixed with black coloration along termen; cilia pale ochreous, marked with grey, whitish before tornus, a dark grey sub-basal line. Hindwing white, suffused and strigulate with grey; cilia whitish, a grey sub-basal line.

Female (fig. 19). Forewing pattern similar to that of male but often with a more pronounced pink flush; hindwing almost uniform dark grey.

Variation. In both sexes only minor variation is found in the intensity of the forewing coloration and markings; in the male the pink flush may be very weak or absent, the general coloration then appearing lighter.

COMMENTS

The dark coloration of the head and thorax distinguish this species from *C. dubitana* and *C. hybridella*.

BIOLOGY

Larva. Head light brown, epistomal region blackish brown; prothoracic plate light brown, posterior margin variably marked with dark brown, sometimes with a solid transverse bar; abdomen strongly shagreened, dull pale yellow, tinged with pink dorsally; pinacula inconspicuous, very light brown; peritreme of spiracles dark brown; anal plate brown with darker spots or mottling; anal comb absent. On *Senecio jacobaea*; larvae from the first generation of moths in July, feeding on the flowers and flower stalks, burrowing into the stems and retarding the growth of the central shoot, causing it to thicken and the leaves to bunch; larvae from the second generation feed in the lower part of the stem and in the rootstock in September and October, overwintering in the burrow until the following spring. The illustration (pl. 8, fig. 2) shows the larval borings in a stem collected in early May.

Pupa. Yellowish brown. April and July, in a white silken cocoon in the larval habitation.

Imago. Bivoltine, the first generation emerging in May and June, the second in late July and August. Frequenting especially coastal areas and to a lesser extent inland chalk downs and rough pastureland. The moth rests during the day on the foodplant or amongst

88

vegetation nearby and may sometimes be disturbed and netted. The male flies at sunset, the female a little later, and both sexes may be taken at light.

DISTRIBUTION

Widely distributed in the British Isles as far north as Inverness-shire; most common in the southern counties of England and Wales, and around the coasts elsewhere. In Ireland this species appears to be mainly coastal.

Europe and North Africa to southern Russia.

Cochylis pallidana Zeller

Pl. 25, fig. 20

[*Eupoecilia pallidana*; Barrett, 1905: 283. *Cochylis pallidana*; Pierce & Metcalfe, 1922: 36, pl. 13 (♂, ♀ genitalia). *Phalonia pallidana*; Meyrick, 1928: 492; Ford, 1949: 46]

DESCRIPTION

♂ ♀ 10–14 mm. Sexes similar; antenna of male shortly ciliate.

Male (fig. 20). Forewing ground colour dull white, costa and dorsum sparsely strigulate with blackish grey; markings pale greyish ochreous, variably suffused with black, the black scales being white-tipped, producing a greyish effect; basal fasciae diffuse, strongly suffused; sub-basal fascia developed on costa, becoming obsolescent towards dorsum; median fascia slightly outward-oblique and well developed from dorsum to beyond middle, dorsally suffused with black, often with an admixture of ferruginous and more or less thickly bordered with shining plumbeous, fractured and obsolescent beyond fold towards costa, outer margin partially represented by a heavy blackish patch on costa with a smaller patch directly below; a small postmedian streak on dorsum; an irregular, elongate patch above tornus in discal area; pre-apical spot represented by blackish costal striae, emitting a fasciate stria curving along termen to tornus; a moderately distinct, black pre-tornal dot; cilia pale ochreous, diffusedly marked with dark grey, with a blackish sub-basal line. Hindwing light grey; cilia paler, with a dark sub-basal line.

Female. Similar to male.

Variation. Only minor variation is found in the strength and development of the forewing markings.

COMMENTS

Distinguished from *C. nana* by the absence of ochreous suffusion in the ground colour of the forewing, and the narrower median fascia which is separate from the sub-basal fascia on the costa.

BIOLOGY

Larva. Head and prothoracic plate varying from brown to blackish brown, the latter darker posteriorly; abdomen rose-pink dorsally and laterally, yellowish white ventrally, the whole tinged with greenish; pinacula inconspicuous; anal plate brown, darker medially; thoracic legs and prolegs yellowish white. July and August, living in the flowers and seedheads of *Jasione montana*, feeding on the seeds. When full-grown at the end of August or in September the larva leaves the seedhead and spins up amongst debris, usually on the ground, hibernating until the following spring before pupating *in situ*.

Pupa. Light brown. April and May, spun up amongst leaf litter or other debris.

Imago. June. A mainly coastal species, frequenting grassland on light sandy or stony soils where sheep's-bit is found, also on inland chalk downs. Both sexes fly freely at dusk but are seldom seen during the day when resting amongst the foodplant or concealed in gorse and other thick cover.

DISTRIBUTION

Locally common in the south-eastern, southern and western coastal districts of Britain as far north as the Isle of Man, Ayrshire and Lanarkshire; also occurring inland on the chalk downs of southern England, where it is scarce and local. The distribution in Ireland appears to be almost entirely coastal; records are known from Co.'s Dublin, Cork, Kerry, Galway and Londonderry.

North and central Europe; Asia Minor; southern Russia.

Cochylis nana (Haworth)

Pl. 25, fig. 21

[*Eupoecilia nana*; Barrett, 1905: 315. — *nana*; Pierce & Metcalfe, 1922: 28, pl. 10 (♂, ♀ genitalia) *Phalonia nana*; Meyrick, 1928: 492; Ford, 1949: 46]

DESCRIPTION

♂ ♀ 9–13 mm. Sexual dimorphism not pronounced; antenna of male shortly ciliate; antenna of female filiform, hindwing darker.

Male (fig. 21). Forewing ground colour white suffused with ochreous, especially distally, strigulate with grey; markings grey mixed with black; basal and sub-basal fasciae indeterminate, represented by dense diffuse strigulation; median fascia diffuse, broad, margins poorly defined, outer margin outward-oblique from costa, a concentration of black on costa and in median fold, often a sprinkling of ferruginous-ochreous scales at middle; pre-apical spot small, produced as a diffuse ochreous-grey stria reaching to tornal area; cilia pale ochreous sparsely irrorate with black, a darker sub-basal line. Hindwing light grey; cilia paler, with a faint sub-basal line.

Female. Similar to male; hindwing darker grey.

Variation. Some variation is found in the strength of the forewing markings which vary from light grey to blackish grey; the ochreous suffusion may be weak, the ground colour then appearing lighter.

COMMENTS

A distinctive species, at once recognized by its small size and characteristic grey and ochreous general coloration, and the broad median fascia of the forewing with its outer margin outward-oblique from the costa and the inner margin diffuse and often confluent with the ill-defined basal fasciae.

BIOLOGY

Ovum. Deposited on the catkin bud of the foodplant in June.

Larva. Head brown; prothoracic plate brownish, an irregular group of dark brown dots above the lateral margin, posterior margin marked with dark brown medially; abdomen whitish tinged with yellowish grey, brownish dorsally. July to September or October, feeding in the catkins of *Betula*. In the autumn the full-grown larva leaves the catkins and spins a silken cocoon in a crevice in the bark, pupating the following spring.

Pupa. Light brown. May and early June, in a silken cocoon on the tree trunk.

Imago. June. Frequenting birch woods, preferring large trees, resting on the trunks or amongst the foliage by day but readily disturbed. At dusk both sexes fly actively around the birches.

DISTRIBUTION

Widely distributed and generally common wherever established birch woods are found at lower elevations in England and Wales. Locally common in Scotland, being generally distributed in Argyll and Perthshire, extending as far north as east Inverness-shire. In Ireland it is recorded by Beirne (1941: 78) as widely distributed but scarce in the southern half of the country, with records from Co.'s Killarney, Cork and Wicklow; it is apparently unknown in the north.

North and central Europe to Asia Minor; North America.

Plates 1–21

COCHYLIDAE (PHALONIIDAE)

Plates 1–8

TORTRICIDAE: TORTRICINAE

Plates 9–21

COCHYLIDAE (PHALONIIDAE)

PLATE I

1 *Hysterosia sodaliana* (Haworth): larval spinning on berries of *Rhamnus catharticus* (Addington, Surrey)

2 *Hysterophora maculosana* (Haworth): larva in seed capsules of *Endymion non-scriptus* (Dodnash Wood, East Suffolk)

3 *Phtheochroa rugosana* (Hübner): larval spinning on flowers of *Bryonia dioica* (Thorpeness, Suffolk)

4 *Phtheochroa rugosana* (Hübner): larval spinning on berries of *Bryonia dioica* (Riddlesdown, Surrey)

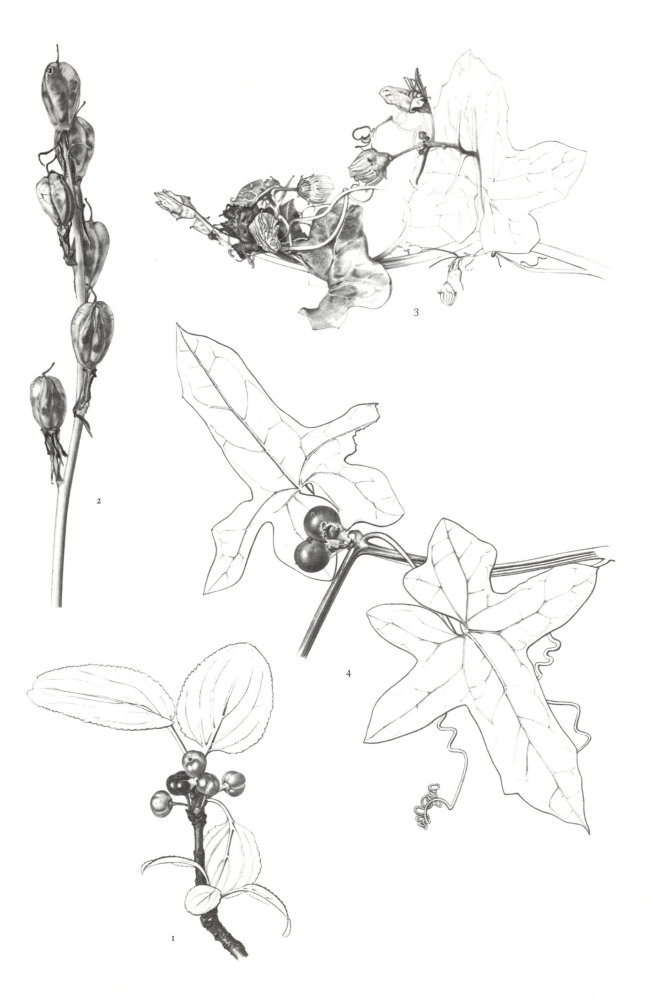

Plate 1

PLATE 2

1 *Phalonidia vectisana* (Humphreys & Westwood): larval spinning in flowerhead of *Triglochin maritima* (Benfleet, Essex)

2 *Phalonidia alismana* (Ragonot): larva in stem of *Alisma plantago-aquatica* (East Horsley, Surrey)

1

2

Plate 2

COCHYLIDAE (PHALONIIDAE)

PLATE 3

1, 2   *Phalonidia gilvicomana* (Zeller): larva in seedheads of
      Mycelis muralis (Mickleham, Surrey)

3   *Stenodes alternana* (Stephens): larva in flowerhead of
      Centaurea scabiosa (St. Margaret's Bay, Kent)

4   *Stenodes straminea* (Haworth): larval habitation in
      stem of *Centaurea nigra* (Wheal Rose, Cornwall)

4

3

1

2

PLATE 4

1 *Juniperus communis*: foodplant of *Aethes rutilana* (Hübner)

2 *Aethes rubigana* (Treitschke): larval borings in seedhead of *Arctium lappa* (Ashtead, Surrey)

3 *Aethes smeathmanniana* (Fabricius): larval spinnings in seedhead of *Achillea millefolium* (Benfleet, Essex)

COCHYLIDAE (PHALONIIDAE)

PLATE 5

1   *Aethes dilucidana* (Stephens): stem of *Pastinaca sativa* showing presence of larva (Mickleham, Surrey)

2   *Aethes francillana* (Fabricius): larval habitation in seedhead of *Daucus carota*

3   *Aethes francillana* (Fabricius): larva burrowing in stem of *Daucus carota* (Riddlesdown, Surrey)

1 *Aethes beatricella* (Walsingham): larva burrowing in stem of *Conium maculatum* (Stanford-le-Hope, Essex)

2 *Commophila aeneana* (Hübner): larva in rootstock of *Senecio jacobaea* (Benfleet, Essex)

3 *Eupoecilia ambiguella* (Hübner): larval spinning on *Frangula alnus* (Holmwood, Surrey)

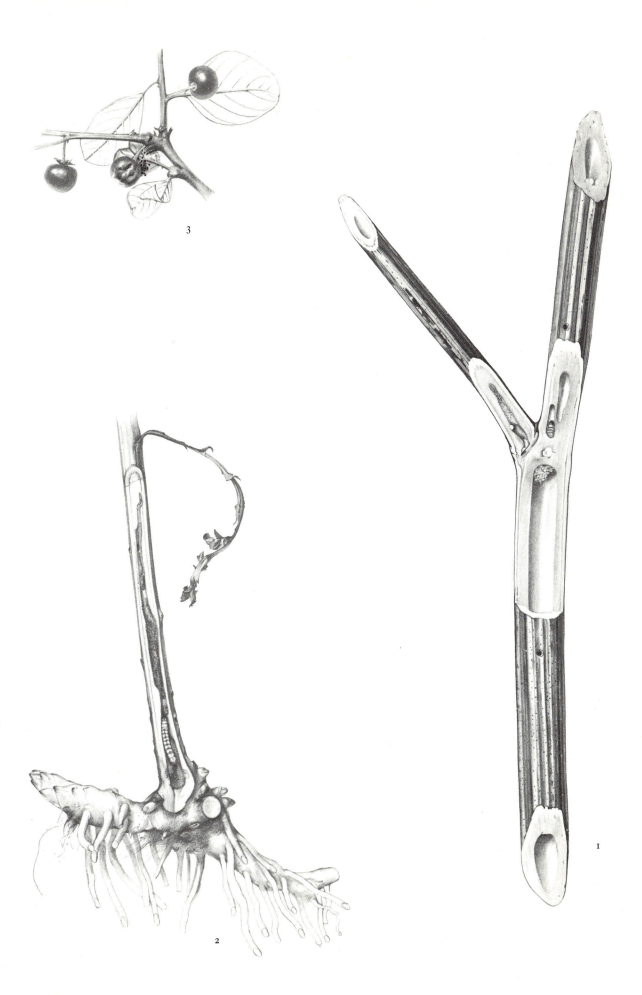

Plate 6

1 *Cochylidia heydeniana* (Herrich-Schäffer): seedhead of
Erigeron acer showing presence of larva (Swanley, Kent)

2 *Falseuncaria degreyana* (McLachlan): larval habitation
in seedhead of *Plantago lanceolata* (Thetford, Norfolk)

3 *Cochylis roseana* (Haworth): seedhead of *Dipsacus
fullonum* showing larval borings and pupa (Wye, Kent)

4 *Cochylis flaviciliana* (Westwood): larval spinning in
seedhead of *Knautia arvensis* (Addington, Surrey)

Plate 7

1   *Cochylis hybridella* (Hübner): larval habitation in
    seedheads of *Picris echioides* (Benfleet, Essex)

2   *Cochylis atricapitana* (Stephens): larval boring in dead
    stem of *Senecio jacobaea* (Brook, Kent)

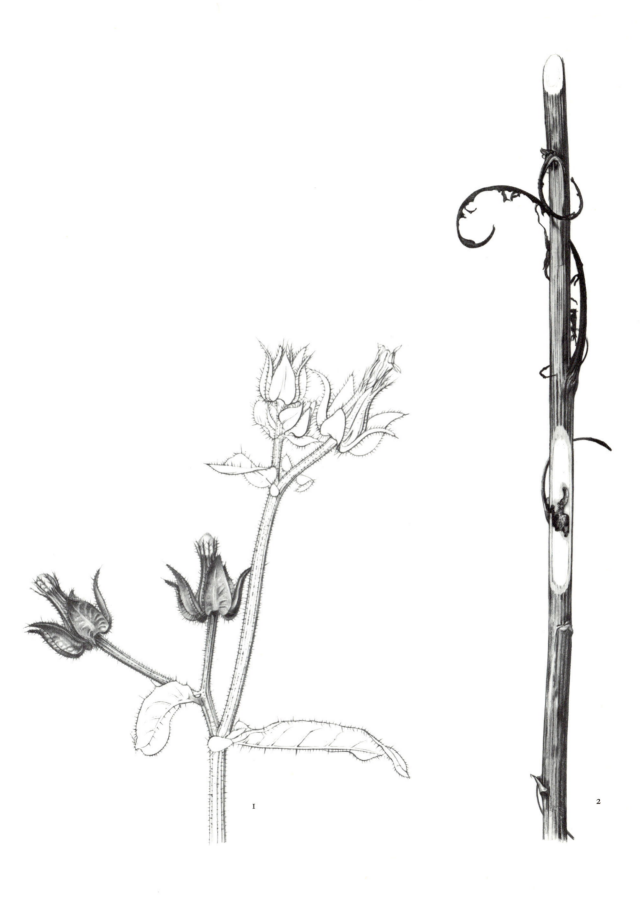

1

2

Plate 8

PLATE 9

1 *Pandemis corylana* (Fabricius): larval spinning on
   Quercus (New Forest, Hants)

2 *Pandemis heparana* (Denis & Schiffermüller): larval
   spinning on *Vaccinium* (Horsley, Surrey)

3 *Archips podana* (Scopoli): larval spinning on
   Euonymus japonicus (Porthtowan, Cornwall)

4 *Archips podana* (Scopoli): apple pitted by early instar
   larva (Cambs)

Plate 9

PLATE 10

1 *Archips crataegana* (Hübner): larva feeding in folded leaf of *Tilia* (Buckingham Palace Garden, London)

2 *Archips xylosteana* (Linnaeus): larval spinning on *Quercus* (Hyde Park, London)

3 *Archips xylosteana* (Linnaeus): larval spinning on *Tilia* (South Kensington, London)

4 *Archips xylosteana* (Linnaeus): larval spinning on *Salix* (Monks Wood, Hunts)

5 *Archips rosana* (Linnaeus): larval spinning on *Corylus* (Burren, Co. Clare)

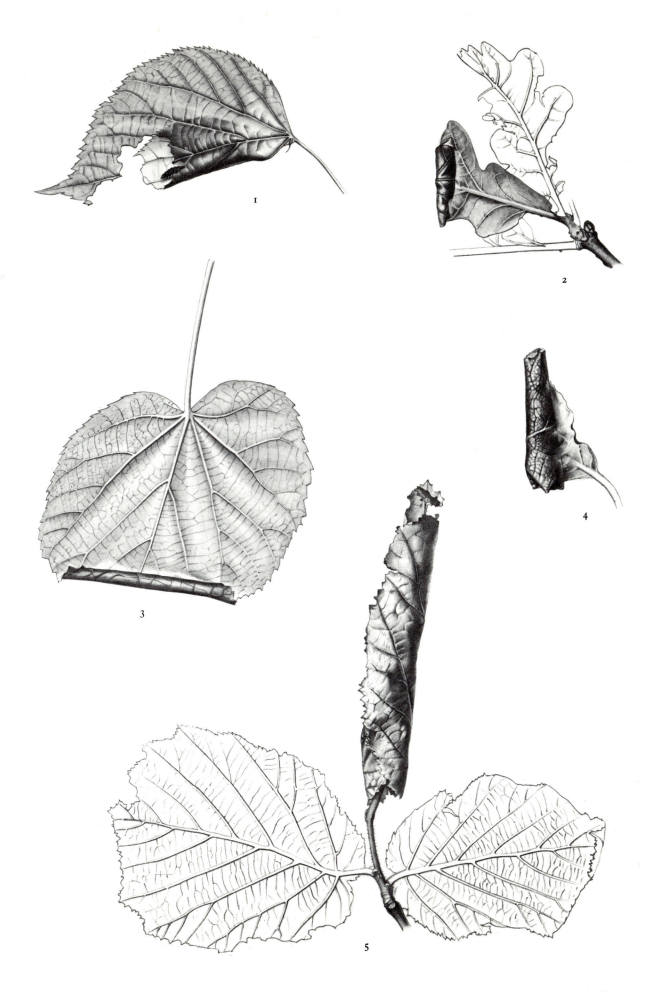

1

2

3

4

5

Plate 10

PLATE II

1 *Choristoneura lafauryana* (Ragonot): larval spinning
   on *Myrica gale* (King's Lynn, Norfolk)

2 *Cacoecimorpha pronubana* (Hübner): larval spinning
   on *Ligustrum* (Byfleet, Surrey)

3 *Cacoecimorpha pronubana* (Hübner): larval spinning
   on *Hippophae rhamnoides* (Weston-super-Mare,
   Somerset)

1

2

3

PLATE 12

1 *Aphelia paleana* (Hübner): larval spinning on
  Arrhenatherum elatius (Swaffham, Norfolk)

2 *Aphelia viburnana* (Denis & Schiffermüller): seedheads
  of *Dryas octopetala* spun together by larva (Burren,
  Co. Clare)

3 *Clepsis spectrana* (Treitschke): larval spinning in shoot
  of *Epilobium* (Whippendell Wood, Herts)

Plate 12

1 *Clepsis consimilana* (Hübner): larval spinning amongst
   withered leaves of *Ligustrum* (Dulwich, London)

2 *Epiphyas postvittana* (Walker): larval spinning on
   Montbretia (Porthtowan, Cornwall)

3 *Adoxophyes orana* (Fischer von Röslerstamm): apple
   scarred by larva (East Malling, Kent)

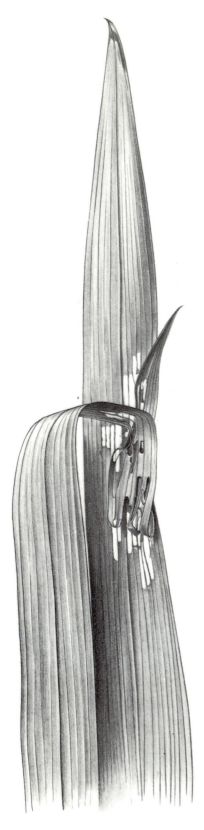

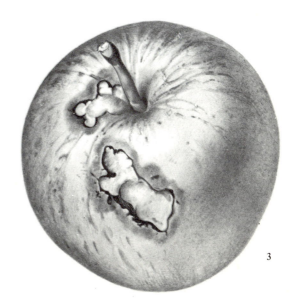

PLATE 14

1   *Lozotaenia forsterana* (Fabricius): larval spinning on
    Hedera (Horsley, Surrey)

2   *Olindia schumacherana* (Fabricius): cocoon in spun
    leaf of *Vaccinium* (Tintern, Mons)

3   *Ditula angustiorana* (Haworth): larval spinning on
    Viscum (Wye Valley, Hereford)

4   *Ditula angustiorana* (Haworth): larval spinning on
    Larix (Farnborough, Kent)

5   *Cnephasia conspersana* Douglas: larval spinning in
    flowerhead of *Chrysanthemum leucanthemum*
    (Sydenham, Kent)

1

2

3

5

4

Plate 14

1 *Cnephasia stephensiana* (Doubleday): larval spinning on *Plantago lanceolata* (Effingham, Surrey)

2 *Cnephasia stephensiana* (Doubleday): larval spinning on *Chrysanthemum leucanthemum* (Effingham, Surrey)

3 *Cnephasia pasiuana* (Hübner): larval spinning in flowerhead of *Chrysanthemum leucanthemum* (Sydenham, Kent)

4, 5 *Cnephasia interjectana* (Haworth): larval spinnings on *Chrysanthemum leucanthemum* (Cudham, Kent)

6 *Cnephasia genitalana* Pierce & Metcalfe: larval spinning in flowerheads of *Ranunculus* (Derby Fen, Norfolk)

Plate 15

PLATE 16

1   *Tortricodes alternella* (Denis & Schiffermüller): imago
     at rest on twig of *Quercus* (Wisley Common, Surrey)

2, 3 *Eana incanana* (Stephens): larval spinnings in
      flowerheads of *Endymion non-scriptus* (Otford, Kent)

4, 7 *Tortrix viridana* (Linnaeus): cocoons spun in leaves of
      Salix (Monks Wood, Hunts)

5, 6 *Tortrix viridana* (Linnaeus): larval spinnings on
      Quercus (New Forest, Hants)

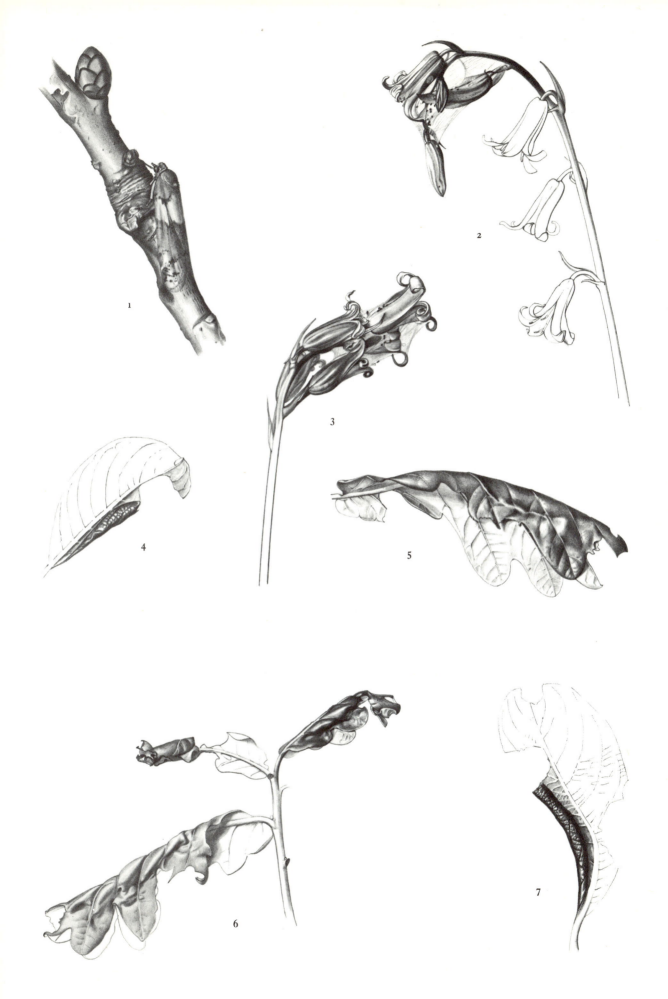

Plate 16

PLATE 17

1 *Croesia bergmanniana* (Linnaeus): larval spinnings on *Rosa* (Benfleet, Essex)

2 *Croesia forsskaleana* (Linnaeus): larval spinning on *Acer campestre* (Wilmington, Kent)

3 *Acleris rhombana* (Denis & Schiffermüller): larval spinning on *Crataegus* (Redhill, Surrey)

4 *Acleris aspersana* (Hübner): larval spinning on *Potentilla* (Swaffham, Norfolk)

5 *Acleris aspersana* (Hübner): larval spinning on *Sanguisorba minor* (Bletchley, Herts)

Plate 17

1 *Acleris sparsana* (Denis & Schiffermüller): larval
   spinning on *Fagus* (Hyde Park, London)

2 *Acleris sparsana* (Denis & Schiffermüller): larval
   spinning on *Acer pseudoplatanus* (Porthtowan,
   Cornwall)

1

2

Plate 18

PLATE 19

1 *Acleris tripunctana* (Hübner): larval spinning on
   Betula (Wisley Common, Surrey)

2 *Acleris ferrugana* (Denis & Schiffermüller): larval
   spinning on *Quercus* (Bookham, Surrey)

3 *Acleris shepherdana* (Stephens): larval spinning on
   Filipendula ulmaria (Woodwalton Fen, Hunts)

PLATE 20

1   *Acleris schalleriana* (Linnaeus): pupal exuviae
    protruding from old larval habitation on *Viburnum
    lantana* (Swanage, Dorset)

2   *Acleris boscana* (Fabricius): larval spinning on *Ulmus*
    (Clandon, Surrey)

1

2

Plate 20

PLATE 21

1 *Acleris hastiana* (Linnaeus): larval spinning on *Salix aurita* (Ashdown Forest, Sussex)

2 *Acleris rufana* (Denis & Schiffermüller): larval spinning on *Myrica gale* (Lochearnhead, Perthshire)

3 *Acleris hyemana* (Haworth): larval spinning on *Erica* (Co. Kerry)

4, 5 *Acleris emargana* (Fabricius): imagines at rest

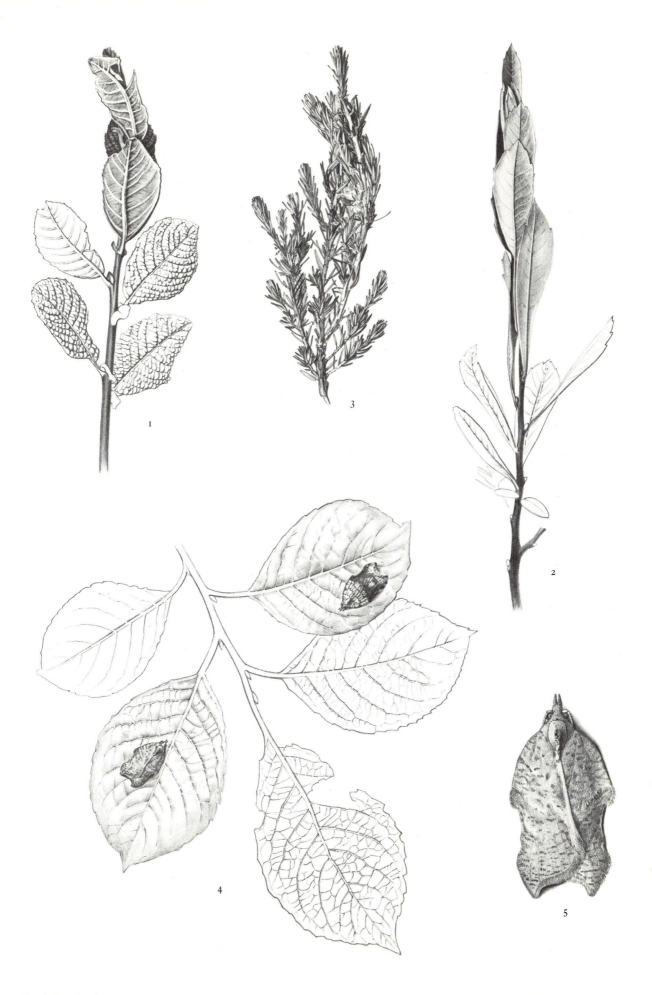

1 2 3 4 5

TORTRICIDAE: TORTRICINAE

ARCHIPINI

Pandemis corylana (Fabricius)

Pl. 26, figs. 1–4

[*Pandemis corylana*; Barrett, 1904: 172; Pierce & Metcalfe, 1922: 3, pl. 1 (♂, ♀ genitalia); Meyrick, 1928: 505; Ford, 1949: 51]

DESCRIPTION

♂ 18–21 mm, ♀ 18–24 mm. Sexual dimorphism not pronounced; antenna of male weakly dentate-ciliate, with deep sinus near base (text-fig. 40), scape and flagellum posteriorly whitish ochreous, segments sometimes diffusely marked with transverse grey bars; antenna of female sparsely ciliate, without sinus (text-fig. 41), flagellum whitish posteriorly, segments with diffuse brown bars.

Male (fig. 1). Frons and dorsal and inner side of labial palpus whitish. Forewing ground colour pale ochreous-yellow, strigulate with ochreous-brown or dark brown; veins, fasciae and termen edged or outlined with dark brown or ferruginous; outer margin of basal fasciae oblique, median fascia almost parallel. Hindwing light grey-brown, weakly suffused with whitish ochreous apically.

Female (fig. 2). Forewing coloration usually deeper and markings stronger than in the male; apical area of hindwing more strongly suffused with whitish ochreous.

Variation. General coloration and markings varying mainly in intensity and strength; in strongly coloured specimens the forewing may approach a rich red-brown (fig. 3) and

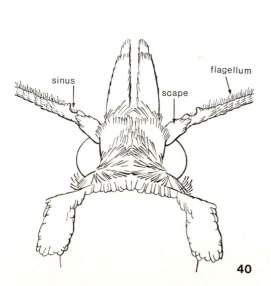
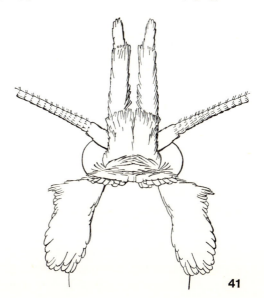

40   **41**

Antennae of *Pandemis corylana* (Fabricius)
Fig. 40, ♂. Fig. 41, ♀

often the cilia are deep ferruginous; more rarely the markings are reduced or obsolescent (fig. 4) and in extreme forms the forewing is unicolorous ochreous-yellow with no trace of markings except for ferruginous coloration at the base, along the termen and in the cilia.

COMMENTS

Similar to *P. cerasana* but distinguished by the characteristic reticulate forewing pattern and the more strongly oblique outer margin of the basal fasciae, which are parallel with the median fascia, and the paler apex of the hindwing.

BIOLOGY

Larva. Rather slender; head and prothoracic and anal plates emerald green, head shaded with yellowish brown, prothoracic plate lightly marked with black; thoracic legs green; abdomen green, paler laterally; anal comb well developed, with up to 9 prongs. May to July, on *Thelycrania sanguinea*, *Corylus*, *Fraxinus*, *Prunus*, *Rubus*, *Quercus* and various other trees and shrubs, spinning leaves together or folding a leaf longitudinally (pl. 9, fig. 1).

Pupa. Light brown, abdominal segments rather slender. June, in the larval habitation.

Imago. July to September; hiding amongst foliage during the day, usually in the more open parts of woods, but readily beaten out and netted; it flies freely at dusk and later frequently comes to light and occasionally to sugar.

DISTRIBUTION

A common woodland species in England, especially in the south, and found throughout Wales. Apparently scarce in Scotland; records from as far north as Ross and Cromarty are unconfirmed. In Ireland it has been recorded from Co.'s Cork, Kerry, Clare, Wicklow and Fermanagh.

North and central Europe to Siberia and Japan.

Pandemis cerasana (Hübner)

Pl. 26, figs. 5–9

[*Pandemis ribeana*; Barrett, 1904: 173; Pierce & Metcalfe, 1922: 4, pl. 1 (♂, ♀ genitalia); Meyrick, 1928: 506; Ford, 1949: 51]

DESCRIPTION

♂ 16–22 mm, ♀ 16–24 mm. Sexual dimorphism not pronounced; antenna of male weakly dentate-ciliate, with deep sinus near base, flagellum whitish ochreous posteriorly, segments indistinctly barred with brown; antenna of female sparsely ciliate, without sinus, scaling of flagellum as in male.

Male (fig. 6). Frons and labial palpus varying from dark brown to light brown, sometimes with greyish white admixture; labial palpus paler interiorly. Forewing ground colour yellowish brown, weakly strigulate distally; markings light brown, edged with reddish brown; outer margin of basal fasciae sinuous, slightly oblique, median fascia strongly oblique. Hindwing almost uniform greyish brown.

94

Female (fig. 7). Similar to male.

Variation. Ground colour of forewing varying from pale ochreous-yellow (fig. 5) to greyish brown (fig. 8), the latter sometimes with an olive tint. The basal half of the wing may be suffused to varying degrees with drab or brown, usually most heavily along the dorsum (fig. 9), and the suffusion may extend over the markings.

COMMENTS

Similar to *P. corylana* but differing in the smoother general appearance and the reduced reticulation, the usually strongly divergent margins of the basal and median fasciae, and the uniformly greyish brown hindwing.

BIOLOGY

Ovum. Deposited in small batches on the leaves or branches, some hatching in late summer, others overwintering and hatching in the spring (Ford, 1949: 51).

Larva. Head light green varying to brownish green, with a conspicuous black postero-lateral stripe; prothoracic plate light green or yellowish green, marked with black laterally; abdomen light green, darker dorsally; pinacula light green, setae yellowish white; anal plate light green or yellowish green, spotted with black; anal comb yellowish, with 6–8 prongs; thoracic legs light green, terminal segments brownish or black; spiracles of prothorax and eighth abdominal segment 2–3 times the diameter of other abdominal spiracles. September to May; some larvae hatch in the autumn and over-winter in an early instar, usually in silken hibernacula on the twigs, recommencing to feed as the buds open in the spring; others do not hatch until the spring. Foodplants include fruit trees of all kinds, also *Acer, Alnus, Corylus, Quercus, Salix, Vaccinium, Tilia, Betula, Sorbus, Ulmus* and *Lysimachia*, the larva feeding in a rolled or folded leaf.

Pupa. Brown to brownish black, darker dorsally. May and June, in a whitish cocoon in a folded leaf or in the larval habitation.

Imago. June to August; frequenting gardens, orchards and open woodland, resting amongst the foliage during the day, flying actively at dusk, later coming to light.

DISTRIBUTION

Widely distributed and locally common throughout the British Isles as far north as Sutherland. In Scotland it is scarce in treeless areas but is often abundant in woodland, the larva usually feeding on *Betula*.

Europe to Siberia, China and Japan; north India.

Pandemis cinnamomeana (Treitschke)

Pl. 26, figs. 10–11

[*Pandemis cinnamomeana*; Barrett, 1904: 175; Pierce & Metcalfe, 1922: 4, pl. 2 (♂, ♀ genitalia); Meyrick, 1928: 506; Ford, 1949: 51]

DESCRIPTION

♂ ♀ 18–24 mm. Sexual dimorphism moderate; antenna of male weakly dentate-ciliate, with deep sinus near base, scape and flagellum white posteriorly; antenna of female sparsely ciliate, scape and flagellum cinnamon-brown.

Male (fig. 10). Frons and upper margin and inner surface of labial palpus white, crown and exterior of labial palpus greyish brown. Forewing ground colour reddish ochreous; markings darker, chestnut or cinnamon-brown; outer margin of basal fasciae and inner and outer margins of median fascia thinly edged with ochreous. Hindwing grey-brown.

Female (fig. 11). Frons, crown of head and labial palpus cinnamon-brown, palpus paler interiorly. Coloration and wing markings similar to those of male, excepting apex of hindwing which is tinged with reddish ochreous.

Variation. Forewing ground colour and markings very constant, showing only slight variation in depth and clarity; median fascia usually constricted near costa but sometimes of equal width throughout, and in some examples the normally conspicuous edging may be subdued and indistinct.

COMMENTS

Weakly marked specimens, especially those with drab coloration, may be confused with forms of *P. heparana*, from which the male may be distinguished by the white head and the female by the presence of reddish ochreous suffusion in the apical area of the hindwing.

BIOLOGY

Larva. Head pale yellow or orange-yellow, a conspicuous black postero-lateral stripe; prothoracic plate similar, edged with dark brown posteriorly, with scattered blackish markings; thoracic legs dark brown or black; abdomen shagreened, grass-green dorsally, whitish green ventrally; pinacula inconspicuous, paler than integument; anal plate light brown, darker anteriorly; anal comb with 5 or 6 slender prongs. May and June; recorded on *Acer*, *Betula*, *Abies alba*, *Larix*, *Prunus*, *Sorbus aucuparia* and *Vaccinium*, living amongst spun leaves. Reported also on *Picea sitchensis* in Wales; on the Continent *Pyrus* and *Quercus* are among the recorded foodplants (Hannemann, 1961: 5).

Pupa. Glossy chestnut-brown. June and July, in the larval habitation.

Imago. End of June and July; hiding during the day, usually amongst the foliage of large trees but sometimes showing a preference for tall bracken; readily disturbed by beating, darting to the ground with a rapid zig-zag flight. At night it may be taken at light, especially in dense woodland.

DISTRIBUTION

Local and seldom abundant, occurring in heavily wooded districts of England north to Westmorland and Durham; unknown in Scotland. In Wales known from Cardiganshire, Caernarvonshire (Michaelis, 1966: 109) and Flintshire, and in Monmouthshire the moth has been taken commonly amongst stands of *Tilia*, which is possibly also a foodplant. Barrett (1861: 13) records it from Powerscourt, Co. Wicklow, but this appears to be the only record from Ireland and is considered dubious by Beirne (1941: 81).

Widely distributed in Europe, except in the south; not recorded from Holland.

Pandemis heparana (Denis & Schiffermüller)

Pl. 27, figs. 1–5

[*Pandemis heparana*; Barrett, 1904: 176; Pierce & Metcalfe, 1922: 4, pl. 2 (♂, ♀ genitalia); Meyrick, 1928: 506; Ford, 1949: 51]

DESCRIPTION

♂ ♀ 16–24 mm. Sexual dimorphism not pronounced; antenna of male weakly dentate-ciliate, with deep sinus near base, scape and flagellum ochreous or ochreous-brown posteriorly, with dark brown segmental bars; female usually larger, apex of forewing more produced, scaling of flagellum similar to that of male.

Male (fig. 1). Head reddish brown varying to purple-brown or fuscous; labial palpus similar exteriorly, comparatively long and conspicuous. Forewing ground colour deep reddish ochreous or reddish brown, weakly reticulate; markings darker, median fascia well-defined, inner margin sharply angulate basad near middle, outer margin curved outward below costa; outer margin of basal fasciae more or less directly transverse, well defined at costa, becoming indistinct towards dorsum. Hindwing rather dark brownish grey.

Female (fig. 2). Similar to male.

Variation. Ground colour of forewing often lighter, the markings contrasting sharply (fig. 3); or ground colour and markings lighter, with reticulation more pronounced (fig. 4). Occasionally the general coloration of the forewing may have a distinct grey tinge; exceptionally the inner margin of the median fascia is not produced (fig. 5).

COMMENTS

Similar in general coloration to *P. cinnamomeana* but lacking the white scaling on the frons in the male and the reddish suffusion in the apical area of the hindwing in the female. The angular projection of the median fascia distinguishes *heparana* from *cinnamomeana* and from brownish forms of *P. cerasana*.

BIOLOGY

Ovum. Deposited in batches of 30–50 eggs on the upper surface of leaves of the foodplants in July, taking about two to three weeks to develop, the young larva overwintering in an early instar until the following spring.

Larva. Head variable in colour, light green, greenish brown or pale yellowish brown, with black lateral markings; prothoracic plate green or yellowish brown, with black postero-lateral markings; thoracic legs green shaded with brown; abdomen emerald-green, paler laterally, sometimes with a bluish green subdorsal line; pinacula concolorous with integument, setae long, whitish; anal plate green, variable, sometimes spotted with blackish; anal comb whitish, normally with 6–8 prongs. Polyphagous, feeding in May and June on various trees and shrubs, especially *Malus*, *Prunus*, *Pyrus*, *Tilia*, *Salix*, *Lonicera*, *Betula*, *Ribes*, *Vaccinium* (pl. 9, fig. 2) and *Myrica*, usually in a rolled leaf. When feeding on *Myrica* the spinning resembles that of *Choristoneura lafauryana*. Occasionally the larva is injurious to the flowers of fruit trees, and also causes superficial damage to the fruit (Adkin, 1924: 188; Balachowsky, 1966: 500).

Pupa. Brownish black, thoracic and dorsal regions darker. June and July, in the larval habitation, or in spun leaves, usually near the tip of a shoot.

Imago. Emerging in late June to August in southern parts of the British Isles, appearing somewhat later northwards, and in Scotland more frequent in August and September. During the day the moth rests amongst foliage of trees and bushes but is easily disturbed; it flies freely at dusk and readily comes to light.

DISTRIBUTION

A common and generally distributed species in England, extending into Scotland as far north as Perthshire and Argyll; in Scotland it is uncommon away from gardens. In Ireland frequent and generally distributed (Beirne, 1941: 81).

Throughout Europe and the Middle East to Siberia, China, Korea and Japan.

Pandemis dumetana (Treitschke)

Pl. 27, figs. 6, 7

[*Pandemis dumetana*; Barrett, 1904: 177; Pierce & Metcalfe, 1922: 4, pl. 2 (♂, ♀ genitalia). *Tortrix dumetana*; Meyrick, 1928: 509; Ford, 1949: 53]

DESCRIPTION

♂ ♀ 18–22 mm. Sexes similar; antenna of male weakly dentate–ciliate, without basal sinus, scape dull brown, flagellum light brown posteriorly with indistinct dark segmental bars; female slightly larger.

Male (fig. 6). Frons, vertex and labial palpus dull brown, palpus paler interiorly. Forewing ground colour characteristically dull brown strigulate with dark grey-brown, veins lined with dark grey-brown forming a reticulate pattern; markings dark grey-brown, margins of basal and median fasciae well defined, edged with dark brown; usually a distinct dark stria from triangular pre-apical spot to tornal area. Hindwing white lightly suffused with grey, weakly strigulate apically.

Female (fig. 7). Similar to male.

Variation. Markings and general coloration very constant; the extent of the strigulation in the forewing shows slight variation, and in some examples the stria from the pre-apical spot is interrupted or terminates near the tornus.

COMMENTS

The relatively broader forewing and weakly strigulate whitish hindwing are characteristic of this species.

BIOLOGY

Ovum. Deposited in batches of various sizes on the underside of the leaves of the foodplants during July and August; the incubation period lasts from 7–10 days (Balachowsky, 1966: 504–505, fig. 219).

Larva. Slender; head green with ocellar region and posterior margin blackish;

prothoracic and anal plates green; abdomen green, with paler sub-dorsal lines, sometimes with a dark green dorsal line; pinacula green. September to June. The second instar larva constructs a silken hibernaculum in the fold of a leaf and hibernates until the following spring (Balachowsky, 1966: 505). Polyphagous on herbaceous and other plants, including *Fragaria*, *Lysimachia*, *Centaurea*, *Mentha*, *Lathyrus palustris*, *Rubus*, *Urtica*, *Thalictrum*, *Quercus* and *Hedera*, living within rolled leaves and occasionally flowerheads.

Pupa. Blackish brown. June and July, in the larval habitation; usually developing in 8–10 days.

Imago. July and August; flying freely from sunset onwards and coming to light.

DISTRIBUTION

Very local and apparently known in the British Isles only from Kent, Sussex, Cambridgeshire, Huntingdonshire, Norfolk and Lancashire. It is found in chalk habitats, such as Barham Downs and Wye in Kent, and near Lewes, Eastbourne, Worthing and Shoreham in Sussex. In marked contrast to the chalkland biotope, it occurs in the comparatively low-lying and less well-drained areas of Monkswood, Huntingdonshire. It also occurs in Wicken Fen, Cambridgeshire, where in the past it was reputed to be common although apparently not found in other fens in East Anglia. Two specimens in the L. T. Ford collection were taken at King's Lynn, Norfolk, by C. Fenn in 1911. The only known record from Lancashire is of specimens taken by Mansbridge (1905: 115) on Kirby Moss, near Liverpool.

Throughout Europe and the Middle East to Siberia, Korea and Japan.

Argyrotaenia pulchellana (Haworth)

Pl. 27, figs. 8, 9

[*Lophoderus politana*; Barrett, 1904: 187. *Argyrotaenia politana*; Pierce & Metcalfe, 1922: 1, pl. 1 (♂, ♀ genitalia). *Eulia politana*; Meyrick, 1928: 510; Ford, 1949: 53]

DESCRIPTION

♂ ♀ 12–16 mm. Sexes similar except for the usually slightly more pronounced apex of the forewing in the female; antenna of male shortly ciliate.

Male (fig. 8). Forewing ground colour silver-white, strigulate with grey; markings dark reddish brown, sprinkled with black; margins of basal and median fasciae irregular. Hindwing grey; cilia whitish.

Female (fig. 9). Similar to male.

Variation. The normally conspicuous silver-white ground colour of the forewing may be obscured to varying degrees by greyish suffusion; the reddish brown markings are sometimes more liberally sprinkled with black, the general coloration then appearing comparatively sombre. The markings are very irregular and variable, and frequently the median fascia is broader, reducing the whitish ground colour.

COMMENTS

The silver-white ground coloration of the forewing is characteristic of this species.

BIOLOGY

Ovum. Deposited in batches of 40–50 eggs, usually on the upper surface of the leaves of the foodplant; hatching in 10–21 days.

Larva. Head light green or yellowish green, epicranium darker posteriorly; prothoracic plate yellowish green with blackish postero-lateral stripe, posterior margin narrowly edged with black; abdomen clear green, sometimes tinged with brown; pinacula indistinct, paler than integument; anal comb present, bearing 6–8 prongs. In the south, June and September; polyphagous, feeding especially on *Erica, Calluna, Myrica* and *Vaccinium;* also on *Betula, Prunus, Abies* and *Pinus* (Styles, 1959: 43), *Larix, Picea* and various herbaceous plants including *Gentiana pneumonanthe, Dryas octopetala, Centaurea* and *Aster.* Recently this species has become a minor pest of *Vitis* and *Malus* on the Continent, attacking both the foliage and fruit (Balachowsky, 1966: 507). The newly hatched larvae skeletonize the underside of the leaf along the midrib, later dispersing and feeding in spun leaves.

Pupa. Pale brown. October to April, overwintering in a cocoon in debris on the ground; June, in a silken cocoon in spun leaves, developing in 7–10 days.

Imago. Bivoltine in the south of the British Isles, the first generation occurring from the end of April to May, the second from the end of June to July; univoltine in the north, occurring from the end of April to June. A common heathland and moorland species, often found in numbers resting by day on heather and ling, its cryptic coloration making detection difficult; in the late afternoon the moth flies freely about its habitat.

DISTRIBUTION

Occurring on heaths and moors throughout the British Isles as far north as Sutherland, but apparently absent from the Outer Hebrides, Orkneys and Shetlands.
Europe to Asia Minor; North America.

Homona menciana (Walker)

Three examples of this species were reared in Sussex from *Camellia* plants imported from Japan in 1964 (Baker, 1964: 275). The plants were imported during the winter and the moths emerged in mid-April.
This species has a tropical distribution and is unlikely to become established in the British Isles. It is closely related to and possibly a form of *H. coffearia* (Nietner), which is a pest of tea in the Far East.

Archips oporana (Linnaeus)

Pl. 27, figs. 10, 11

[*Tortrix piceana*; Barrett, 1904: 155. *Archips piceana*; Pierce & Metcalfe, 1922: 1, pl. 1 (♂, ♀ genitalia). *Cacoecia piceana*; Meyrick, 1928: 502; Ford, 1949: 49]

DESCRIPTION

♂ 19–21 mm, ♀ 22–28 mm. Sexual dimorphism pronounced; male with strong costal fold from base to near middle, broad throughout; female without costal fold, apex of forewing produced, hindwing suffused with orange apically.

Male (fig. 10). Forewing ground colour purple-brown; markings velvety chestnut-red or purplish red, thinly edged with white; inner margin of sub-basal fascia roughened with cream scales. Hindwing dark grey-brown, sometimes tinged with cupreous apically.

Female (fig. 11). Ground colour of forewing purplish ochreous, weakly strigulate; markings chestnut-brown. Hindwing fuscous, apical half with cupreous suffusion.

Variation. In both sexes the markings of the forewing show little variation, but the general coloration varies in depth and intensity.

COMMENTS

The conspicuous purple-brown ground colour of the forewing and the dark grey-brown hindwing distinguish the male of this species from those of *A. podana* and *A. betulana*; the chestnut-brown markings and characteristic weakly strigulate pattern of the forewing distinguish the female.

BIOLOGY

Larva. Head light brown varying to black; prothoracic plate yellowish brown or brown, thoracic legs black; abdomen yellowish green, spiracles black, pinacula brown· September to early June; on *Pinus sylvestris*, *Abies alba*, *Picea glauca* and *Juniperus*. In the autumn the larva lives in a silken tube spun among the needles, overwintering within; in the spring it feeds on the new shoots, and also bores into the stems, causing them to shrivel and break off. On juniper it spins a thick white web among the leaves. *Larix* is recorded as an additional foodplant on the Continent (Swatschek, 1958: 39; Hannemann, 1961: 13).

Pupa. Reddish brown. June and July, in the larval habitation.

Imago. End of June and July; during the early part of the day it may be disturbed from rest by tapping the branches of large fir trees, when it will flutter to another branch or to the ground. From mid-afternoon to early evening and at dusk it flies actively around the conifers.

DISTRIBUTION

Very local, recorded only from conifer woods in Surrey, Hampshire, Dorset, Berkshire and Oxfordshire. Prior to 1888 it was known only from a few specimens taken in the New Forest and Woolmer Forest, Hampshire (Adkin, 1888: 160), but subsequently was found to be not uncommon at Esher and Hindhead, Surrey. In recent years it appears to have become scarcer.

Europe to China and Japan, sometimes sufficiently abundant to be destructive to pine forests; North America.

Archips podana (Scopoli)

Pl. 28, figs. 1–5

[*Tortrix podana*; Barrett, 1904: 157. *Archips podana*; Pierce & Metcalfe, 1922: 2, pl. 1 (♂, ♀ genitalia). *Cacoecia podana*; Meyrick, 1928: 503; Ford, 1949: 49]

DESCRIPTION

♂ 19–23 mm, ♀ 20–28 mm. Sexual dimorphism pronounced; male with costal fold from base to near middle, broad at base, narrowing rather abruptly beyond; female without costal fold, apex of forewing strongly produced, prominent.

Male (fig. 1). Forewing ground colour purplish ochreous, suffused with purple in the basal two-thirds, weakly strigulate distally; markings velvety dark red-brown; inner margin of sub-basal fascia roughened with whitish ochreous scales; inner margin of median fascia edged with whitish; subterminal stria thickened to form an oblique fascia. Hindwing grey, mixed with orange apically.

Female (fig. 2). Forewing ground colour almost uniform purplish ochreous, extensively strigulate with brown, radial veins lined with brown and forming a reticulate pattern; markings weak, often represented in outline only; subterminal stria linear. Hindwing fuscous grey, apical half bright orange weakly strigulate with fuscous.

Variation. In both sexes the markings and coloration show limited variation. A not uncommon melanic form occurs in which the markings of the forewing are partially obliterate in both sexes, and in the male (fig. 3) the orange coloration is absent on the hindwing, but is usually present though reduced in the female (fig. 4). Rarely a pale-coloured form of the male occurs (fig. 5) lacking the strong purplish suffusion and with the markings very weak.

COMMENTS

The male lacks the purple-brown ground colour of the forewing of *A. oporana*; the female is more similar to *A. betulana* but readily distinguished by the orange coloration in the apical half of the hindwing.

BIOLOGY

Ovum. Deposited in several batches of 50–100 eggs on the upper surface of the leaves during June and July. The egg mass is covered with a wax-like secretion and so closely matches the green colour of the leaf that it can only be detected with difficulty; hatching in 17–23 days (Massee, 1946: 59; Balachowsky, 1966: 530, fig. 230).

Larva. Head chestnut-brown, sometimes black; prothoracic plate chestnut-brown with posterior margin darker, or black, divided medially by a narrow whitish or brown line, anterior margin of prothorax whitish; abdomen light yellowish green or green; pinacula paler than integument; thoracic legs brownish black or black. July to early May; polyphagous on a wide range of deciduous trees, also on shrubs and occasionally conifers, feeding usually in spun leaves (pl. 9, fig. 3) or flowers. The larva is common in fruit orchards and sometimes a pest of apple. Larvae hatching from an egg mass on apple feed on both surfaces of the leaf for a few days before dispersing; they then spin a fine web on the underside of a leaf, where they continue to feed until the first ecdysis (Massee, 1946: 60). In the second and third instars the larvae may feed on the skin of the fruit (pl. 9, fig. 4), spinning a leaf to the surface for shelter, and continue to feed in this way until the end of October or until the fruit are picked. At this time they normally spin a silken hibernaculum on a twig or spur (Balachowsky, 1966: 531, fig. 231) and hibernate as third instar larvae until spring, when they enter the fourth instar and feed on the opening buds, often boring into them and causing serious damage, and later on the leaves, usually spinning two or more together.

Pupa. Dark yellowish brown, brownish black dorsally. June and July, in the larval habitation.

Imago. End of June and July; resting openly during the day on foliage of trees and bushes; flying freely in the late afternoon and at dusk and coming readily to light after dark. Males may often be observed assembling around a freshly emerged female (South, 1898: 91).

DISTRIBUTION

Common and widely distributed in England and Wales except in mountainous regions; occurring plentifully in most gardens and orchards in the southern counties and East Anglia. Rare in Scotland and apparently mainly of south-western distribution; recorded from the Clyde area and also from Berwickshire and Roxburghshire. In Ireland widely distributed and generally common in gardens.

Europe north to Lapland, Asia Minor, southern Russia, and eastern Siberia to Japan; recorded as a probable introduction to North America (Freeman, 1958: 16).

Archips betulana (Hübner)

Pl. 28, figs. 6, 7

[*Tortrix decretana*; Barrett, 1904: 159. *Archips decretana*; Pierce & Metcalfe, 1922: 2, pl. 1 (♂, ♀ genitalia). *Cacoecia decretana*; Meyrick, 1928: 503; Ford, 1949: 50]

DESCRIPTION

♂ 19–21 mm, ♀ 20–24 mm. Sexual dimorphism pronounced; male with narrow costal fold reaching to about one-third; female without costal fold, apex strongly produced, prominent.

Male (fig. 6). Forewing ground colour ochreous suffused with purplish, suffusion weaker distally; markings velvety reddish chestnut; inner margin of sub-basal fascia overlaid with coarse greyish ochreous scales; median fascia reaching to costa, inner margin and anterior of outer margin edged with whitish. Hindwing grey, costal half suffused with whitish ochreous.

Female (fig. 7). Forewing ground colour ochreous-brown, reticulate with dark brown; markings dull brown or reddish brown, basal area reticulate; median fascia weak, usually interrupted below costa. Hindwing as in male but costal half more broadly suffused with whitish ochreous, apex with weak orange tinge.

Variation. In both sexes there is considerable variation in the coloration and development of the markings of the forewing. In the male the markings sometimes have a heavier purplish suffusion and stand out strongly, or may be weak and diffuse and show very little contrast with the ground colour. In the female the general coloration of the forewing varies from dull brown to reddish ochreous; the triangular pre-apical spot and costal part of the median fascia are sometimes deeper brown and more conspicuous; very rarely almost unicolorous light ochreous-brown forms occur.

COMMENTS

Both sexes differ from *A. podana* in the whitish ochreous suffusion in the costal half of the hindwing, and the male differs further in lacking orange coloration in the apical area.

BIOLOGY

Larva. Head and prothoracic plate shining brown or dark brown, anterior margin of prothoracic plate whitish, posterior margin sometimes edged with whitish; abdomen yellowish brown or greenish brown, darker dorsally, with a pale subspiracular line; pinacula paler than integument, generally small and inconspicuous; anal plate green or yellowish green (Atmore, 1889: 244). Feeding from May to the beginning of July on *Myrica gale*, spinning together the leaves of a terminal shoot, very often including an old seedhead. On the Continent recorded also on *Betula*, *Corylus* and *Vaccinium*, but in the British Isles it seems to be restricted to *Myrica*.

Pupa. Blackish, wings tinged with brown. July, in a slight web in the larval habitation or in leaf litter on the ground.

Imago. August; hiding during the day amongst foliage and not easily disturbed. The male becomes active at sunset, the female at dusk, and both sexes fly with great swiftness over the foodplant.

DISTRIBUTION

In the British Isles this species has been recorded only from Norfolk in the neighbourhood of King's Lynn on boggy heaths where *Myrica* is found. It was first taken about 1881 by A. E. Atmore (Warren, 1887a: 125; Barrett, 1904: 160) and was not uncommon for some years subsequently, but then became scarce and was last seen about 1900 (H. C. Huggins, *in litt.*). The recent rediscovery of *Choristoneura lafauryana*, which was also only known from the King's Lynn area, suggests that *betulana* may still occur there.

Europe to north-west and south-east Russia; Korea.

Archips crataegana (Hübner)
Pl. 28, figs. 8–11

[*Tortrix crataegana*; Barrett, 1904: 160. *Archips crataegana*; Pierce & Metcalfe, 1922: 2, pl. 1 (♂, ♀ genitalia). *Cacoecia crataegana*; Meyrick, 1928: 503; Ford, 1949: 50]

DESCRIPTION

♂ 19–22 mm, ♀ 23–27 mm. Sexual dimorphism moderately pronounced; male with costal fold extending beyond middle; female without costal fold, apex more pronounced.

Male (fig. 8). Forewing ground colour light brown; markings velvety chocolate-brown; median fascia not reaching costa, inner margin edged with whitish, angulate near middle; pre-apical spot produced nearly to termen. Hindwing brownish grey.

Female (fig. 9). Forewing ground colour usually darker brown than in male, reticulate with dark fuscous; median fascia reaching costa, not edged inwardly with whitish. Hindwing brownish grey, apex sometimes tinged with yellow.

Variation. In the male the forewing ground colour varies from dark grey to whitish ochreous (fig. 10), the markings usually remaining dark. In the female the ground colour varies from dark brown to light ochreous (fig. 11); the reticulation and markings vary in intensity and may contrast strongly with a light ground colour. When present, the yellowish suffusion at the apex of the hindwing varies in intensity and sometimes the apex may appear distinctly yellow.

COMMENTS

Usually larger than *A. xylosteana* and differing in the broader forewing which in general appearance is browner and less variegated. The male is further separated by the median fascia which is obsolete on the costa; the female by the fuscous reticulation and the more prominent apex of the forewing.

BIOLOGY

Ovum. Deposited during June–August on the bark, usually high up the tree, in batches of about 30 eggs laid close to each other in a white cement-like substance, which gives the egg mass the appearance of a bird-dropping (Balachowsky, 1966: 535, fig. 234). The ova overwinter and hatch the following spring (Massee, 1946: 63).

Larva. Head and prothoracic plate shining black, prothoracic plate large, less shining than head; abdomen dark olive-green or black, translucent, integument shagreened; pinacula black, setae long, whitish; anal plate shining black, strongly convex anteriorly; anal comb black, with 6–8 prongs. April and May, feeding on the foliage of *Quercus*, *Ulmus*, *Fraxinus*, *Tilia*, *Salix*, and orchard and other deciduous trees and shrubs. In the first instar the larva feeds mainly on the under surface of the leaf; in later instars the edge of the leaf is rolled into a tight tube (pl. 10, fig. 1).

Pupa. Head and thoracic region shining black, abdominal segments dull blackish brown (Sheldon, 1922: 195). June and July, in the larval habitation.

Imago. Late June to August; frequenting rides and borders of woods and forests, and orchards and wooded areas generally. During the day the moth may be disturbed from rest among the foliage, dashing swiftly away to another hiding place, often at a good height. At sunset it may be glimpsed flying erratically about the trees and is difficult to net.

DISTRIBUTION

A typically woodland species, never very common but generally distributed in the southern counties of England, occurring in mixed woodland westwards to Monmouth-shire, and more sparingly northwards to Yorkshire, Durham and Northumberland. Unknown from Cheshire and Lancashire, but occurs in Flintshire, North Wales. In Scotland it has been recorded from Berwickshire (Bolam, 1929: 115). In Ireland recorded from Co.'s Kerry, Wicklow and Antrim (Beirne, 1941: 80), and evidently uncommon.

North and central Europe, Asia Minor, Transcaucasia to China and Japan.

Archips xylosteana (Linnaeus)
Pl. 29, figs. 1–4

[*Tortrix xylosteana*; Barrett, 1904: 162. *Archips xylosteana*; Pierce & Metcalfe, 1922: 2, pl. 1 (♂, ♀ genitalia). *Cacoecia xylosteana*; Meyrick, 1928: 503; Ford, 1949: 50]

DESCRIPTION

♂ 15–21 mm, ♀ 16–23 mm. Sexual dimorphism not pronounced; male with narrow costal fold to beyond middle; female without costal fold, apex slightly produced.

Male (fig. 1). Forewing ground colour whitish ochreous, partially suffused with olive-grey; markings reddish brown, thinly edged with clear ground colour; inner margin of median fascia sinuate, pre-apical spot semi-ovate, usually contiguous with stria-like marking to tornus.

Female (fig. 2). Forewing ground colour as in male; markings less reddish, often darker. Hindwing grey, apical area sometimes tinged with yellow or cupreous.

Variation. Considerable minor variation is found in the forewing markings. The inner margin of the median fascia may be very shallowly sinuous or almost angulate at the middle; when well developed the tornal marking is strongly connected to the pre-apical spot, or it may be reduced to a small irregularly shaped spot. In some specimens the ground colour and markings may be extremely pale (fig. 3) and in others abnormally dark (fig. 4).

COMMENTS

Most similar to *A. crataegana* but differing in the more variegated appearance and smaller size.

BIOLOGY

Ovum. Deposited in small batches on the trunk and branches of the foodplant during July. The egg mass is covered with a brownish secretion which blends in colour with the bark and conceals its contour (Balachowsky, 1966: 539).

Larva. Head shining black; prothoracic plate dark brown or black, edged with white anteriorly and sometimes posteriorly, divided medially by a narrow whitish line; abdomen whitish grey varying to dark bluish grey, paler or whitish laterally; pinacula light grey; setae whitish; anal plate black or blackish brown; anal comb present; thoracic legs black, prolegs green dotted with black. April to early June; polyphagous on deciduous trees and shrubs, particularly *Quercus, Ulmus, Tilia, Corylus, Acer,*

Fraxinus and fruit trees, living within a rolled leaf (pl. 10, figs. 2–4). *Hypericum*, *Rubus* and *Lonicera* are also known foodplants and the larva is occasionally found on *Abies*.

Pupa. Dark brown or black. June, in the larval habitation.

Imago. July; resting during the day amongst the foliage of trees and shrubs, but readily disturbed and flying quickly to another hiding place. At dusk the moths fly freely and are often to be seen flying wildly about hedgerows and gardens.

DISTRIBUTION

Occurring in woodlands and gardens in most parts of Britain as far north as Perthshire, but in Scotland more frequently in the south. In Ireland widely distributed and frequent in woodland.

Europe to Asia Minor, east Siberia, China and Japan.

Archips rosana (Linnaeus)

Pl. 29, figs. 5–7

[*Tortrix rosana*; Barrett, 1904: 163. *Archips rosana*; Pierce & Metcalfe, 1922: 2, pl. 1 (♂, ♀ genitalia). *Cacoecia rosana*; Meyrick, 1928: 504; Ford, 1949: 50]

DESCRIPTION

♂ 15–18 mm, ♀ 17–24 mm. Sexual dimorphism pronounced; male with costal fold to beyond middle; female without costal fold, apex more pointed, apex of hindwing usually distinctly tinged with orange-yellow.

Male (fig. 5). Forewing ground colour varying from light brown to purplish brown, weakly strigulate; markings dark brown, often reddish tinged; a dark stria from pre-apical spot to tornal area. Hindwing grey; cilia paler.

Female (fig. 6). Forewing ground colour more strongly strigulate with dark brown; markings weaker, outer margin of median fascia diffuse, stria from pre-apical spot weakly developed. Hindwing grey, apical area suffused with orange-yellow.

Variation. In both sexes the forewing pattern shows considerable minor variation and the general coloration tends to be darker rather than lighter. Occasionally specimens are found with the ground colour light brown, the markings contrasting strongly; more frequent are dark forms in which the markings are obscure (fig. 7).

COMMENTS

A generally smaller species than *A. crataegana*; the male is readily distinguished by the unbroken median fascia; the more uniform coloration of the forewing and the relatively fine subterminal stria separate the female.

BIOLOGY

Ovum. Deposited in greenish masses, usually on the bark; August, overwintering to April (Balachowsky, 1966: 519, fig. 227).

Larva. Head light brown varying to shining black; prothoracic plate light brown or black, sometimes greenish anteriorly, posterior and lateral margins usually darker; thoracic legs blackish; abdomen light green varying to dark green dorsally, paler ventrally; pinacula paler than ground colour; anal plate light brown or green. April to early June; polyphagous, mainly on fruit trees and other deciduous trees and shrubs including raspberry and blackcurrant; occasionally on conifers. On hatching the larva feeds in a bud, later in a rolled leaf (pl. 10, fig. 5) or spun leaves, and also on the flowers and young fruits.

Pupa. Rich dark brown, wings and abdomen paler laterally. June and July, in a cocoon in the larval habitation.

Imago. July to early September; during the day the moth rests amongst foliage of trees and bushes but is readily disturbed, flying freely at dusk and coming to light.

DISTRIBUTION

Generally distributed throughout the British Isles as far north as the Outer Hebrides. It is often abundant in the southern counties of England, defoliating apple and other fruit trees in orchards (Massee, 1946: 64). In Scotland it is common in the south, particularly in gardens, but is rare in the Highlands. In Ireland widely distributed in wooded areas and sometimes common in orchards and gardens; abundant in the Burren, Co. Clare, and also on the bogs of West Galway where the foodplant is *Myrica*.

Throughout Europe, including Scandinavia to the Arctic Circle, Asia Minor and Transcaucasia; coastal regions of North America.

Choristoneura diversana (Hübner)

Pl. 29, figs. 8, 9

[*Pandemis diversana*; Barrett, 1904: 178. *Archips diversana*; Pierce & Metcalfe, 1922: 3, pl. 1 (♂, ♀ genitalia). *Tortrix diversana*; Meyrick, 1928: 509; Ford, 1949: 53]

DESCRIPTION

♂ 15–20 mm, ♀ 19–23 mm. Sexes similar; male without costal fold; female usually slightly larger, coloration darker.

Male (fig. 8). Forewing ground colour light brown, weakly strigulate distally; markings deeper brown, outer margin of basal fasciae oblique, slightly convex; inner margin of median fascia slightly irregular at middle. Hindwing grey varying to brownish grey.

Female (fig. 9). Similar to male but forewing usually more extensively strigulate with darker brown.

Variation. In the male the forewing ground colour varies slightly in intensity but is more often lighter than darker; in light specimens the markings tend to be stronger and more pronounced and may have the outer margin of the basal fasciae and inner margin of the median fascia weakly edged with orange-brown; rarely the median fascia is interrupted above the middle. Occasionally the ground colour is dark brown but the markings remain moderately distinct; in the female it is more often darker, the markings then becoming diffuse and obscure.

COMMENTS

Occasionally confused with *Syndemis musculana* but distinguished by the absence of white coloration in the forewing.

BIOLOGY

Ovum. Flattened, green; deposited in large batches during July (Fenn, 1890: 216).

Larva. Head and prothoracic plate dark brown or reddish brown; abdomen normally green or greyish green, but varying to grey or greyish white; pinacula yellow. Polyphagous on various trees, shrubs and herbaceous plants, including *Betula*, *Malus*, *Pyrus*, *Prunus*, *Salix*, *Quercus*, *Rhamnus*, *Lonicera*, *Syringa*, *Trifolium* and *Achillea*; in spun leaves. The ova hatch in August and September, the young larvae hibernating until spring, becoming full-grown by the end of May.

Pupa. Reddish brown, slender. June, in the larval habitation or in spun leaves.

Imago. July; flying during the evening and at dusk over the tops of young trees and bushes, later coming to light. During the day the moth rests amongst the foliage, dropping to the ground when disturbed.

DISTRIBUTION

Local, restricted mainly to the southern counties of England, ranging from Kent, where it has been observed in great abundance in plum orchards near Sittingbourne (Huggins, 1959: 121), to Somerset (Emmet, 1969: 96), Devon, Hereford and Norfolk; sometimes common in Wicken Fen (Cambridgeshire) where the main foodplant appears to be *Rhamnus*; also recorded from Yorkshire. The only Irish record is from the Belfast district (Watts, 1894: 129), which Beirne (1941: 82) considers doubtful.

North and central Europe to the Ukraine, Caucasus and Asia Minor, and east Siberia to Japan; sometimes occurring as a defoliator of minor importance in apple, pear and plum orchards.

Choristoneura hebenstreitella (Müller)

Pl. 29, figs. 10–11

[*Tortrix sorbiana*; Barrett, 1904: 165. *Archips sorbiana*; Pierce & Metcalfe, 1922: 2, pl. 1 (♂, ♀ genitalia). *Cacoecia sorbiana*; Meyrick, 1928: 504; Ford, 1949: 50]

DESCRIPTION

♂ 19–25 mm, ♀ 24–30 mm. Sexual dimorphism not pronounced; male with short costal fold from above base to near middle; female larger, without costal fold, apex more produced.

Male (fig. 10). Forewing ground colour pale yellowish brown, weakly reticulate with light brown; markings olive-brown; basal fasciae complete, outer margin oblique, slightly sinuous; outer margin of median fascia indented below costa, inner margin usually unbroken. Hindwing dark grey; cilia paler.

Female (fig. 11). Similar to male except for larger size, and the slightly heavier reticulation and the stronger median fascia of the forewing. Hindwing with apical cilia often tinged with yellow.

Variation. Slight variation of the forewing ground colour occurs in both sexes and some specimens, especially females, have a distinct yellowish appearance. In the markings the greatest variation is found in the median fascia which may be completely broken below the costa.

COMMENTS

The large size and distinctive olive-brown markings are characteristic of this species.

BIOLOGY

Larva. Head dark brown or black; prothoracic plate brown, postero-lateral margins black, anterior margin straight, edged with whitish; abdomen greyish dark green; pinacula small, greenish white; thoracic legs black; anal plate yellowish brown. September to May, feeding on *Quercus, Betula, Corylus, Salix caprea, Myrica, Vaccinium* and *Hedera*; also recorded on *Sorbus, Sambucus, Prunus, Pyrus, Malus* and *Ulmus*. The first instar larva, which overwinters in a round or oval silken hibernaculum spun in a crevice or at the base of a twig (Pato˘ka, 1958: 194), occasionally attacks apple fruits (Hey & Massee, 1934: 228) in the autumn. In the spring the larva feeds in spun or rolled leaves.

Pupa. Brownish black. June, in the larval habitation.

Imago. June and July; hiding during the day amongst foliage of trees and bushes and not easily disturbed. At dusk it flies freely and may often be seen flying wildly about open spaces, rides and borders of woods.

DISTRIBUTION

A typically woodland species, occurring as far north as Northumberland and Berwickshire (Bolam, 1929: 115); most common in large woods in the south of England. On the Pennines it occurs up to 1500 ft on open moors where it feeds on *Vaccinium*, and is common at Arnside, Westmorland; also recorded on *Vaccinium* from similar localities in the mountains of North Wales (Michaelis, 1969b: 188). In Ireland it is uncommon but has been recorded from Co.'s Wicklow, Dublin, Sligo, Tipperary and Galway (Beirne, 1941: 80), and has been taken at Glengarriff, Co. Cork (Huggins, 1953b: 252). Europe to Asia Minor; Japan.

Choristoneura lafauryana (Ragonot)

Pl. 30, figs. 1–3

[*Tortrix lafauryana*; Barrett, 1904: 166. *Pandemis lafauryana*; Pierce & Metcalfe, 1922: 4, pl. 2 (♂, ♀ genitalia). *Cacoecia lafauryana*; Meyrick, 1928: 504; Ford, 1949: 50]

DESCRIPTION

♂ 18–22 mm, ♀ 20–24 mm. Sexes moderately dimorphic; male with short costal fold from beyond base to near middle; anal tuft large and prominent; female without costal fold. In both sexes the forewings have a strong silky sheen.

Male (fig. 1). Ground colour of forewing deep yellow-ochreous or yellow-brown, indistinctly strigulate; markings reddish brown, usually discernible though often rather weak and obscure (fig. 2); outer margin of basal fascia very oblique; outer margin of median fascia interrupted below costa. Hindwing light grey.

Female (fig. 3). Similar to male but markings weaker and usually indicated only by faint lines and striae.

Variation. Ground colour of forewing varying from light yellow-ochre to red-brown; almost unicolorous fuscous-brown forms occur rarely. Males with light yellow-ochre ground colour have the markings more pronounced, especially on the costa when the median fascia is interrupted.

COMMENTS

The male of this species may be confused with well-marked forms of the male of *Sparganothis pilleriana*, and the almost unicolorous fuscous-brown form of the female with similar dark females of that species. The exceptionally long labial palpi of *pilleriana* readily separate the two species.

BIOLOGY

Ovum. Deposited in elongate batches of 70–100 eggs on the upper surface of the leaves of the foodplant during July and August (Balachowsky, 1966: 515, fig. 225).

Larva. Head dull yellowish green, ocellar region black; prothoracic plate dark green; abdomen pea-green, a darker green dorsal line; pinacula paler than integument, rather inconspicuous; anal plate dark green. August to June and early July, overwintering in an early instar; on *Myrica gale*, spinning the upper terminal leaves into a vertical tube (pl. 11, fig. 1). On the Continent it has been reported as a minor pest on *Malus*, *Pyrus*, *Ribes*, *Fragaria* and *Boehmeria nivea* (Balachowsky, 1966: 515), and also on *Salix* and *Forsythia* (Zangheri, 1965: 8).

Pupa. Shining black. July, in the larval habitation or spun up among dead leaves or litter on the ground, the moth emerging in 8–10 days.

Imago. July and August; resting during the day amongst the foliage of the foodplant, the male flying when disturbed, the female dropping to the ground. The flight period begins at sunset, the male appearing first and flying in a lively manner about the foodplant; the female flies a little later towards dusk and is swifter on the wing. This species has been taken at light.

DISTRIBUTION

In the British Isles known only from Norfolk on boggy heaths near King's Lynn. The first two specimens were taken in July, 1880, by A. E. Atmore (1881a: 17; 1881b: 153) and the species occurred commonly up to about 1900, but apparently became scarce after that time. According to H. C. Huggins (*in litt.*), Atmore informed him in 1923 that he had not bred the species since 1900. However, the species is still to be found in the

King's Lynn district, but is apparently rare. A male was bred on 24th July, 1962, from a larva collected after several days searching at the beginning of the month. The spinning of this larva is illustrated.

Spain, France, Belgium, Netherlands, Denmark, Germany, Switzerland, north Italy, Transcaucasia, Siberia, Amurland and Japan.

Cacoecimorpha pronubana (Hübner)
Pl. 30, figs. 4–6

[*Tortrix pronubana*; Barrett, 1907: 268.—*pronubana*; Pierce & Metcalfe, 1922: 8, pl. 3 (♂, ♀ genitalia). *Cacoecia pronubana*; Meyrick, 1928: 505; Ford, 1949: 51]

DESCRIPTION

♂ 14–18 mm, ♀ 16–24 mm. Sexes moderately dimorphic; male without costal fold; female usually larger, hindwing without pronounced black scaling at apex and along termen.

Male (fig. 4). Forewing ground colour deep ochreous-brown, weakly reticulate with darker brown distally; markings dark brown or purplish brown, suffused with dull plumbeous; outer margin of basal fasciae usually indicated in outline only; inner margin of median fascia straight; pre-apical spot extended as a broad terminal fascia. Hindwing bright orange; apex, terminal and inner margins suffused with black scaling; cilia orange-yellow.

Female (fig. 6). Forewing coloration similar to that of male, reticulation more pronounced; median fascia weaker towards costa; pre-apical spot reduced, emitting a usually distinct stria from inner margin to tornal area. Hindwing orange; black scaling present at inner margin, sparse or absent along termen and at apex.

Variation. The coloration of the male varies in intensity; melanistic forms occur (fig. 5), but are less frequent in the female.

COMMENTS

The bright orange coloration of the hindwing is characteristic of this species.

BIOLOGY

Ovum. Deposited in small batches, greenish in colour, on the upperside of a leaf, hatching in about two to three weeks (Adkin, 1908: 49, pl. 2, fig. 7; Smart, 1917: 280; Ford, 1949: 51).

Larva. Head varying from grey-green or green to yellowish brown, ocellar region blackish brown, a usually heavy black or blackish brown marking on postero-lateral margin, posterior margin variably edged with blackish brown, epicranium sometimes marked with brown; prothoracic plate yellowish brown, greenish brown or grey-green, medial sulcus line paler, a usually conspicuous elliptical or triangular black or blackish brown marking on the postero-lateral margin; several other smaller and less conspicuous markings also usually present, one rather elongate and situated on the lateral margin anteriorly, another as an irregular dot in the angle formed by the posterior margin and mid-dorsal line, and several which may appear as irregular brownish specks or blemishes in central area of each half of the plate; abdomen very variable in colour, ranging from various shades of green—bright green, yellow-green, olive-green, grey-green—to greyish brown, paler ventrally; pinacula lighter than ground colour, setae whitish; peritreme of spiracles dark brown or black; anal plate light green marked with brown or blackish brown; anal comb green, usually with 4 long and 2 short (outer) prongs; thoracic legs green marked with brown. Fisher (1924: 418) has shown that the colour of the larva varies according to the plant on which it is feeding, a phenomenon which probably occurs in many species of Tortricidae. May to August, a few overwintering; September to April, the broods overlapping. The majority of larvae from the early summer brood of moths reach maturity and produce imagines from the end of August to September, but a few overwinter and produce moths in the spring. Larvae from the late summer brood of imagines overwinter in an early instar, recommencing to feed in the spring and producing moths from May onwards. Polyphagous, feeding on a wide variety of plants, including *Hippophae rhamnoides* (pl. 11, fig. 3), but especially common on *Euonymus japonicus* (pl. 11, fig. 2), spinning the terminal or lateral leaves of a shoot together and living within; often common on garden plants, including strawberry (Vernon, 1971: 75), tomatoes, fuchsias, carnations, bay (*Laurus*) and *Robinia*.

Pupa. Blackish brown, thoracic region and wings black. End of April to June and August to October, in a folded leaf or spun leaves. The structure of the pupa of this species is described in detail by Fisher (1924: 419).

Imago. Bivoltine, the first generation emerging from May to July and the second from late August to October and occasionally November, the broods overlapping. Frequenting gardens and hedgerows, flying during the day in the sunshine with a characteristic erratic flight.

DISTRIBUTION

The earliest record of this species in the British Isles is of two male specimens, one captured on 23rd October, 1905, at Bognor, Sussex by W. H. B. Fletcher (1905: 276), the other in early October the same year by Image (1906: 13). In 1906 it was recorded as breeding at Eastbourne, Sussex by Adkin (1906: 274). Sich (1914: 250) recorded that it had occurred in the Chiswick area of London since 1907, the larva feeding on *Laurus nobilis*. The species had evidently become locally established in the south of England, and Adkin (1908: 49) gives a detailed account of its life history. In 1932 Adkin (1932: 32) reported it as having spread over the greater part of Sussex and that it had become a pest in greenhouses; the latter observation was subsequently confirmed by Fryer (1934b: 7). It has since spread throughout most of the southern counties of England and is often locally abundant; it is common at Portland, Dorset, where the chief foodplant is *Euphorbia amygdaloides*. Its range extends into Wales: Glamorgan (Cardiff) (Cox, 1940a: 236), Denbighshire (Deganwy, Glan Conway, Llandudno) (Michaelis, 1969a: 2) and Caernarvonshire. The most northerly record in the British Isles is of a specimen from Cheshire (Hoylake) taken by C. M. Jones in 1936 (Michaelis, 1953: 74; 1954a: 60).

Europe to Asia Minor; North and South Africa; North America.

Syndemis musculana (Hübner)

Pl. 30, figs. 7–12

[*Lozotaenia musculana*; Barrett, 1904: 181. *Archips musculana*; Pierce & Metcalfe, 1922: 3, pl. 1 (♂, ♀ genitalia). *Tortrix musculana*; Meyrick, 1928: 509; Ford, 1949: 53]

DESCRIPTION

♂ ♀ 15–22 mm. Sexual dimorphism not pronounced; forewing of male with narrow costal fold from near base to about one-third; female usually larger, forewing broader.

Male (fig. 7). Forewing ground colour white suffused with grey, sparsely irrorate with blackish; markings dark brown, often with a slight ferruginous admixture; basal fasciae indistinct, outer margin almost vertical, sinuous, angulate at middle; median fascia well developed. Hindwing grey.

Female (fig. 8). Similar to male.

Variation. This species shows considerable individual variation in the forewing ground colour, which ranges from white to grey-brown, and in the intensity of markings. Towards the north of England and in Scotland the ground colour tends to be whiter or silver-grey (figs. 9, 10). In the Outer Hebrides, Orkneys and Shetlands an extreme form predominates in which the silver-grey ground colour contrasts sharply with the dark markings (figs. 11, 12). This form has been described from the Shetlands as subspecies *musculinana* (Kennel).

COMMENTS

The greyish white or white ground colour and the contrasting median fascia are characteristic of this species.

BIOLOGY

Larva. Head yellowish brown varying to orange-brown, ocellar region and posterior margin dark brown; prothoracic plate greyish brown varying to yellowish brown, marked with black posteriorly; abdomen varying from olive-green or yellowish green to grey-brown or blackish brown, paler ventrally, a broad dorsal line slightly darker than ground colour; pinacula rather prominent, lighter than ground colour; anal plate greenish or yellowish brown, sometimes with darker mottling. In the British Isles the principal foodplants are *Rubus*, *Quercus* and *Betula*, but the larva is polyphagous and will feed on various trees and shrubs and occasionally on herbaceous plants and grasses. July to September or October, living in a compact tube of spun leaves or in a folded leaf, hibernating in the spinning and pupating in the spring. Styles (1960: 146) records it on *Picea* and *Larix*, which were previously unknown as foodplants of this species in the British Isles, and states that the species is now widely distributed in larch plantations and also causes damage to conifer seedlings in forest nurseries.

Pupa. Blackish brown. April and May, in the larval habitation or in a loosely spun cocoon amongst leaf litter on the ground.

Imago. May and June, sometimes early July, frequenting especially lanes, open woodland and the borders of forests, also mountain moorland. It may be readily disturbed from rest amongst vegetation during the day, flying from late afternoon until well after dusk, occasionally coming to light and sugar.

DISTRIBUTION

Common and generally distributed throughout the British Isles. Found up to 1,800 ft on the Pennines and to 2,500 ft in Snowdonia. In Derbyshire it is plentiful on acid soils but uncommon on limestone. In the Shetlands, Orkneys and Outer Hebrides it is represented by the subspecies *S. musculana musculinana* (Kennel).

Europe to Siberia and Japan. In North America it is represented by the vicarious species *S. afflictana* (Walker), which may be no more than a geographical race (Freeman, 1958: 43; Powell, 1964: 160).

Ptycholomoides aeriferanus (Herrich-Schäffer)

Pl. 31, fig. 1

[*Cacoecia aeriferana*; Scott, 1952: 170. *Ptycholomoides aeriferana*; Ford, 1958: 3]

DESCRIPTION

♂♀ 17–21 mm. Sexes superficially similar; male with well-developed costal fold from base to near middle.

Male (fig. 1). Forewing with characteristic pale golden-yellow ground colour, diffusely strigulate with dark brown; markings blackish brown, often with weakly iridescent plumbeous mottling. Hindwing dark chocolate brown.

Female. Usually slightly larger and similar in coloration and markings, but sometimes with slightly paler golden-yellow ground colour.

Variation. The forewing ground colour and pattern are very constant; the blackish brown markings vary in intensity but, even in strongly marked specimens in which the strigulation may be heavier, the characteristic golden-yellow ground colour is evident.

COMMENTS

The yellowish ground colour of the forewing and the dark hindwing are characteristic of this species.

BIOLOGY

Larva. Head light brown, with darker mottling dorsally; prothoracic plate yellowish brown, blackish medially; abdomen green, weakly shagreened; anal plate yellowish green. May and June, living between spun needles of *Larix*.

Pupa. June, in the larval habitation.

Imago. Late June, July and August; the moth is seldom seen by day and most specimens have generally been taken at light. It is apparently migratory and has been taken in localities where its foodplant is not found.

DISTRIBUTION

This species was first recorded from the British Isles by Scott (1952: 170; 1953: 139) who took a series of specimens at light in July and August, 1951, at Westwell, near Ashford, Kent. It occurred in the same locality in 1952 and 1953; in the latter year a specimen was also taken on the 8th July at Elham Park Woods, Kent by S. Wakely (1953: 302). Specimens have subsequently been recorded from a number of scattered localities, including Harling Forest, Norfolk (Styles, 1955: 82), Balcombe, Sussex (Fairclough, 1955: 34), Westcliff-on-Sea, Essex (Huggins, 1961a: 150; 1962: 41), Carshalton, Surrey (1957), Hampshire (1960), Welwyn, Hertfordshire (1952), Buckinghamshire, and Buckingham Palace Garden, London (July, 1963) (Bradley & Mere, 1964: 68).

Central and south-east Europe; east Russia; south-east Siberia and Japan.

Aphelia viburnana (Denis & Schiffermüller)
Pl. 31, figs. 2–6

[*Heterognomon viburniana*; Barrett, 1905: 196. *Amelia viburniana*; Pierce & Metcalfe, 1922: 8, pl. 3 (♂, ♀ genitalia). *Tortrix viburniana*; Meyrick, 1928: 508. *Tortrix viburnana*; Ford, 1949: 52]

DESCRIPTION

♂ ♀ 15–22 mm. Sexes moderately dimorphic; male without costal fold, antenna weakly dentate-ciliate; forewing of female narrower, with costa strongly sinuate and apex more pronounced.

Male (figs. 2, 4, 5). Forewing ground colour varying from grey-brown and olive-brown to reddish brown, occasionally greyish ochreous; when present, markings dark brown or reddish brown. Hindwing dark grey or grey-brown; cilia whitish.

Female (figs. 3, 6). Forewing colour ground as in male; in well-marked forms the median fascia is usually more pronounced; unicolorous forms are common, often with very rich red-brown coloration.

Variation. This species shows minor variation in forewing coloration and markings which appear to be influenced by foodplant and biotope.

COMMENTS

Pale unicolorous forms may be confused with *A. paleana* but the brownish content of the forewing coloration distinguishes *viburnana* from that species. The males of these

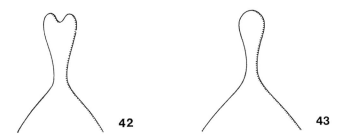

Male genitalia of *Aphelia* species
Fig. 42, *A. viburnana* (Denis & Schiffermüller), uncus. Fig. 43, *A. paleana* (Hübner), uncus

two species may be readily distinguished by examining the uncus with the abdomen *in situ*; the uncus is easily exposed by brushing away a few scales of the anal tuft. In *viburnana* (text-fig. 42) the uncus is deeply emarginate apically, in *paleana* (text-fig. 43) it is broadly rounded.

BIOLOGY

Larva. Head light yellowish brown, marked with black postero-laterally; prothoracic plate variable, light brown or greenish brown, variably marked with black; abdomen dark green varying to greenish black, subspiracular line paler; pinacula whitish green, rather conspicuous, setae whitish; peritreme of spiracles black; anal plate light green or brown, spotted with black; anal comb brownish; thoracic legs greenish or brownish, marked with black (Barrett, 1883: 132; Prest, 1877: 49). May and June, in spun leaves of various herbaceous and other plants, including *Vaccinium, Erica, Myrica, Salix, Helianthemum, Potentilla, Lythrum, Lonicera, Spiraea, Pastinaca, Teucrium* and *Centaurea*; also on *Abies* and *Pinus* (Huggins, 1964b: 230). On the Continent *Viburnum, Andromeda, Alisma, Ledum, Sanguisorba, Artemisia, Coronilla, Lysimachia, Scrophularia* and *Lotus* are among the foodplants recorded for this species.

On moorlands, where the chief foodplant is *Vaccinium vitis-idaea*, half-grown larvae may be found in late April, indicating that the ova hatch the previous year and that the larvae hibernate. The larva is very active and moves from shoot to shoot. When feeding on *Vaccinium myrtillus* it favours the tips of the shoots, eating the upper cuticle and parenchyma, leaving the lower cuticle intact; in some years when the larvae are plentiful on the Derbyshire moors, the attacked shoots appear pale brown over large areas of moorland. In the limestone area of the Burren in western Ireland, the larva has been found on *Dryas octopetala*, feeding in spun shoots and seedheads (pl. 12, fig. 2), and also on *Geranium sanguineum* and *Filipendula ulmaria*; in the Thames estuary, it is common on the salt marshes, feeding in spun shoots of *Aster tripolium* and *Artemisia maritima* (Huggins, 1961a: 151; 1964b: 231).

Pupa. Black. June and July, in a white silken cocoon spun in the larval habitation.

Imago. July; easily put up during the day in most of its habitats. Frequenting moors and mosses where it often occurs in numbers, flying actively on hot sunny days. In the west of Ireland it has been taken on the wing in September.

DISTRIBUTION

A locally common species throughout the British Isles, including the Outer Hebrides, St. Kilda and Orkneys; most plentiful on heaths, mosses and blanket bogs.

Europe to east Siberia.

Aphelia paleana (Hübner)

Pl. 31, figs. 7–10

[*Heterognomon icterana*; Barrett, 1905: 194. *Aphelia palleana*; Pierce & Metcalfe, 1922: 8, pl. 3 (♂, ♀ genitalia). *Tortrix paleana*; Meyrick, 1928: 507; Ford, 1949: 52]

DESCRIPTION

♂♀ 17–24 mm. Sexual dimorphism moderately pronounced; male with comparatively broad wings without costal fold, antenna weakly dentate-ciliate; female usually smaller, wings attenuated.

Male (figs. 7, 9). Forewing pale yellow-ochreous or whitish ochreous without markings; cilia paler; base of forewing, thorax, tegula and head suffused with orange-yellow. Hindwing dark grey; cilia paler.

Female (figs. 8, 10). Forewing coloration similar to that of male but usually paler, base of wing, thorax, tegula and head concolorous with rest of wing or only slightly darker. Hindwing dark grey; cilia paler.

Variation. In both sexes the coloration of the forewing varies from whitish or greyish ochreous, found in the more common forms, to pale clear yellow.

COMMENTS

Whitish or greyish ochreous forms of this species can be confused with *A. unitana*, but may usually be distinguished by the darker cilia of the forewing in both sexes, and by the orange-yellow suffusion on the head and thoracic region and on the basal area of the forewing, especially in the male. As mentioned under *A. viburnana*, the males of that species and *paleana* can be readily separated by differences in the shape of the uncus (text-figs. 42, 43).

BIOLOGY

Ovum. Deposited during July and August in small batches on the upper surface of a leaf of the foodplant, each batch consisting of about 10–30 eggs arranged in a double row longitudinally along the leaf (Balachowsky, 1966: fig. 236); hatching in about 6 days.

Larva. Head and prothoracic plate brownish black or jet black, plate divided medially by a light brown line, anterior margin broadly edged with grey or brown; abdomen varying from dark grey-green to black; pinacula prominent, shining white; anal plate shining black; anal comb present; thoracic legs black. September to May and June, hibernating during the winter and continuing to feed in the spring; polyphagous, feeding in spun leaves of various Gramineae, including *Agropyron repens* and *Phragmites*, and *Centaurea*, *Tussilago*, *Plantago*, *Scabiosa*, *Caltha palustris*, *Filipendula* and other herbaceous plants; occasionally on *Quercus* and *Fagus*, and on *Picea sitchensis* in Scotland. At Swaffham, Norfolk, it has been found in spun leaves of *Arrhenatherum elatius* (pl. 12, fig. 1), and in the Burren, Co. Clare, on *Dactylis glomerata* (Bradley & Pelham-Clinton, 1967: 134). Although not normally a pest in the British Isles, in Lincolnshire in June, 1951, larvae of this species caused considerable damage to crops of *Phleum pratense* (Timothy grass), feeding in spun leaves and sometimes in the seedheads (Gair, 1959: 95). In Scandinavia and Finland the larvae occasionally cause serious damage to crops of Timothy grass.

Pupa. Dark brown or black, abdomen paler. June, in a folded edge of a leaf.

Imago. July and August; during the day the moth may be disturbed from rest amongst rough herbage in meadows and pastures, and flies leisurely about its habitat in the evening.

DISTRIBUTION

Generally distributed in coastal areas, including sandhills, as far north as the Inner and Outer Hebrides and Aberdeenshire. More local inland and recorded from Middlesex, Hertfordshire, Cambridgeshire (Wicken Fen), Staffordshire, Shropshire, Cheshire and Lancashire (Manchester district). In Derbyshire it is common in the limestone dales, and at Malham (Yorkshire) it occurs amongst damp carr entirely surrounded by heather and birch-mossland. In Scotland its distribution is almost entirely coastal and it is common only in the south. In Ireland it is generally distributed in fields and meadows (Beirne, 1941: 82) and is common in the limestone area of the Burren, Co. Clare.

North and central Europe; Iran; Buchara; Siberia.

Aphelia unitana (Hübner)
Pl. 31, figs. 11–13
[*Aphelia unitana*; Bradley, 1964: 75, figs. 3–6 (♂, ♀ genitalia)]

The inclusion of this species in the British list (Bradley, 1964: 75) followed the work of Holst (1962: 303), who reinstated *unitana* as a distinct species in continental Europe. However, *unitana* and *A. paleana* are very similar superficially and, although showing slight comparative differences in shape and coloration of the forewing, are difficult to separate with certainty because of individual variation and intergradation and the close similarity in genitalic structure. The specific status of *unitana* will perhaps be clarified on further investigation into the biology.

DESCRIPTION

♂ ♀ 19–24 mm. Sexual dimorphism moderately pronounced; male with comparatively broad wings; female usually smaller, wings attenuate.

Male (figs. 11, 13). Forewing ground colour grey, sometimes overlaid with whitish, giving the wing a silver-grey appearance, occasionally with a slight yellowish tinge; cilia whitish. Hindwing grey; cilia whitish.

Female (fig. 12). Forewing whitish yellow varying to clear pale yellow; cilia concolorous. Hindwing dark grey; cilia whitish.

Variation. Males from Northumberland and Yorkshire are darker silver-grey in appearance compared with those from western Ireland.

COMMENTS

This species is very difficult to distinguish from *paleana*, both superficially and morphologically. The male of *unitana* can almost invariably be recognized by the characteristic silver-grey coloration of the forewing and the whitish cilia, and the absence of orange-yellow suffusion in the basal area of the forewing and on the thorax.

The female is usually more robust than that of *paleana*, the forewing being relatively broad and less attenuated, showing less marked sexual dimorphism.

BIOLOGY

Larva. A detailed description of the larva is not available, but so far as is known it resembles that of *paleana* in appearance. May and June, in spun leaves of *Heracleum* and *Rubus* (Bradley, 1964: 75), *Angelica*, *Allium* and probably other herbaceous plants. Very little is known about the biology of *unitana*, but it appears that the larvae of both this species and *paleana* feed on various low-growing plants and grasses and are full-grown in spring, and that both may occur in the same locality.

Pupa. June; in the larval habitation.

Imago. June and July; frequenting high moorland and limestone dales in England, but occurring at lower elevations on limestone in western Ireland.

DISTRIBUTION

Owing to the confusion of this species with *paleana* because of the similarity of the imagines, its distribution is at present incompletely known. Material examined shows that in England it occurs in Northumberland (Kielder, Morpeth), Yorkshire (Malham Tarn) (Michaelis, 1965: 16), Derbyshire (Dovedale), and Staffordshire (Manifold Valley); and in Ireland in the Burren, Co. Clare.

On the Continent this species is recorded from Denmark by Holst (1962: 303), and also from France, Sweden, Austria, Germany, Yugoslavia and Russia (Leningrad). In Denmark it occurs on moorland and in southern Europe in mountain districts. Opheim (1965: 23) records it as widely distributed in Norway, occurring as far as 67°N, while *paleana* is apparently restricted to the Oslo-fjord district and is the less common species.

Clepsis senecionana (Hübner)

Pl. 32, figs. 1–3

[*Clepsis rusticana*; sensu Barrett, 1905: 208; sensu Pierce & Metcalfe, 1922: 5, pl. 2 (♂, ♀ genitalia). *Tortrix rusticana*; sensu Meyrick, 1928: 508; sensu Ford, 1949: 52]

DESCRIPTION

♂ ♀ 13–16 mm. Sexual dimorphism not pronounced; forewing of male without costal fold, antenna dentate-ciliate, flagellum rough-scaled; forewing of female narrower, apex more pointed, coloration paler, more yellowish ochreous, antenna simple, smooth-scaled.

Male (fig. 1). Forewing ground colour greyish ochreous, weakly irrorate with olive-yellow or olive-brown, with darker transverse oblique striae. Hindwing grey.

Female (figs. 2, 3). Forewing pale ochreous or whitish ochreous, irroration and striae often weak.

Variation. In the male the forewing ground colour varies from light greyish ochreous to comparatively dark ochreous-brown, the irroration varying in intensity. The female is less variable. In both sexes a very light almost whitish form occasionally occurs.

COMMENTS

Distinguishing features of this species are its small size, the pale ochreous general coloration of the forewing and lack of any strong markings.

BIOLOGY

Larva. Head and prothoracic plate pale brown or yellow-brown; ocelli black; thoracic legs yellow-brown; abdomen pale greenish yellow, dorsal and lateral lines broad, olive-green; pinacula paler than integument; peritreme of spiracles black; anal plate pale yellowish brown, mottled with brown. July to April; in a rolled leaf or spun terminal leaves of *Myrica* and *Vaccinium myrtillus*; also known on *Picea*, *Pinus* and *Larix* in Scotland. On the Continent, *Lotus*, *Gentianella amarella*, *Polygonatum*, *Potentilla*, *Lysimachia*, *Dorycnium*, *Onobrychis viciifolia* and *Convallaria* are recorded as additional foodplants. The larva overwinters in a white silken cocoon constructed in the middle of a small cluster of shoot leaves, the tips of which are often eaten off evenly all round (Bankes, 1909: 151).

Pupa. Head and thorax blackish, abdomen lighter. April, in a white silken cocoon in a spun shoot.

Imago. May and June; frequenting damp grassy clearings in woods, and rough fields, marshes and fens, moorlands and mosses. It is of rather retiring habits, hiding during the day amongst grass and other herbage, creeping rather than flying to another hiding place when disturbed, but flies freely in warm sunshine during the afternoon and evening, disappearing instantly the sky becomes cloudy.

DISTRIBUTION

Local but widely distributed, occurring in suitable localities throughout the British Isles. In north and central Wales, and on the southern Pennines and similar northern localities in England, it may be found on *Vaccinium* on the mountains up to 2,000 ft, while in Cheshire, Flintshire and Westmorland it occurs locally on *Myrica* on low-level mosses. In Scotland it is widely distributed on moorlands as far north as the Shetlands. In Ireland locally common in woodlands (Beirne, 1941: 82) and in hazel scrub in the Burren, Co. Clare.

Europe to east Siberia.

Clepsis rurinana (Linnaeus)

Pl. 32, fig. 4

[*Tortrix semialbana*; Barrett, 1904: 168. *Lozotaenia semialbana*; Pierce & Metcalfe, 1922: 6, pl. 2 (♂, ♀ genitalia). *Tortrix semialbana*; Meyrick, 1928: 509; Ford, 1949: 52]

DESCRIPTION

♂ 16–19 mm, ♀ 17–20 mm. Sexual dimorphism not pronounced; male with short costal fold in basal third, antenna weakly dentate-ciliate.

Male (fig. 4). Forewing ground colour pale yellow-brown, sparsely strigulate with brownish in distal half; markings ochreous-brown, basal fasciae indeterminate, represented by a diffuse patch on dorsal half of base; median fascia with inner margin well defined, nearly straight; stria from pre-apical spot usually reaching tornus. Hindwing whitish grey, tornal area to cubital vein strongly infuscate.

Female. Forewing coloration and pattern similar to that of male.

Variation. Both sexes show only slight variation; the median fascia may be weak and sometimes interrupted a little below the costa and the stria from the pre-apical spot may be slightly irregular or obsolescent.

COMMENTS

The whitish apical area of the hindwing readily distinguishes *rurinana* from *Pandemis* species, and the relatively complete median fascia, which is well developed on the dorsum, separates it from *C. spectrana*.

BIOLOGY

Larva. Head and prothoracic plate rather glossy pale brown, plate with black median and lateral markings; abdomen grey-green, pinacula paler; anal plate pale brown, marked with black. The larva may be found from April to June in rolled leaves of *Lonicera, Acer, Fagus, Quercus* and *Rosa.* On the Continent *Chelidonium, Urtica, Convolvulus, Euphorbia, Pulicaria, Anthriscus cerefolium* and *Lilium* have also been recorded as foodplants.

Pupa. June; in the larval habitation.

Imago. End of June and July; very retiring in its habits, frequenting lanes and open woodland, hiding during the day amongst the foliage of trees and bushes, dropping when disturbed either to the upper surface of a leaf or to the ground.

DISTRIBUTION

Local and scarce, occurring in England in Essex, Kent (Darenth, Folkestone, Greenhithe, Lewisham), Surrey, Hampshire, Gloucestershire (Cranham, Nailsworth, Stroud), Lancashire (Grange-over-Sands) and Yorkshire (Doncaster). The only confirmed records from Wales are from Caernarvonshire (Glan Conway, Great Ormes Head) (Michaelis, 1969a: 2); other alleged Welsh specimens examined have proved to be *Pandemis cerasana.* A record from Argyll (Lochgoilhead) was confirmed by Chapman (1907: 258) but records from Ayrshire, Renfrewshire and Arran are unconfirmed. In Ireland known only from the Burren, Co. Clare, where several specimens were taken in 1970.

Europe to Siberia, India, China, Korea and Japan.

Clepsis spectrana (Treitschke)

Pl. 32, figs. 5–9

[*Tortrix costana*; sensu Barrett, 1904: 169. *Clepsis costana*; sensu Pierce & Metcalfe, 1922: 5, pl. 2 (♂, ♀ genitalia). *Tortrix costana*; sensu Meyrick, 1928: 508; sensu Ford, 1949: 52]

DESCRIPTION

♂ 15–22 mm, ♀ 17–24 mm. Sexual dimorphism moderately pronounced; male with costal fold from base to about one-third, costa strongly arched, antenna dentate-ciliate; female without costal fold, costa sinuate, apex slightly produced.

Male (fig. 5). Forewing ground colour pale ochreous, sparsely irrorate with brown; markings dark brown, strongly developed in costal half of wing; median fascia oblique, diffuse beyond middle and becoming obsolescent towards tornus. Hindwing whitish grey, darker terminally.

Female (fig. 6). Forewing coloration and pattern as in male.

Variation. An extremely variable species; the forewing ground colour varies from whitish ochreous to yellowish or reddish brown (fig. 8) and the brown irroration may be either obsolescent or heavy. Melanic forms (fig. 9, f. *liverana* Mansbridge) are apparently uncommon. Mansbridge (1914: 18) reported an example bred from a larva on *Ilex* at Allerton in the Liverpool district, and subsequently obtained similar specimens from Burnley (Lancashire) and Peterborough (Northamptonshire) from larvae feeding on *Iris*, and a further specimen from Allerton from a larva on oak. This melanic form was named f. *liverana* by Mansbridge and has also been found at Blackpool (Lancashire) and in Derbyshire. The forewing markings vary in strength and in well-marked specimens the median fascia extends to the tornus (fig. 7); extreme forms occur, especially in the Thames estuary, with the markings obsolescent, the forewing appearing unicolorous.

COMMENTS

Yellowish forms of this species may be distinguished from *C. rurinana* by the stronger markings on the costa of the forewing and the more uniform whitish grey hindwing.

BIOLOGY

Ovum. Deposited in small batches on the foodplant.

Larva. Head and prothoracic plate shining black or blackish brown; abdomen greyish olive-green varying to brown, paler dorsally, a whitish subspiracular line; pinacula cream-white; anal plate cream-white mottled with black or brown; anal comb with 6–8 long prongs. The full-grown larva is most frequently found in May and June, in spun leaves and flowers of *Epilobium* (pl. 12, fig. 3), *Spiraea*—usually in the tops— *Potentilla*, *Aster tripolium*, *Iris*, *Scirpus*, *Urtica* and most herbaceous plants in fens and marshes, and *Limonium vulgare*, *Artemisia maritima* and other maritime plants; also on cultivated *Cyclamen*, feeding on the leaves, fruits and seeds, and occasionally on cultivated strawberry, hops and *Pelargonium* (Vernon, 1971: 74).

Pupa. Matt black. May and June, spun up in a white silken cocoon in the larval habitation.

Imago. Early May to July, the period of peak emergence occurring about mid-June, the early moths apparently producing a small second generation in August and September (Barrett, 1904: 171). The species has also been recorded in September from western Ireland (Bradley & Pelham-Clinton, 1967: 134). During the day the moth hides amongst tall foliage in marshes, fens, bogs and damp woods and is very difficult to flush, slipping away deeper into the vegetation if disturbed. Towards dusk it is active and flies gently about its habitat, and after dusk comes freely to light, the females flying throughout the night.

DISTRIBUTION

Common in suitable habitats in England, Wales and Ireland, especially in fens near the coast; in Scotland it is less common and appears to be restricted to the south-western and south-eastern counties.

Northern and central Europe, south-east Russia.

Clepsis consimilana (Hübner)

Pl. 32, figs. 10–12

[*Lozotaenia unifasciana*; Barrett, 1904: 180. — *unifasciana*; Pierce & Metcalfe, 1922: 5, pl. 2 (♂, ♀ genitalia). *Tortrix unifasciana*; Meyrick 1928: 509; Ford, 1949: 52]

DESCRIPTION

♂ 13–17 mm, ♀ 15–19 mm. Sexes moderately dimorphic; forewing of male with well-developed costal fold to middle or slightly beyond, antenna very weakly dentate-ciliate; female larger, costa of forewing less strongly arched, without costal fold, markings obsolescent.

Male (fig. 10). Forewing ground colour ochreous; markings poorly developed, purplish brown; basal fasciae scarcely indicated, a comparatively distinct fuscous spot at outer margin on dorsum; median fascia oblique, an ill-defined fuscous spot at inner margin on dorsum; cilia concolorous. Hindwing grey.

Female (fig. 11). Forewing ground colour darker than in male, diffusely irrorate with reddish brown; markings obsolescent except for the two fuscous spots on dorsum. Hindwing as in male, sometimes slightly darker.

Variation. The forewing markings of the male vary in strength; not infrequently specimens are found with comparatively well-developed fasciae (fig. 12), sometimes with the ground colour irrorate with reddish brown as in the female. The forewing ground colour of the female may be paler and without the reddish brown irroration; more rarely the forewing is suffused with plumbeous, and the reddish brown irroration is intensified, the cilia remaining yellowish brown.

COMMENTS

Distinguished by its small size, indistinct forewing markings and the two small fuscous spots on the dorsum.

BIOLOGY

Ovum. Deposited in small, silver-grey batches of 6–12 eggs on the upper surface of the leaves along the midrib; hatching in about 12 days.

Larva. Head light brown; prothoracic plate brown, edged with black posteriorly; abdomen violet-green or violet-grey, integument shagreened; pinacula brown, not very prominent. August to May, overwintering in the third instar in a hibernaculum spun in dead leaves on the foodplant; feeding on *Ligustrum*, living in a dense untidy spinning (pl. 13, fig. 1), *Syringa*, *Hedera* (Huggins, 1923: 15), *Lonicera*, *Polygonum* and *Malus*, also on *Carpinus* and *Crataegus* (Thurnall, 1902: 130; 1920: 93), usually on dead or withered leaves. Sheldon (1920a: 49) describes the life history and found that in captivity larvae would eat fresh leaves.

Pupa. Light brown. June, in the larval habitation.

Imago. June to August and September; often abundant in suburban gardens and other favoured situations, flying at the slightest disturbance, dashing off erratically to another resting place. During the afternoon and evening it may be seen flying actively around its foodplants. At Dungeness, Kent, it has been observed swarming around *Ulex* bushes (Edwards & Wakely, 1958: 93), and in Cheshire and Lancashire has been taken frequently on heather and birch mosses, which indicates that the larva has other foodplants in addition to those mentioned. In Scotland it emerges a little later than in the south, and is on the wing until early September.

DISTRIBUTION

Widely distributed and generally common in lanes and hedges, and especially gardens containing privet hedges, throughout the British Isles as far north as the Inner Hebrides, the Clyde area, Aberdeenshire and Morayshire, but rare in mountain districts.

Europe to Asia Minor and Syria; North Africa; east Siberia; also recorded from the eastern United States (Klots, 1941: 126) and Madagascar.

Clepsis trileucana (Doubleday)

Stephens (1834: 81) records two specimens of this North American species from Cornwall, tentatively determined as *Lozotaenia schreberiana* (Linnaeus). The specimen from Stephens' collection is figured by Wood (1839: 133, pl. 30, fig. 875). Later, Doubleday (1847: 1729) stated that the two specimens bore no resemblance to the true *schreberiana*, except for the white costal spot, and that they represented a common American species, which he named "*Sericoris trileucana* (Gm.)". The abbreviation "Gm." possibly refers to Gmelin, but no description by that author of a species under this name has been traced, consequently the name should be attributed to Doubleday. Barrett (1887: 35) confirmed that *trileucana* is conspecific with *persicana* Fitch, under which name the species is known by American authors.

There are no other known records of this species from the British Isles.

Clepsis melaleucanus (Walker)

This species was added to the British list by Stephens (1834: 78), who described it under the name *Lozotaenia biustulana* Stephens from two specimens reputed to have been captured in the west of England. Wood (1839: 132, pl. 30, fig. 867) and Humphreys & Westwood (1845: 113, pl. 80, fig. 10) illustrate one of these specimens. Doubleday (1847: 1729) and Barrett (1887: 35) considered the two Stephens specimens to have originated from North America and the species is discounted as British.

Epichoristodes acerbella (Walker)

Several instances have occurred in recent years of larvae of this species being intercepted at Heathrow Airport, London, on carnations (*Dianthus*) imported from East Africa. A moth was bred in Edinburgh from a larva feeding on chincherinchee (*Ornithogalum thyrsoides*) originating from South Africa (Pelham-Clinton, 1969: 72).

The larva feeds on various herbaceous plants, but so far as is known there is no record of it occurring under natural conditions in the British Isles.

Epiphyas postvittana (Walker)
Pl. 32, figs. 13–16
[*Tortrix postvittana*; Meyrick, 1937: 256; Ford, 1949: 53]

DESCRIPTION

♂ 16–21 mm, ♀ 17–25 mm. Sexual dimorphism pronounced; male usually smaller, antenna weakly dentate-ciliate, length of cilia approximately equal to width of flagellum, basal half of forewing usually sharply demarcated, well-developed costal fold from base to about two-fifths; antenna of female minutely ciliate, forewing longer, apex produced.

Male (fig. 13). Basal half of forewing light buff or pale yellow, contrasting strongly with the dark brown and ferruginous coloration of the distal half, the demarcation often emphasized by the deeper coloration of the oblique narrow median fascia, the inner edge of which is sharply defined and usually straight, but sometimes is slightly sinuate at the middle; pre-apical spot obscure, its inner margin usually defined by ferruginous ground coloration separating it from the median fascia. Hindwing grey.

Female (fig. 14). General coloration of the forewing more uniform, with less contrast between the basal and distal halves; median fascia usually reduced.

Variation. An extremely variable species with numerous recurring forms. In strongly marked forms of the male the distal half of the forewing may vary from reddish brown to blackish, often with purplish mottling; the contrasting pale basal half may be sparsely speckled with black. Lightly marked forms resembling the female in appearance occur; an extreme form in which the usually dark outer half of the forewing is light and the pre-apical spot discernible is uncommon (fig. 15). Only minor variation is found in the female; often the forewing is irrorate with black in both the basal and distal halves of the wing (fig. 16).

COMMENTS

The male is usually distinguished by the abrupt division of the forewing medially into a pale basal area and darker apical area, and the female by its large size and relatively elongate forewing, often with greatly reduced markings.

BIOLOGY

Ovum. June, and from August to September; deposited in batches of 15–30 eggs on the upperside of the leaves, hatching in about ten days.

Larva. Head pale brown; prothoracic plate light greenish brown, divided medially by a pale greenish line; thoracic legs brown; abdomen pale yellowish green, darker dorsally; pinacula paler than integument, setae whitish; anal plate light brownish green; anal comb light green. June and July, and September to April, feeding on *Euonymus japonicus*, loosely spinning the shoots in a manner similar to that of *Archips podana* and *Cacoecimorpha pronubana*. In Devon and Cornwall it originally showed a preference for *Euonymus*, but has since been recorded on various other plants, including *Baccharis*, *Rubus*, *Buddleia*, *Centranthus*, *Chrysanthemum*, *Escallonia*, *Crataegus*, *Hebe*, *Hypericum*, *Hedera*, *Jasminum*, *Lavandula*, *Mesembryanthemum*, *Mentha*, Montbretia (pl. 13, fig. 2), *Urtica*, *Ligustrum*, *Pyracantha* and *Pulicaria* (Baker, 1968: 169). In Australia, where this species is native, it attacks *Malus* and is a pest, but in the British Isles it has only been recorded on this foodplant in captivity (Huggins, 1958a: 53).

Pupa. Dark reddish brown. May and June, August, in the larval habitation.

Imago. Bivoltine. May to October, the two generations overlapping. The moth is easily disturbed during the day from rest amongst *Euonymus* and other shrubs. It flies freely after dusk and comes readily to light.

DISTRIBUTION

This indigenous Australian species is apparently adventitious to the British Isles. It was first found to be established at Newquay, Cornwall, in 1936 by F. C. Woodbridge (Meyrick, 1937: 256), and is now widespread in west Cornwall and has been recorded from Penzance, Falmouth, Camborne, Redruth, Portreath (Tremewan, 1957: 76), Porthtowan, Ellbridge, Gerrans and Portscatho. It is also known from Devon (Torquay) (Bradley & Martin, 1956: 151) and Hampshire (Alverstoke) (Sadler, 1967: 87). A full account of its history and distribution in England has been published by Baker (1968). Prior to 1936, there were several records of specimens bred from larvae imported on New Zealand apples (Fletcher, 1933: 165; Jacobs, 1933: 70; Fryer, 1934a: 7).

Australia, where it is an important pest in fruit orchards and commonly known as the light-brown apple moth or apple leaf-roller; introduced to Tasmania, New Zealand, New Caledonia and the Hawaiian Islands.

Adoxophyes orana (Fischer von Röslerstamm)

Pl. 33, figs. 1, 2

[*Adoxophyes orana*; Bradley, 1952a: 1; Ford, 1958: 5]

DESCRIPTION

♂ 15–19 mm, ♀ 18–22 mm. Sexual dimorphism pronounced; antenna of male shortly ciliate, forewing with broad costal fold from base to about one-third, markings usually conspicuous, contrasting with paler ground colour; female usually larger, antenna minutely ciliate, forewing without costal fold, with darker general coloration and less contrasting markings.

Male (fig. 1). Ground colour of forewing light greyish brown; markings dark brown suffused with ochreous; outer margin of basal fasciae poorly defined, oblique to middle; median fascia narrow, margins irregular, usually constricted at middle before emitting strong tornal spur; pre-apical spot broken and reduced, emitting a strong stria extending to the tornal area, and a second much thinner stria parallel with termen. Hindwing grey.

Female (fig. 2). Forewing ground colour greyish brown; markings essentially as in male but more subdued and often partially obsolete. Hindwing grey.

Variation. This species shows little variation. In the male the forewing ground colour is sometimes dull and the markings show less contrast. The female is seldom strongly marked, but occasionally has a rather conspicuous reticulate pattern in the forewing, especially in the distal half.

COMMENTS

This species is the only representative of the genus *Adoxophyes* known in the British Isles. In the wing venation, the stalking of veins 7 (R_5) and 8 (R_4) of the forewing, with vein 3 (CuA_1) rising well before the angle of the cell, and the separation of veins 3 (CuA_1), 4 (M_3) and 5 (M_2) in the hindwing, are characteristic.

BIOLOGY

Ovum. June, and in the autumn; deposited in batches of 4–150 or more eggs, lemon-yellow in colour, usually on the upper and under surfaces of leaves and, in the autumn, occasionally on the fruit. In total a female may lay 200–400 eggs; incubation lasts from 8–20 days (Balachowsky, 1966: 555; Groves, 1952: 152; Hardman, 1953: 267).

Larva. Head and prothoracic plate yellowish brown, black in earlier instars; abdomen variable in colour, yellow-green, olive-green or dark green; pinacula small, yellowish. June and July, and another generation from September to May; polyphagous, feeding on leaves and fruit of *Malus*, *Pyrus* and *Prunus*; also on *Betula*, *Salix*, *Populus*, *Alnus*, *Corylus*, *Rosa*, *Rubus*, *Ribes*, *Humulus*, *Lonicera*, *Solanum dulcamara* and various other plants. Larvae feeding in June and July, producing the summer generation of moths, cause considerable damage to apple and scar the fruit (pl. 13, fig. 3), spinning a white silken web along the midrib on the underside of the leaf and gnawing the under surface without perforating the leaf. In a later instar the larvae disperse and live in spun leaves. Sometimes the larva spins a leaf to a fruit and feeds on the skin, thus damaging the developing fruit which is scarred when mature. Larvae from the summer generation of moths hatch in September and normally overwinter in the second or third instar. In the autumn, the young larvae feed on the foliage or on the surface of the fruit before hibernating in a silken hibernaculum spun either between a dead leaf and a twig, two shoots on a spur, a mummified apple and a branch, or any similar situation; in the spring they recommence to feed, eating into the buds and developing leaves from late

March to early May. Detailed accounts of the biology of this species in Britain have been published by Groves (1952) and Hardman (1953).

Pupa. Dark brown, wings paler. May and June, in the larval habitation in buds or spun leaves; August and September, in the larval habitation; 10–20 days duration.

Imago. Bivoltine, the first generation of moths emerging in June, a second from mid-August to October; the emergence periods may vary according to climatic conditions and sometimes slightly overlap. Frequenting orchards and gardens, not very active by day, resting amongst the foliage but flying if the branches are shaken; both sexes fly towards dusk and later come to light.

DISTRIBUTION

First discovered in the British Isles in 1950 in apple orchards at Teynham, Kent (Groves, 1951: 259). The appearance of *orana* in these islands followed the extension of its range in continental Europe. It quickly became established as a pest in apple and cherry orchards in south-east England (Bradley, 1952a: 1; 1952b: 16), and has been recorded from widely separated localities in Kent, north of the North Downs from the Medway estuary to the east coast at Sandwich, and at Gravesend (Hardman, 1953: 265), High Halstow (Chalmers-Hunt, 1968b: 179), East Malling (Bradley, 1952b: 17) and Bromley (Jacobs, 1952: 86). Other records include London at Blackheath (Allen, 1959: 153) and Camberwell (Wakely, 1969: 95); Essex at Bradwell-on-Sea and Chelmsford (Hardman, 1953: 265), and Westcliff-on-Sea (Huggins, 1953a: 189; 1954: 128); Suffolk at Risely and Walberswick (Chipperfield, 1970: 87); Cambridgeshire at Upwell and Wisbech, and Norfolk at Walton Highway (Hardman, 1953: 266). There is an unconfirmed record from Cheshire (Barritt, 1952: 235).

Europe to China and Japan.

Ptycholoma lecheana (Linnaeus)

Pl. 33, figs. 3–5

[*Ptycholoma lecheana*; Barrett, 1904: 183; Pierce & Metcalfe, 1922: 7, pl. 3 (♂, ♀ genitalia). *Cacoecia lecheana*; Meyrick, 1928: 504; Ford, 1949: 50]

DESCRIPTION

♂♀ 16–23 mm. Sexes moderately dimorphic; antenna of male dentate-ciliate, forewing with strong costal fold from base to middle; female usually larger, antenna minutely ciliate, forewing broader, without costal fold.

Male (fig. 3). Forewing ground colour fuscous, variably suffused with greenish yellow from base to median fascia, except costal fold; distal half of wing sparsely suffused with yellow-ochreous or ferruginous-ochreous; median fascia diffuse and obscure, only slightly darker than ground colour, edged on both sides with shining metallic plumbeous; pre-apical spot obsolescent, its apex indicated by metallic plumbeous scales. Hindwing blackish brown; cilia whitish, with a dark brown sub-basal line.

Female (fig. 4). Similar to the male but with stronger yellowish suffusion in the distal part of the forewing, the median fascia very weak, usually indicated only by darker coloration towards the dorsum, partially edged on both sides with metallic plumbeous which on the outer margin is sometimes connected to the plumbeous scaling in the apical area.

Variation. In the male the fuscous ground colour of the forewing is very constant, but the yellowish suffusion in the basal half varies in strength and may be weak; when the suffusion is strong the basal half may be distinctly yellow. Occasionally the metallic plumbeous edging of the median fascia is interrupted and reduced or entirely absent (fig. 5). Female similarly variable in intensity of yellowish suffusion and development of metallic plumbeous markings.

COMMENTS

Distinguished by the generally dark coloration of the fore- and hindwings, the shining metallic edging of the median fascia and the metallic scaling in the apical area of the forewing.

BIOLOGY

Larva. Head yellowish brown, edged with black posteriorly; prothoracic plate yellow, with black lateral margins; sometimes head and prothoracic plate wholly black; abdomen grey-green or yellow-green dorsally, with a darker dorsal line, pale yellowish green ventrally, with a darker lateral line; pinacula weakly sclerotized, yellowish; anal plate light green; a small anal comb present; thoracic legs black. August to May; overwintering as an early instar larva in a small silken hibernaculum on the bark of twigs, recommencing to feed early in the spring on the opening buds and young shoots. Polyphagous, feeding in the rolled leaves of *Malus*, *Prunus* and various other fruit trees, also on *Acer*, *Populus*, *Quercus*, *Salix*, *Abies*, *Picea*, *Larix* and other woodland trees and shrubs. The larva is frequently found in fruit orchards but seldom in sufficient numbers to cause serious defoliation.

Pupa. Black. June, in a dense, white cocoon in a rolled leaf or the larval habitation.

Imago. June and July; frequenting wooded areas, particularly orchards. When at rest during the day on bushes and trees the moth closely resembles the fallen capsule of a leaf bud; in the afternoon and evening it flies wildly over the trees and bushes.

DISTRIBUTION

Occurring throughout England and Wales, sometimes abundantly in the southern counties of England, becoming scarcer northwards. In Scotland it is comparatively uncommon, occurring in oak woods in the south, and recorded as far north as Perthshire and Argyll. In Ireland widely distributed and common in woods.

Throughout Europe to Asia Minor and Transcaucasia.

Lozotaeniodes formosanus (Geyer)

Pl. 33, fig. 6

[*Eulia formosana*; Parfitt, 1947: 225; Ford, 1949: 54]

DESCRIPTION

♂ ♀ 20–26 mm. Sexes similar; male without costal fold on forewing, antenna dentate-ciliate; female usually larger, antenna minutely ciliate.

Male. Forewing ground colour pale yellow-ochre overlaid with greyish purple; numerous irregularly interconnected narrow chestnut-brown fasciae, often interrupted, extensively edged with black, forming a characteristic bold reticulate pattern. Hindwing light grey, with pale ochreous or cupreous suffusion in apical area.

Female (fig. 6). Similar to male.

Variation. In both sexes the formation of the transverse fasciae shows considerable minor variation; asymmetrically marked specimens are not uncommon.

COMMENTS

The bold reticulate forewing pattern is very characteristic.

BIOLOGY

Ovum. Deposited in a row on the concave side of a pine needle during July and August; hatching in 14–21 days.

Larva. Head reddish brown, clypeal and ocellar regions darker; prothoracic plate reddish brown, darker posteriorly, edged laterally with brownish yellow; abdomen grey, with reddish suffusion, stronger dorsally; pinacula whitish; anal plate reddish brown, sometimes with darker markings; thoracic legs brownish yellow, shaded with blackish brown (de Joannis, 1900: 299). September to June; on *Pinus sylvestris*, feeding first on the brown scaling of the buds and, after hibernation, on the needles, constructing a silken tube 2–3 cm long, open at both ends and spun along a twig near the terminal shoot. In captivity the larva will feed on *Picea*.

Pupa. Reddish brown. June and July.

Imago. Late June to August; seldom seen during the day, but on the wing at dusk and coming readily to light.

DISTRIBUTION

First discovered in the British Isles in 1945 when a specimen was taken at light in a wooded locality in north-west Surrey; in 1946 a further specimen was captured in the same locality in July and another in August (Parfitt, 1947: 225). It has subsequently been taken at light in other localities in Surrey, including Horsell (de Worms, 1952: 215) and Weybridge (Messenger, 1951: 272); and at Whitstable (Huggins, 1951: 120), Folkestone, Ham Street and Sandwich, Kent (Chalmers-Hunt, 1951: 272); Hastings, Sussex; Sway (Brown, 1953: 83) and Winchester, Hampshire; the Isle of Wight; Canford Heath, Dorset (Brown, 1953: 83); Middlesex; Camberwell, London (Wakely, 1965: 91; 1968a: 164); Essex; Needham Market, Reydon, Walberswick (Chipperfield, 1967: 319; 1968: 61) and Thorpeness (Wakely, 1968b: 313), Suffolk. It has been observed in localities in Essex where *Pinus sylvestris* does not grow, such as at Westcliff-on-Sea, Hockley and Bradwell-on-Sea, indicating that it is a migratory species (Huggins, 1958c: 269; 1961b: 181; 1962: 41).

Europe to southern Russia.

Lozotaenia forsterana (Fabricius)

Pl. 33, figs. 7–9

[*Heterognomon forsterana*; Barrett, 1904: 192. *Lozotaenia forsterana*; Pierce & Metcalfe, 1922: 6, pl. 2 (♂, ♀ genitalia). *Tortrix forsterana*; Meyrick, 1928: 508; Ford, 1949: 52]

DESCRIPTION

♂ ♀ 20–29 mm. Sexes moderately dimorphic; forewing of male without costal fold but sometimes with edge of costa slightly raised from near base to near middle, antenna weakly dentate-ciliate; female usually larger, antenna minutely ciliate, apex of forewing more prominent.

Male (fig. 7). Forewing ground colour light greyish brown, sparsely strigulate with brownish; markings dark brown; basal fasciae obsolete; median fascia present on costa as a strong quadrate spot and represented in the pre-tornal area by a broad suffusion; pre-apical spot well developed. Hindwing grey; cilia paler, with an inconspicuous pre-apical fuscous spot on costal margin.

Female (fig. 9). Similar to male except in wing shape and larger size.

Variation. In both sexes the ground colour of the forewing varies from brownish grey to greyish brown. The markings vary in strength, being somewhat darker in heavily marked specimens; the median fascia is often obsolescent on the dorsum, but on the costa the markings, although reduced, usually remain prominent. Specimens lacking the strigulation occasionally occur, the forewing ground colour being almost unicolorous.

Specimens of this species from the Highlands of Scotland (fig. 8) apparently represent a subspecies in which the forewing coloration is distinctly grey in general appearance and the markings are weak and reduced.

COMMENTS

Distinguished by its relatively large size and the reticulate greyish brown forewing with strong costal markings.

BIOLOGY

Larva. Head black, or brown with blackish margins; prothoracic plate green or light brown, with a large black marking laterally; abdomen dull grey-green, darker dorsally; pinacula indistinct, shining green or whitish; anal plate greenish or yellowish grey, heavily marked with black laterally; anal comb present; thoracic legs brown. The larva hatches in September and after hibernation recommences to feed in April and May, living between two leaves, spun one above the other, on *Hedera* (pl. 14, fig. 1), *Ligustrum*, *Lonicera*, *Stachys*, *Vaccinium*, *Ribes*, *Larix*, *Abies*, *Picea* and *Pinus*. It is often responsible for scarring ivy leaves, eating the parenchyma and causing unsightly brown patches; in Cambridgeshire and Gloucestershire it has been found attacking leaves of cultivated strawberry (Vernon, 1971: 74).

Pupa. Blackish brown. June, in the larval habitation or in spun leaves (Tutt, 1890: 65).

Imago. Late June and July; frequenting copses, hedgerows, gardens and the borders of woods.

DISTRIBUTION

Occurring throughout most of Britain as far north as Ross and Cromarty, but nowhere abundant. In Ireland common and widely distributed (Beirne, 1941: 82); recently recorded from the Aran Islands (Dunn, 1969: 277). In the Scottish Highlands it is apparently represented by a distinct subspecies and occurs chiefly amongst *Vaccinium myrtillus*, its presumed foodplant in this region. The exact distribution of the two races is unknown, but the southern form extends as far north as Argyll.

Widely distributed throughout Europe; south-east Siberia.

Lozotaenia subocellana (Stephens)

The original description of this species was based on two specimens which, according to Stephens (1834: 75), were taken either in Devon or Cornwall. Although Wood (1839: 131, pl. 29, fig. 859) illustrates one of the original specimens, both of which are now apparently lost, the species has subsequently remained unrecognized.

Paramesia gnomana (Clerck)

Pl. 33, fig. 10

[*Dichelia gnomana*; Barrett, 1905: 214. *Capua gnomana*; Meyrick, 1928: 501. *Paramesia gnomana*; Ford, 1958: 3]

DESCRIPTION

♂ ♀ 16–20 mm. Sexes similar; antenna of male weakly dentate-ciliate, forewing without costal fold; female usually larger.

Male (fig. 10). Forewing ground colour pale ochreous-yellow, faintly strigulate with ochreous-brown; basal fasciae obsolete; median fascia ochreous-brown, darker on costa, represented by a very oblique, slender wedge-like streak from costa to about middle of wing; a dark sometimes blackish suffusion at tornus; pre-apical spot dark brown, well developed, usually interrupted towards apex of wing. Hindwing ochreous-white, weakly suffused with grey dorsally.

Female. Similar to male.

Variation. Continental specimens examined show little variation in general coloration and only minor variation in the strength of the markings.

COMMENTS

Superficially this species resembles lightly marked 'yellowish' forms of *Clepsis spectrana*. It can, however, be readily distinguished by venational differences in the forewing; in *gnomana* veins 7(R_5) and 8(R_4) are stalked from near the middle, whereas in *spectrana* veins 7(R_5) and 8(R_4) are separate.

BIOLOGY

Larva. Head brownish yellow; prothoracic and anal plates brown; abdomen yellow-green or grey-green; thoracic pinacula brown; abdominal pinacula yellowish green. The life history of this species in the British Isles is unknown. On the Continent, May and June; polyphagous, feeding on the leaves of deciduous trees and on *Vaccinium myrtillus*, *Stachys* and *Iris*, also on *Plantago*, *Taraxacum* and grasses (Swatschek, 1958: 54; Hannemann, 1961: 27).

Imago. On the Continent the moth is on the wing from June to August, frequenting damp places, hiding amongst the herbage by day and very difficult to disturb.

DISTRIBUTION

This species can only be regarded as doubtfully British. Barrett (1872: 129) recorded three specimens which had been given by a local collector to J. B. Hodgkinson, who lived in Lancashire. These specimens are now in the British Museum (Natural History) and are without exact locality data. Subsequently, Barrett (1905: 214) considered that the species should be removed from the British list, since the origin of the specimens was uncertain.

However, a further reputed British specimen was recorded by Huggins (1932: 104) who detected it among the series of *Clepsis spectrana* in J. C. Melvill's collection. This specimen likewise lacks locality data. Nevertheless, there appears to be a strong possibility that *gnomana* occurs in the British Isles but has been overlooked because of its superficial resemblance to *spectrana*.

North, central and southern Europe to Asia Minor, Iran, central and eastern Asia, occurring from low-lying areas up to 5,000 ft in mountains.

Paraclepsis cinctana (Denis & Schiffermüller)

Pl. 33, fig. 11

[*Lophoderus cinctana*; Barrett, 1904: 189. — *cinctana*; Pierce & Metcalfe, 1922: 6, pl. 3 (♂, ♀ genitalia). *Eulia cinctana*; Meyrick, 1928: 511; Ford, 1949: 53]

DESCRIPTION

♂ ♀ 14–18 mm. Sexes similar; male with antenna dentate-ciliate, a well-developed but very short costal fold at base of forewing; antenna of female sparsely ciliate.

Male (fig. 11). Forewing ground colour white, strigulate with plumbeous, most pronounced along costal and dorsal margins and in distal area of wing; markings ferruginous-brown extensively overlaid with plumbeous and sparsely irrorate with black; basal fasciae well developed, outer margin directly transverse; median fascia complete; pre-apical spot well developed, with heavier black irroration. Hindwing white, sparsely stippled with brownish grey.

Female. Similar to male.

Variation. Only minor variation is found in the forewing markings and in the stippling of the hindwing of this distinctive species.

COMMENTS

The white ground coloration and contrasting markings are distinctive of this species.

BIOLOGY

Larva. Head and prothoracic plate blackish brown; abdomen greenish brown, integument shagreened; anal plate light brown, spotted with black (Swatschek, 1958: 54). The life history of this species in the British Isles appears to be unknown. On the Continent, September to June; feeding on *Anthyllis*, *Lotus*, *Genista*, *Sarothamnus* and *Artemisia campestris*, living in a tubular silken gallery.

Imago. July; frequenting dry grassland on chalk downs inland and on the coast. The moth flies in sunshine, making short flights over the tops of the grasses, appearing conspicuously white while on the wing.

DISTRIBUTION

Known only from Kent, where it was first collected in 1857 by S. C. Tress Beale (Stainton, 1858: 88; Beale, 1858: 35), who took a few specimens on the sloping bank of a field at Alkham, near Dover. It also occurs elsewhere in the Dover district and inland at Barham Downs, near Canterbury.

Europe to Asia Minor.

Epagoge grotiana (Fabricius)

Pl. 33, figs. 12, 13

[*Dichelia grotiana*; Barrett, 1905: 212. *Epagoge grotiana*; Pierce & Metcalfe, 1922: 7, pl. 3 (♂, ♀ genitalia). *Capua grotiana*; Meyrick, 1928: 501; Ford, 1949: 49]

DESCRIPTION

♂ ♀ 13–17 mm. Sexes similar; antenna of male dentate-ciliate, forewing without costal fold but sometimes slightly upturned at middle; female usually larger, antenna minutely ciliate, forewing comparatively narrow.

Male (fig. 12). Forewing ground colour bright ochreous-yellow, strigulate with fuscous or ferruginous-brown; markings ferruginous-brown with fuscous admixture; basal fasciae weak, formed of numerous striae, outer margin diffuse and ill-defined, convex; median fascia complete, broadest on dorsum; pre-apical spot large, usually with several ferruginous-brown strigulae emanating from apex. Hindwing greyish brown.

Female (fig. 13). Similar to male.

Variation. The characteristic rich ochreous-yellow coloration and subtle ferruginous-brown markings of the forewing show slight variation in tone and intensity. The basal fasciae, never very strong, are sometimes obsolescent; variation is found in the width of the median fascia which may be slightly constricted or indented on the inner margin near the middle; more rarely the pre-apical spot is broadly connected to the middle of the median fascia by a brownish suffusion.

COMMENTS

The bright ochreous-yellow ground colour of the forewing and the ferruginous-brown markings are characteristic.

BIOLOGY

Ovum. Deposited in batches of 2 or 3 eggs, brilliant white in appearance; hatching in 10 days (Chrétien, 1897: 259).

Larva. Head brown with black ocellar and epistomal regions, or entirely black; abdomen green or greenish brown; pinacula small, black; thoracic legs black. August to May, overwintering in a narrow fold of a leaf; feeding on the leaves of *Quercus*, *Crataegus* and *Rubus*. On the Continent it is recorded as polyphagous on low plants, including *Petasites albus*, *Rosa* and *Vaccinium*; also on fallen leaves and fruit.

Pupa. Dark brown. June, spun up in or near the larval habitation.

Imago. June and July; during the day the moth may be beaten from its resting place on trees and bushes and will usually fly to the ground, although at times it can be very active. Most often found in the vicinity of woods, usually at rest on leaves of oak, but also occurring on sandhills where it rests on bramble bushes. It flies actively at dusk and again at dawn.

DISTRIBUTION

Generally distributed and sometimes locally common in wooded areas in England as far north as Westmorland, Northumberland and Durham, but scarce in the Midlands and northern counties mentioned. On the Isles of Scilly it occurs plentifully on sandhills on Tresco and St. Mary's (Richardson & Mere, 1958: 142). In Wales it is known from Flintshire, Denbighshire, Caernarvonshire, Merionethshire and Cardiganshire. It is apparently rare in Scotland and is known only from Glen Moriston (Pelham-Clinton, 1959: 68) and Aviemore, Inverness-shire. In Ireland widely distributed and locally common in oak woods.

Europe to Asia Minor; China, Korea and Japan.

Capua vulgana (Frölich)

Pl. 34, figs. 1, 2

[*Capua favillaceana*; Barrett, 1905: 251; Pierce & Metcalfe, 1922: 7, pl. 3 (♂, ♀ genitalia); Meyrick, 1928: 500; Ford, 1949: 49]

DESCRIPTION

♂ ♀ 13–19 mm. Sexes moderately dimorphic; antenna of male serrate-ciliate, forewing with a conspicuous costal fold extending from near base to middle; female usually larger, antenna filiform, minutely ciliate.

Male (fig. 1). Forewing ground colour greyish buff, obscurely strigulate with greyish olive-drab; basal fasciae, median fascia and pre-apical spot ill-defined, represented by brownish or fuscous suffusion; median fascia darker towards costa; a rather distinctive pale buff basal area posterior to costal fold; head and labial palpi also pale buff, enhancing the light basal area; thorax greyish buff. Hindwing brownish grey, paler basally.

Female (fig. 2). Forewing comparatively uniform brownish buff, lacking the pronounced costal fold and pale buff basal area of the male. Head and labial palpi concolorous with the forewing. Hindwing brownish grey.

136

Variation. In both sexes the intensity of the greyish or brownish buff coloration of the forewing varies. In the male the markings are sometimes intensified with an admixture of fuscous scales, but the markings are never very clearly defined; the female may also have stronger markings but seldom as dark as in the male. Exceptionally the markings are obsolescent and the forewing is almost unicolorous greyish buff except for a weak fuscous suffusion in the basal and median areas.

COMMENTS

The male may be distinguished by the paler basal area of the forewing and the pale buff head and labial palpi; the female somewhat resembles an obscurely marked *Cnephasia* but differs in having the forewing distinctly buff-coloured.

BIOLOGY

Larva. Head yellowish; prothoracic plate light brown; abdomen greyish green, integument shagreened (Swatschek, 1958: 50). July to September, overwintering either as a larva or pupa; feeding on the leaves of *Carpinus*, *Alnus*, *Sorbus* and *Vaccinium*; on the Continent *Rubus* and *Quercus* are recorded as additional foodplants.

Pupa. Apparently undescribed.

Imago. May and June; the male may often be seen flying about trees in woods, or over *Vaccinium* in open country; both sexes are on the wing towards dusk, when the flight is generally much lower down. In the Burren, Co. Clare, it is common amongst scrub hazel (*Corylus*), which is possibly the foodplant in that locality.

DISTRIBUTION

Widely distributed and locally common; most plentiful in the south of England, extending as far north as Argyll. Less common in the east of Scotland but recorded from East Lothian and Perthshire. In Wales it occurs in Flintshire, Denbighshire and Caernarvonshire, and according to Barrett (1905: 252) is common in the south. Locally common in woodlands in most parts of Ireland except the north and north-west (Beirne, 1941: 79); common in hazel scrub in the Burren, Co. Clare.

Europe to Asia Minor; east Siberia; Japan.

Philedone gerningana (Denis & Schiffermüller)

Pl. 34, figs. 3, 4

[*Amphysa gerningana*; Barrett, 1905: 210. *Philedone gerningana*; Pierce & Metcalfe, 1922: 6, pl. 2 (♂, ♀ genitalia); Meyrick, 1928: 501; Ford, 1949: 49]

DESCRIPTION

♂ ♀ 12–17 mm. Sexes moderately dimorphic; antenna of male strongly bipectinate-ciliate, forewing without costal fold, anal tuft pale yellow with orange-brown admixture; female with antenna simple, forewing comparatively narrow, apex produced, apex of hindwing with weak cupreous suffusion.

Male (fig. 3). Forewing ground colour light ochreous-yellow, basal area adjacent to median fascia weakly strigulate with olive-brown, distal half of wing strigulate with reddish brown; markings reddish brown; outer margin of basal fascia diffuse; sub-basal fascia indeterminate; median fascia very oblique, with a characteristic, sharply defined inner margin, the outer margin diffuse and irregular; pre-apical spot ill-defined, sometimes darkened with blackish purple irroration. Hindwing dark grey; cilia paler, with a dark grey sub-basal line.

Female (fig. 4). Forewing ground colour extensively suffused with dull purplish brown, especially in basal area, obliterating strigulae. Hindwing grey, suffused apically with cupreous mottled with grey.

Variation. Minor individual variation in general coloration and markings is frequent but no extreme forms are known.

The British populations of this species apparently represent a distinct subspecies which is smaller than the European nominate form and has more sharply demarcated basal and distal halves of the forewing pattern, generally darker coloration, and weak strigulation in place of strong reticulation.

COMMENTS

Distinguished by the contrasting basal and distal areas of the forewing in both sexes, and the strongly bipectinate antennae of the male.

BIOLOGY

Larva. Head light brownish yellow or reddish brown, darker mottled; prothoracic plate light brown, weakly mottled; abdomen varying from light to greyish green, integument somewhat roughened dorsally; pinacula brown, setae dark brown; anal plate yellowish, edged with black; anal comb whitish. May and June, in spun leaves and flowerheads of *Potentilla*, *Helianthemum*, *Vaccinium* and *Armeria maritima*; also known on *Picea sitchensis* in Scotland. On the Continent, *Lotus*, *Plantago*, *Scabiosa columbaria* and *Peucedanum* are also recorded as foodplants (Hannemann, 1961: 30).

Pupa. Yellowish brown. June, in the larval habitation.

Imago. July and August, sometimes extending into September; frequenting heaths and limestone areas. The male is active at any time of the day when the sun is shining, even as early as sunrise, but the female is much more lethargic and flies in the evening; occasionally coming to light.

DISTRIBUTION

Typically a mountain heathland species, but also occurring at lower levels, and recorded in England from Kent, Somerset, Cornwall, Herefordshire, Cheshire, Derbyshire, south Westmorland, north Lancashire, Yorkshire, Cumberland, Durham and Northumberland. In Wales it occurs on the mountains of Flintshire and Caernarvonshire southwards to Pembrokeshire. In Scotland it is very local but widely distributed, its range extending to the Outer Hebrides, St. Kilda and the Shetlands. Recorded from a few localities in the northern half of Ireland, but Beirne (1941: 80) states that it is probably frequent in the mountains and bogs; it occurs not uncommonly in the limestone area of the Burren, Co. Clare (Bradley & Pelham-Clinton, 1967: 135), and in bogs in west Galway.

Europe to Asia Minor, Transcaucasia, Central Asia to south-east Siberia.

Philedonides lunana (Thunberg)
Pl. 34, figs. 5–7

[*Amphysa prodromana*; Barrett, 1905: 211. *Amelia prodromana*; Pierce & Metcalfe, 1922: 9, pl. 3 (♂, ♀ genitalia). *Philedone prodromana*; Meyrick, 1928: 501; Ford, 1949: 49]

DESCRIPTION

♂ ♀ 12–16 mm. Sexual dimorphism pronounced; antenna of male strongly pectinate, pectinations short, ending in long tufts of cilia, forewing without costal fold; female usually smaller, antenna simple, forewing somewhat attenuate.

Male (figs. 5, 6). Forewing ground colour whitish grey, coarsely irrorate with grey; markings dark brown or ferruginous-brown, sparsely irrorate with black; basal fasciae diffuse; median fascia well defined, dilated in dorsal area; pre-apical spot diffuse, inner margin produced towards tornus. Hindwing dark grey, with faint darker mottling.

Female (fig. 7). Forewing ground colour greyish white, sparsely irrorate with ferruginous-brown; basal fasciae more clearly defined than in male. Hindwing greyish white, basal two-thirds suffused with greyish, distal area strigulate with grey.

Variation. In the male minor variation occurs in the coloration of the forewing markings; in specimens in which the ferruginous-brown content is intensified the markings show more contrast with the ground colour. In both sexes the pre-apical spot extends to the tornal area and may be confluent with the median fascia.

COMMENTS

The dark coloration of the fore- and hindwings, and the strongly pectinate antennae, are characteristic of the male; in the female the forewing is attenuated and the markings contrast with the greyish white ground colour.

BIOLOGY

Ovum. Deposited in small batches, covered in a yellowish glutinous substance, on the leaves of the foodplant at the end of March and April (Gregson, 1880b: 90).

Larva. Rather slender; head light yellowish brown, ocellar region dark brown, a blackish spot laterally; prothoracic plate broad, dark green or greenish yellow; thoracic legs brownish; abdomen dull green or olive-green, integument shagreened, pinacula paler; anal plate greenish yellow. May and June; feeding in spun or folded leaves and flowers of *Potentilla*, *Smyrnium olusatrum*, *Mentha*, *Myrica*, *Vaccinium*, *Erica*, *Calluna*, *Valeriana* and other herbaceous plants. In the Burren district in western Ireland the larva has been found on *Arctostaphylos*, and on the Isle of Rhum, Inner Hebrides, it has been recorded on young *Pinus contorta* (Wormell, 1969: 112); on the mainland of Scotland and in Wales the larva has been found on *P. contorta* and *Picea sitchensis*.

Pupa. June to March, spun up in leaves of the foodplant and in debris on the ground.
Imago. Mid-March to May; favouring open country near mountain heaths and dry moorland. The males fly freely in calm, sunny weather, especially at sunrise, midday and evening, the females sitting on the tops of *Calluna*, *Vaccinium* and other vegetation nearby, where pairing takes place. At other times they hide amongst the dead leaves on the ground and are then very difficult to find (White, 1880: 114).

DISTRIBUTION

Locally common on heaths and moorland from Cheshire northwards to Ross and Cromarty and the Orkneys; records from the Shetlands require confirmation. In Cheshire and Derbyshire it apparently does not occur below 600 ft, but in Northumberland and Durham it is found on coastal sandhills. Little is known of its occurrence in Wales, but it has been recorded from the Denbighshire moors (Michaelis, 1969b: 189) and there are unconfirmed records from Caernarvonshire and Glamorgan. In Ireland, Beirne (1941: 80) records it as locally common on the mountains and bogs and apparently most frequent in the northern half of the country. Bradley & Pelham-Clinton (1967: 135) record it as common on the lower slopes of Mt. Doughbranneen, in the Burren, Co. Clare.

Northern and central Europe.

Ditula angustiorana (Haworth)

Pl. 34, figs. 8, 9

[*Batodes angustiorana*; Barrett, 1905: 206. *Capua angustiorana*; Pierce & Metcalfe, 1922: 7, pl. 3 (♂, ♀ genitalia). *Batodes angustiorana*; Meyrick, 1928: 500; Ford, 1949: 49]

DESCRIPTION

♂ ♀ 12–18 mm. Sexual dimorphism moderately pronounced; antenna of male strongly dentate-ciliate, forewing with broad costal fold from base to one-third; female larger, antenna simple, forewing without costal fold, coloration paler.

Male (fig. 8). Forewing ground colour brown with ochreous admixture, most apparent as a characteristic semi-elliptical patch on the dorsum; basal fasciae indistinct, sub-basal fascia forming a rectangular, outward-oblique, dark ferruginous-brown patch on dorsum from near base to middle; costal fold and basal area anterior to sub-basal fascia dark greyish brown; median fascia inconspicuous, pale ferruginous-brown, thinly edged with ochreous ground colour, narrow towards costa and concealed by costal fold; pre-apical spot ill-defined, ferruginous-brown coarsely irrorate with black; postmedian area with variable plumbeous suffusion. Hindwing dark chocolate-brown.

Female (fig. 9). Forewing ground colour pale tawny, mixed with whitish ochreous distally, and forming a distinctive patch near middle of costa; markings pale ferruginous-brown or fulvous; sub-basal fascia more distinct than in male; median fascia extending to costa. Hindwing coloration paler than in male.

Variation. Extreme forms appear to be unknown but minor variation frequently occurs in the intensity of the coloration and markings.

COMMENTS

The forewing pattern is reminiscent of *Archips xylosteana*, but the two species differ in wing shape in both sexes, and in the length of the costal fold in the male, that of *angustiorana* extending only to one-third, that of *xylosteana* to two-thirds. The dark hindwings of *angustiorana* are also characteristic.

BIOLOGY

Ovum. The pale yellow egg mass is deposited on the leaves during June and July; it is very similar to that of *Archips podana* but is somewhat smaller and more yellowish (Massee, 1946: 61).

Larva. Slender; head light yellowish brown or greenish yellow, ocellar region and posterior margin marked with blackish brown; prothoracic plate variable, yellowish green, light brown or dark brown, usually darker postero-laterally, anterior margin sometimes paler; thoracic legs light green, with blackish terminal segments; abdomen varying from light yellowish green to brownish green, or grey, weakly shagreened dorsally; pinacula light green, most conspicuous in darker dorsal region; spiracles small, peritreme dark brown, spiracle on eighth segment more than twice diameter of others; anal plate greenish, weakly marked with brown or entirely brown; anal comb yellowish or brown, with 4 prongs. August to April and May; polyphagous, feeding on the foliage of orchard trees such as *Malus*, *Pyrus* and *Prunus*, usually amongst spun leaves (pl. 14, figs. 3, 4) and developing fruit buds, sometimes causing superficial damage to the mature fruits; also on *Vitis*, *Taxus*, *Juniperus*, *Pinus sylvestris*, *Larix*, *Viscum*, *Hedera*, *Rhododendron*, *Quercus*, *Ilex*, *Buxus*, *Laurus*, *Hippophae* and various other trees and shrubs and sometimes herbaceous plants. On fruit trees the newly hatched larva feeds on the leaves and after the second moult it turns its attention to the fruits; in late autumn it spins a silken hibernaculum attached to a bud or spur, recommencing to feed in the early spring on the buds and later on the flowers and developing fruit. When feeding on *Viscum* it eats into the parenchyma, blistering the leaf and usually pulling two or three leaves together with silk; on *Quercus* it hibernates on a withered leaf spun to a twig.

Pupa. Light brown, darker dorsally. June, usually on the foodplant in spun leaves or a folded leaf, or in debris on the ground; sometimes in an old spongy oak gall.

Imago. June and July; frequenting orchards, gardens, hedgerows and the borders of woods. The male is very active in the sunshine, but in dull weather may often be found sitting openly on leaves, tree trunks and palings. The female is less active but readily disturbed. At dusk both sexes are on the wing and may occasionally be taken at light and sugar.

DISTRIBUTION

A common urban and rural species in the southern counties of England; locally common in Wales and the northern counties of England, though according to Barrett (1905: 207) scarce in Durham and Northumberland. In Scotland it is common in the south and has been recorded as far north as the Clyde area and Perthshire. Widely distributed and common in Ireland, particularly in gardens.

Europe to Asia Minor; North Africa; North America (coastal regions).

Pseudargyrotoza conwagana (Fabricius)

Pl. 34, figs. 10–12

[*Dictyopteryx conwayana*; Barrett, 1905: 201. — *conwayana*; Pierce & Metcalfe, 1922: 16, pl. 6 (♂, ♀ genitalia). *Argyrotoxa conwayana*; Meyrick, 1928: 518; Ford, 1949: 57]

DESCRIPTION

♂ ♀ 11–15 mm. Sexual dimorphism not pronounced; antenna of male strongly dentate-ciliate, forewing without costal fold; female usually larger, antenna simple, forewing coloration more yellow.

Male (fig. 10). Forewing ground colour yellow, weakly suffused with light ferruginous; markings poorly defined, fuscous-brown; basal area strigulate, followed by a rather conspicuous subquadrate pale yellow dorsal blotch; median fascia consolidated with blackish suffusion on costa; markings and ground colour coarsely irrorate with metallic bluish plumbeous. Hindwing dark greyish brown; cilia paler, with a brown sub-basal line.

Female (fig. 12). Similar to male but with markings and ferruginous suffusion weaker; general coloration more yellow.

Variation. The markings vary in strength; heavily marked specimens of the male may have a blackish admixture which contrasts with the bright yellow ground colour (fig. 11). In the female, forms are found in which the markings are reduced or obsolescent and the general coloration is ferruginous-orange, with the plumbeous irroration more conspicuous.

COMMENTS

The general yellowish coloration and metallic irroration of the forewing and the usually conspicuous pale yellow dorsal blotch readily distinguish this species.

BIOLOGY

Larva. Head glossy brownish yellow; prothoracic plate similar, rather broad; abdomen yellowish white, with a darker dorsal line, integument shagreened; pinacula weakly sclerotized. August to September or October, in the seeds of *Fraxinus* and berries of *Ligustrum*; on the Continent it is known to feed also on the fruits of *Berberis* and *Syringa*.

Pupa. October to April; in a flattened, white cocoon in the ground (Barrett, 1905: 201).

Imago. May to early July; on warm sunny days the moth may be seen flying actively about the foodplants, particularly large ash trees. Towards dusk it flies farther afield but is seldom seen at light.

DISTRIBUTION

Common and sometimes abundant in the British Isles; probably occurring wherever ash and uncut privet grow. Recorded as far north as Morayshire, and extending to Wester Ross.

Europe to Asia Minor; south-east Siberia; China.

SPARGANOTHINI

Sparganothis pilleriana (Denis & Schiffermüller)
Pl. 34, figs. 13–16
[*Oenectra pilleriana*; Barrett, 1904: 184. *Sparganothis pilleriana*; Pierce & Metcalfe, 1922: 25, pl. 10 (♂, ♀ genitalia); Meyrick, 1928: 519; Ford, 1949: 57]

DESCRIPTION

♂ 15–20 mm, ♀ 17–22 mm. Sexes strongly dimorphic; antenna of male rough-scaled, finely ciliate, forewing truncate apically, with very narrow costal fold from base to about one-fifth, markings developed; female usually slightly larger, antenna minutely ciliate, apex of forewing angulate, wing pattern obsolete. Labial palpi markedly long and porrect in both sexes.

Male (fig. 13). Forewing ground colour ochreous, with a brassy tinge from base to beyond median fascia, sparsely strigulate with brown distally; markings brown with an ochreous admixture; basal fasciae obscurely indicated on costa, sub-basal fascia represented by a small quadrate spot on dorsum; median fascia narrow, very oblique from costa at one-third, outer margin irregular; pre-apical spot quadrate, only a little beyond middle of wing, usually emitting apically several irregular transverse, often confluent, strigulae; a rather thick almost straight subterminal strigula. Hindwing brownish grey; cilia pale ochreous, with a darker sub-basal line.

Female (fig. 14). Forewing coloration almost uniform red-brown with a weak brassy sheen; wing pattern obsolete. Hindwing brownish grey; cilia paler, with a weak sub-basal line.

Variation. The male shows extensive variation in reduction of the forewing markings; in specimens in which they are obsolescent the ground colour is usually paler, varying from greyish ochreous (fig. 15) to whitish ochreous. In the female the forewing ground colour shows similar variation (fig. 16).

The variation in this species appears to be influenced by the biotope. In salt marshes, forms with dark ferruginous or red-brown coloration predominate, and on wet heathland the general coloration is olive-green or brown; ochreous forms intermediate between these two extremes occur on chalk downs.

COMMENTS

This species may be readily recognized by its long, porrect labial palpi (text-fig. 9).

BIOLOGY

Ovum. Deposited on the foodplant in batches of 40–60 eggs during July; emerald-green when first laid, changing to greenish yellow and then pure yellow.

Larva. Head and prothoracic plate black; abdomen varying from pale green to greyish green, with a narrow darker green dorsal line, integument strongly shagreened; pinacula small, paler than integument; anal plate yellowish brown; thoracic legs black. September, overwintering and feeding until the following May and June; polyphagous on various plants, including *Limonium vulgare*, *Narthecium ossifragum*, *Stachys*, *Iris*, *Clematis*, *Humulus*, *Pyrus*, *Origanum*, *Centaurea* and *Plantago*, living in folded or spun leaves, usually folding the leaf from the upperside. On the Continent it is injurious to *Vitis vinifera*, feeding in the autumn on the fruit and leaves, and on the buds and developing leaves in the spring. In Turkey the larva has been found feeding in shoots of *Pinus pinea* and *P. halepensis* var. *brutia*.

Pupa. Dark brown. June, between spun leaves or in a rolled leaf of the foodplant.

Imago. July; frequenting coastal salt marshes and, in August, inland on bogs and wet heaths. The moth is seldom seen during the day, resting amongst the herbage. The male flies at dusk, the female is less active.

DISTRIBUTION

In the British Isles known only from the southern counties of England and in South Wales. Extremely local but often very common where it occurs, in Sussex (Worthing), Hampshire (New Forest), Isle of Wight (Ventnor sands, Yarmouth), Dorset (Bloxworth, Studland), Devon (Torquay), Cornwall, and Glamorgan (Gower Peninsula).

Europe to Siberia; Korea, China and Japan; North America.

CNEPHASIINI

Olindia schumacherana (Fabricius)

Pl. 35, figs. 1, 2

[*Olindia ulmana*; Barrett, 1905: 276; Pierce & Metcalfe, 1922: 15, pl. 6 (♂, ♀ genitalia); Meyrick, 1928: 517; Ford, 1949: 56]

DESCRIPTION

♂ ♀ 11–16 mm. Sexes moderately dimorphic; antenna of male very shortly ciliate, forewing with white antemedian fascia narrow; female usually larger, antenna simple, antemedian fascia broad.

Male (fig. 1). Forewing fuscous-brown, coarsely irrorate with black, mottled with plumbeous in basal and distal areas; a narrow white antemedian fascia, often interrupted a little below costa. Hindwing dark fuscous-brown; cilia paler.

Female (fig. 2). Plumbeous mottling of forewing heavier and forming blotches in apical area; white antemedian fascia conspicuous, broader than in male and almost invariably complete.

Variation. The irregularly margined white antemedian fascia varies in both sexes. In the male the fascia is often constricted or interrupted below the costa and in extreme forms may be reduced to distinct costal and subdorsal spots; occasionally the normally broad fascia in the female is narrow, the margins being more irregular.

COMMENTS

The dark fuscous-brown coloration of the fore- and hindwings and the contrasting white antemedian fascia of the forewing are characteristic of this species.

BIOLOGY

Larva. Head brownish yellow, posterior margin blackish; prothoracic plate and thoracic legs dark brown or black; abdomen yellowish green, integument shagreened; pinacula dark brown or black. The larva may be found from April to June on *Ranunculus ficaria*, folding down the edge of a leaf and living within, eating the cuticle and tissue but leaving the upper surface intact, or gnawing neighbouring leaves; also in spun or folded leaves of *Aquilegia*, *Mercurialis*, *Vaccinium* and other herbaceous plants. On the Continent, *Plantago* and *Chrysosplenium* are also recorded as foodplants.

Pupa. June, in the larval habitation (pl. 14, fig. 2), or in a cocoon on the ground.

Imago. June and July; hiding during the day, preferring the thick cover of yew trees and bushes, especially hazel, alder and blackthorn, often low down near the ground. Although readily disturbed it quickly returns to cover. The male flies in the afternoon sunshine and is joined by the female at dusk.

DISTRIBUTION

Widely distributed but rather local and never very common in the British Isles, occurring as far north as the Clyde area, Perthshire and Kincardineshire; most frequent in the south of England, becoming scarcer northwards and rare in the south of Scotland. Scarce in Cheshire and only occasional in Flintshire, Denbighshire and Caernarvonshire. In Ireland, Beirne (1941: 84) records it as frequent in woodland in Co.'s Kerry, Wicklow, Mayo and Donegal, and Mere & Pelham-Clinton (1966: 180) record it from Cratloe, Co. Clare.

North and central Europe.

Isotrias rectifasciana (Haworth)

Pl. 35, figs. 3, 4

[*Olindia hybridana*; sensu Barrett, 1905: 275. *Isotrias rectifasciana*; Pierce & Metcalfe, 1922: 15, pl. 6 (♂, ♀ genitalia); Meyrick, 1928: 516; Ford, 1949: 56]

DESCRIPTION

♂ 14–16 mm, ♀ 11–14 mm. Sexual dimorphism pronounced; male usually larger than female, forewing broader distally and termen more oblique, with paler general coloration, antenna rough-scaled, minutely ciliate; female with antenna simple, forewing markings heavier and more distinct.

Male (fig. 3). Forewing ground colour cream-white diffusely irrorate with pale ochreous-brown; markings pale ochreous-brown sprinkled with black; outer margin of basal fasciae shallowly convex, irregular; median fascia almost directly transverse, inner margin indented at middle; pre-apical spot represented by two small chevron-like marks on costa, these followed by a larger spot close to apex and extended across wing as a relatively broad irregular subterminal streak; a quadrate pre-tornal spot on dorsum; cilia whitish yellow. Hindwing brownish grey; cilia paler.

Female (fig. 4). Ground colour of forewing clearer and less suffused than in male;

markings dark brown, more heavily sprinkled with black; median fascia slightly broader at middle than in male.

Variation. This species shows considerable minor variation in the forewing markings and in the irroration of the ground colour; no extreme forms are known.

COMMENTS

Resembling a *Cnephasia* in general appearance but differs in having the forewing markings almost directly transverse.

BIOLOGY

The life history of this species is apparently unknown, but the pupa has been beaten from *Crataegus* (Barrett, 1905: 276), which may be the larval foodplant.

Imago. May and June. Frequenting hedgerows in which elm, hawthorn, blackthorn and other bushes grow; easily beaten out during the day but not very active and soon settling again. Both sexes fly at sunset.

DISTRIBUTION

A common species in hedges and lanes, occurring in England northwards to Yorkshire, Westmorland and Cumberland. Known in Wales from Flintshire, Denbighshire, Caernarvonshire and Merionethshire, and in Scotland from Ayrshire, Renfrewshire and Arran; the record from Argyll (Meyrick, 1928: 517) requires confirmation. In Ireland it has been recorded from Co.'s Galway, Dublin, Sligo and Down, but it is evidently rare and records require confirmation (Beirne, 1941: 84).

Europe to Asia Minor.

Eulia ministrana (Linnaeus)

Pl. 35, figs. 5–7

[*Lophoderus ministrana*; Barrett, 1904: 186. *Eulia ministrana*; Pierce & Metcalfe, 1922: 9, pl. 3 (♂, ♀ genitalia); Meyrick, 1928: 510; Ford, 1949: 53]

DESCRIPTION

♂ ♀ 18–22 mm. Sexual dimorphism not pronounced; antenna of male dentate-serrate, shortly ciliate; antenna of female weakly dentate, finely pilose.

Male (fig. 5). Forewing ground colour whitish yellow overlaid with dark ferruginous-brown in basal, median and terminal areas; a usually conspicuous white discocellular spot; cilia ferruginous-brown or dark brown. Hindwing brownish grey; cilia paler.

Female. Similar to male.

Variation. In both sexes considerable variation is found in the extent and intensity of the ferruginous-brown suffusion. Recurring dark forms (figs. 6, 7) are not uncommon in hill and mountain districts from the Midlands northwards; in extreme heavily suffused forms the white discal spot may be obliterated and the whole wing is almost uniform dark reddish brown. In intermediate forms the suffusion of the ground colour of the forewing is reduced and distinctly whitish and the basal area of the hindwing is often weakly infuscate.

COMMENTS

The moderately large size and the whitish yellow ground colour of the forewing with obscure ferruginous-brown markings are characteristic.

BIOLOGY

Ovum. Deposited in small, yellowish brown batches on the foodplant during May and June; hatching in about 18 days (Peers, 1865: 250).

Larva. Head chestnut-brown, ocellar region black; prothoracic plate similar, varying to yellowish green; abdomen green, integument finely shagreened; pinacula and anal plate yellowish green; anal comb with 8–10 large prongs. July to April, on *Corylus*, *Fraxinus*, *Rhamnus*, *Sorbus*, *Betula*, *Prunus*, *Alnus* and *Vaccinium myrtillus*, feeding first at the tip of a leaf, later between two spun leaves in a tubular spinning open at both ends. The larva reaches maturity at the beginning of October and overwinters before pupation in the spring. On the Continent, *Epilobium*, *Rubus chamaemorus* (Benander, 1929: 133), *Rosa*, *Fagus*, *Quercus* and *Tilia* are also recorded as foodplants.

Pupa. Dull brown, abdomen paler. April and May, in the larval habitation, the openings of which have been closed.

Imago. May and June; frequenting wooded areas and lanes, flying fairly high up about the trees at sunset and occasionally coming to light after dark. Seldom seen during the day, possibly dropping unnoticed to the ground when disturbed. It may also be found amongst tall sallows in boggy places on open moorland and mosses. In Scotland it is locally common around birch and alder.

DISTRIBUTION

A common woodland species in most parts of the British Isles, occurring as far north as Sutherland in Scotland. Local in Ireland and apparently most frequent in the north.

Europe to Siberia and Japan; North America.

Cnephasia longana (Haworth)

Pl. 35, figs. 8–10

[*Sphaleroptera ictericana*; Barrett, 1905: 254. *Cnephasia longana*; Pierce & Metcalfe, 1922: 11, pl. 4 (♂, ♀ genitalia); Meyrick, 1928: 513; Ford, 1949: 54]

DESCRIPTION

♂♀ 15–22 mm. Sexual dimorphism pronounced; antenna of male very shortly ciliate, forewing unicolorous, anal tuft well developed; antenna of female simple, forewing with brownish markings.

Male (fig. 8). Forewing unicolorous, pale ochreous to brownish ochreous; cilia concolorous. Hindwing pale translucent grey; cilia paler.

Female (fig. 9). Forewing ground colour whitish ochreous, diffusely irrorate with ochreous-brown; markings ochreous-brown; sub-basal fascia sharply angulate medially; median fascia from beyond middle of costa to before tornus, its inner margin deeply indentate above and below middle; pre-apical spot diffuse; a broad subterminal stria. Hindwing as in male; cilia whitish.

Variation. In the male the forewing varies from whitish grey to dark ochreous-brown; very rarely the darker forms have a sprinkling of blackish scales. The female shows similar variation in ground colour, with the markings varying from light ochreous-brown to dark ochreous-brown with an admixture of fuscous or black; heavily marked specimens are sometimes distinctly grey in general appearance and the markings are diffuse (fig. 10). Weakly marked almost unicolorous pale forms of the female occur rarely. In both sexes the hindwing varies from whitish to dark grey.

COMMENTS

The male differs from other *Cnephasia* species in the unicolorous ochreous forewing; in the female the characteristic markings are ochreous-brown.

BIOLOGY

Ovum. 100–200 eggs per female are laid singly or in small batches on the foodplant during July; hatching in 11–20 days (Balachowsky, 1966: 598). In North America the biology has received considerable attention because of the economic importance of this species. According to Rosenstiel (1941: 255) oviposition occurs at night and the eggs are laid singly on rough surfaces of trees, posts and similar situations. The eggs are deposited in crevices and depressions and are brushed by the ovipositor pads while still sticky, causing scales and particles of debris to adhere.

Larva. Head lustrous light brown, ocellar region and postero-lateral margin dark brown, ocellus III large and prominent; prothoracic plate light brown, or greenish and concolorous with abdomen, usually with a dark brown or blackish marking on the lateral margin and weakly mottled with brown along the posterior margin; abdomen green-grey or yellowish grey, with paler longitudinal subdorsal and lateral lines; pinacula small, paler than integument; anal plate and comb light brown, comb strongly developed with about 6–8 long prongs. The eggs hatch in August and the young larvae immediately enter hibernation, spinning a hibernaculum in the vicinity of the old egg batch (Edwards & Mote, 1936: 1121; Balachowsky, 1966: 598). The larvae commence feeding in early spring, and in May and June are found on *Chrysanthemum*, *Hypochoeris*, *Anthemis*, *Aster tripolium*, *Ranunculus*, *Armeria*, *Lychnis* and various other low-growing plants, including occasionally cultivated field beans. On the Continent the recorded foodplants include *Selinum* and *Ligularia* (Hannemann, 1961: 38; Razowski, 1959: 249). On plants such as *Selinum*, *Sempervivum* and *Ligularia*, the larva mines the leaves in the first and sometimes the second instar (Hering, 1957: 3, pl. 73, fig. 628); the type of mine and parts of the leaves eaten vary with the foodplant, and feeding may occur between the folded parts of a young leaf or shoot rather than within the parenchyma. In later instars the larva lives on the terminal parts of plants, particularly the flowers, drawing together the florets, anthers or petals, and feeding within. According to

Rosenstiel, Ferguson & Mote (1944: 815), in North America the first instar larvae are wind-dispersed from elevated positions to nearby low vegetation after leaving the hibernaculum, putting out a silken thread as an aid to buoyancy.

Pupa. Light brown. June, in a flimsy cocoon amongst surface debris, dead leaves or in empty seed capsules.

Imago. July; favouring rough coastal and chalkland habitats. Readily disturbed during the day from rest amongst gorse, bracken and other herbage; the female has a tendency to sit on tree trunks and fences. Both sexes fly at dusk and come to light.

DISTRIBUTION

Widely distributed and often common on the coast and on chalk downs inland in England north to Westmorland and Cumberland. According to Barrett (1905: 256) it is common throughout the southern and eastern counties to Hertfordshire, Oxfordshire, Derbyshire, Cheshire, Lancashire and Yorkshire, and was formerly common in the southern suburbs of London. In Wales known from Denbighshire, Caernarvonshire and Cardiganshire. Apparently rare in Scotland, and records from Aberdeenshire (Reid, 1893a: 10), Sutherland (Wickham, 1927: 43) and the Outer Hebrides (Weir, 1881: 219) are unconfirmed. Apparently very local in Ireland, recorded from Co.'s Kerry (Killarney), Down, Sligo, Fermanagh and Antrim (Beirne, 1941: 83).

Europe to Asia Minor; north-west Africa and the Canary Islands. Also occurs in western North America where it was first recorded in 1929; it has become a serious pest on cultivated strawberry (boring into the fruit), flax, vetch and other plants (Edwards & Mote, 1936: 1118; Cram & Tonks, 1959: 156; Powell, 1964: 124), and is known as the omnivorous leaf-tier.

Cnephasia gueneana (Duponchel)

A specimen of *C. gueneana* was bred in April, 1960, from a larva found in anemones and jonquils the previous February. The flowers had been purchased from a London florist and were believed to have been imported from the south of France (Manley, 1961: 18).

This species is indigenous to the Mediterranean sub-region, ranging from Palestine and southern Europe to Morocco and the Canary Islands. De Lucca (1951: 205) describes the early stages and life history. In the Mediterranean region the moth is on the wing from March to June, and the larva feeds in the flower buds and spun leaves of various wild and cultivated herbaceous plants during January and February.

C. gueneana is closely related to *C. longana*, which occurs in the British Isles, and the two species are alike in coloration and wing pattern. Compared with the female of *longana*, the markings of the forewing of the female of *gueneana* are normally comparatively bold and contrast more strongly with the ground colour. The male is whitish yellow or whitish brown in general appearance and, like the male of *longana*, is without pronounced markings.

A description of this species, including genitalic differences which distinguish it from *longana*, is given by Razowski (1959: 253).

Cnephasia communana (Herrich-Schäffer)

Pl. 35, figs. 11, 12

[*Sciaphila communana*; Barrett, 1905: 263. *Cnephasia communana*; Pierce & Metcalfe, 1922: 12, pl. 5 (♂, ♀ genitalia); Meyrick, 1928: 514; Ford, 1949: 55]

DESCRIPTION

♂ ♀ 18–22 mm. Sexual dimorphism not pronounced; antenna of male very shortly ciliate, forewing long and narrow; forewing of female comparatively broader.

Male (fig. 11). Forewing ground colour light grey diffusely irrorate and suffused with fuscous; markings fuscous, sprinkled with black, especially on margins; sub-basal fascia from costa, terminating above dorsum, outer margin prominently angulate at middle of wing; inner margin of median fascia sinuous, strongly angulate at middle, outer margin diffuse. Hindwing grey, paler basally.

Female (fig. 12). Similar to male but forewing markings generally darker.

Variation. The coloration and markings of the forewing usually show only minor variation; however, at Gomshall, Surrey, an extreme form occurs in which the ground colour in the basal half of the wing is almost white, the distal half is powdered with whitish grey, and the fasciae are blackish brown and conspicuous.

COMMENTS

One of the earliest *Cnephasia* species to emerge and the only one on the wing during most of May.

BIOLOGY

The immature stages of this species are apparently undescribed. On the Continent the larva feeds on *Rumex*, *Lotus*, *Chrysanthemum*, *Plantago* and other plants (Razowski, 1959: 221).

Imago. May and June; frequenting rough meadows, fens and chalk downland, hiding amongst vegetation or sitting on tree trunks during the day and easily disturbed. It flies in the early morning and at night, occasionally coming to light.

DISTRIBUTION

Very local but often common, occurring in East Anglia and the southern counties of England to Gloucestershire (Tetbury, Avening). Robinson (1967: 276) records this species from Westmorland (Haversham), and it has probably been overlooked in the Midlands and northern counties.

Europe and North Africa to Asia Minor and Siberia.

Cnephasia conspersana Douglas

Pl. 35, figs. 13–15

[*Sciaphila conspersana*; Barrett, 1905: 261. *Cnephasia conspersana*; Pierce & Metcalfe, 1922: 11, pl. 4 ♂, ♀ genitalia); Meyrick, 1928: 515; Ford, 1949: 56]

DESCRIPTION

♂ ♀ 15–22 mm. Sexual dimorphism moderately pronounced; antenna of male very shortly ciliate, forewing whitish grey, unicolorous or only weakly marked; forewing of female strongly marked.

Male (fig. 13). Forewing ground colour chalk-white, mixed or irrorate with grey; markings obsolescent, diffusely indicated by a darker shade of grey mixed with a few black scales; cilia concolorous. Hindwing greyish brown; cilia paler.

Female (fig. 14). Forewing ground colour similar to that of male; markings greyish brown sprinkled with black, heavier and more distinct than in male; sub-basal fascia extending across wing to dorsum, angulate at middle, weak on dorsum; median fascia with margins irregular, inner margin indented at middle and above dorsum; costal and terminal markings diffuse.

Variation. The forewing ground colour in the male varies from unicolorous whitish grey to dark slate-grey with a strong indication of the markings. The ground colour of the female is often whitish grey with contrasting heavy markings (fig. 15). In chalk districts in the south of England distinctly whitish forms occur, usually without markings in the male, while the female is only lightly marked. In limestone districts in the west of Ireland, the male tends to be greyer and the female greyish white with strong markings. Huggins (1964: 159) states that on the coast of Kerry both sexes are pale grey, and that on the Blasket Islands the male is pale grey with almost black markings and the female shining leaden black.

COMMENTS

Similar to *C. stephensiana* but with the forewing comparatively narrow and with the sub-basal fascia usually evident in the dorsal area.

BIOLOGY

Ovum. Approximately 300 eggs are deposited in small batches of about a dozen on the foodplant.

Larva. Head and prothoracic plate light brown or yellowish brown, head with a black lateral marking, plate marked with black; abdomen light yellow, sometimes pale greenish grey dorsally, translucent; pinacula small, black; peritreme of spiracles black; anal plate yellowish brown, marked with black; anal comb black; thoracic legs yellowish, terminal segments brownish. The larva may be found in June, feeding in the flower-heads of *Chrysanthemum*, *Taraxacum*, *Teucrium*, *Hieracium*, *Hypochoeris*, *Leontodon*, *Helianthemum*, *Dryas octopetala* and other composite and rosaceous flowers, drawing together the petals to form a shelter and eating out the immature seeds (pl. 14, fig. 5); also on young shoots of *Silene maritima*, *Teucrium* and *Senecio*, drawing together the terminal leaves and eating out the heart.

Pupa. Blackish brown. June and July, usually spun up in ground debris or some situation other than the larval habitation.

Imago. July; favouring open situations wherever the foodplants grow along the slopes and tops of sea cliffs and adjoining fields, and also on chalk downs. The male may often be found during the day sitting on fence posts and similar exposed situations, and in sunny weather is easily disturbed from rest amongst low-growing herbage. Both sexes are nocturnal and come to light.

DISTRIBUTION

A mainly littoral species, occurring commonly around the coasts of England and Wales, including the Isles of Scilly, and also found inland on chalk downs in Surrey, Berkshire, Oxfordshire, Hertfordshire and Cambridgeshire. Apparently scarce in Scotland, but possibly overlooked; there are unconfirmed records from Aberdeenshire (Reid, 1893a: 10) and Midlothian (Evans, 1897: 103). It has recently been taken in Ardnamurchan, Argyll, and Fife by E. C. Pelham-Clinton, and specimens were recorded from Sanday, Inner Hebrides (Bradley, 1958: 11). Generally distributed in Ireland and common in the Burren, Co. Clare, where the moth has been taken in May and June and in August and September (Bradley & Pelham-Clinton, 1967: 135), indicating that in this area the species is probably bivoltine.

Abroad the species does not appear to be well known; recorded from Spain, North Africa (Razowski, 1959: 242), the Netherlands (Bentinck & Diakonoff, 1968: 48), and France.

Cnephasia stephensiana (Doubleday)

Pl. 36, figs. 1–6

[*Sciaphila octomaculana*; Barrett, 1905: 260. *S. chrysantheana*; sensu Barrett, 1905: 264. *Cnephasia octomaculana*; Pierce & Metcalfe, 1922: 11, pl. 4 (♂, ♀ genitalia). *C. chrysantheana*; sensu Pierce & Metcalfe, 1922: 11, pl. 4 (♂, ♀ genitalia). *C. octomaculana*; Meyrick, 1928: 515. *C. chrysanthemana*; sensu Meyrick, 1928: 515. *C. octomaculana*; Ford, 1949: 55. *C. chrysantheana*; sensu Ford, 1949: 55]

DESCRIPTION

♂ ♀ 18–22 mm. Sexual dimorphism not pronounced; antenna of male very shortly ciliate.

Male (fig. 1). Forewing ground colour whitish grey diffusely irrorate with grey; markings brownish grey irregularly edged and speckled with black; sub-basal fascia terminating well above dorsum; median fascia with inner margin irregularly dentate, strongly edged with black, outer margin diffuse, deeply excavated above and below middle; costal and terminal markings diffuse and ill-defined. Hindwing brownish grey.

Female (fig. 2). Similar to male; forewing comparatively broader, hindwing darker.

Variation. The typical form of this species shows considerable minor variation in the ground colour and markings of the forewing. Melanistic forms occur in which the ground colour of the forewing is dark fuscous and the markings are inconspicuous; in extreme melanistic forms (fig. 3) found commonly in the London area the markings may be obliterated. Light forms are generally rare but an almost unicolorous white form (fig. 4) is locally common on coastal chalkland in south-east Kent.

The form *octomaculana* Curtis (figs. 5, 6) has the forewing ground colour almost clear white, the markings contrasting and well defined. In this form the median fascia is often weak or interrupted a little above the dorsum.

COMMENTS

The largest of the British *Cnephasia* species, with relatively broad forewings; small specimens may be confused with *C. interjectana* but in that species the costal spot of the forewing is stronger and usually extends to the termen.

BIOLOGY

Ovum. The cream-buff eggs are deposited singly or in small batches on the foodplants, 300–400 or more eggs being produced per female; hatching in 18–21 days (Reid, 1941).

Larva. Head variable, light brown marked with dark brown, or entirely brownish black or black; prothoracic plate black, edged anteriorly with grey, a whitish narrow medial sulcus; abdomen shining grey or bluish grey, variable and sometimes darker or lighter, or with a slight greenish tinge; anal plate blackish brown; anal comb absent; thoracic legs black. Barrett (1905: 261) describes the larva of f. *octomaculana* as having the head pale brown, posterior margin blackish, prothoracic and anal plates black; abdomen smoky black, bluish green dorsally; pinacula large, shining black. September to May and June; on hatching the young larva spins a silken hibernaculum and overwinters without feeding until the following April. Polyphagous on herbaceous plants, including *Chrysanthemum*, *Inula*, *Hieracium*, *Taraxacum*, *Centaurea*, *Sonchus*, *Heracleum*, *Vicia*, *Chenopodium*, *Rumex*, *Ranunculus* and *Plantago*; occasionally on cultivated peas and beans. In the first instar the larva mines the leaves, later living externally within spun leaves, the type of spinning varying with the foodplant; on *Plantago* (pl. 15, fig. 1) and *Chrysanthemum leucanthemum* (pl. 15, fig. 2) the leaf is spun into a tube, the frass collecting at the bottom; on the other plants an edge of a leaf may be folded over, or the leaf puckered on the underside, the larva living beneath in a web. On the Continent, the larva has been recorded on *Primula veris*, *Carlina*, *Tussilago*, *Cirsium* and *Serratula* (Hannemann, 1961: 36), and Hering (1957) lists more than 120 foodplants.

Pupa (text-fig. 44). Blackish brown; cremaster spines prominent, curved; caudal bristles filamentous; no distinct transverse ridge ventrad of the cremaster spines (Hering, 1951: 231, fig. 176a). June and July, in the larval habitation or spun up in a nearby leaf or flower.

Imago. July and August; resting during the day but easily disturbed from amongst herbage in gardens, rough fields and meadows, fens and marshes, downland, and similar habitats. Both sexes fly at early dusk and during the night, coming to light and sugar.

DISTRIBUTION

The nominate form is widely distributed and locally common in England north to Westmorland. Known from Caernarvonshire and Flintshire in Wales, but possibly overlooked elsewhere. In Scotland it has been recorded from Perthshire (Meyrick, 1928: 515), Berwickshire (Bolam, 1929: 123) and Fife (Evans, 1905: 157). Widely distributed but somewhat local in Ireland (Beirne, 1941: 84).

The form with the white ground colour, known as f. *octomaculana* Curtis, is apparently most prevalent from Cheshire and Lancashire in England north to Aberdeenshire on the mainland of Scotland. It is scarce in Cheshire and Lancashire, becoming more plentiful northwards and common in the lowlands of southern Scotland and the Isle of

Arran. Elsewhere in Britain it is known from Caernarvonshire in North Wales, the Isle of Man (Chalmers-Hunt, 1968a: 42) and the Shetlands. In Ireland, Beirne (1941: 84) records it as rare, occurring in Co.'s Sligo, Antrim and Kildare. An almost unicolorous white form (fig. 4) occurs along the south-east coast of Kent (Kingsdown, St. Margaret's Bay, Dover).

Generally distributed in the Palaearctic region.

Cnephasia interjectana (Haworth)

Pl. 36, figs. 7–9

[*Sciaphila virgaureana*; Barrett, 1905: 269. *Cnephasia virgaureana*; Pierce & Metcalfe, 1922: 12, pl. 5 (♂, ♀ genitalia); Meyrick, 1928: 514; Ford, 1949: 55]

DESCRIPTION

♂♀ 15–18 mm. Sexual dimorphism not pronounced; male usually smaller, antenna very shortly ciliate.

Male (fig. 7). Forewing ground colour whitish grey coarsely irrorate and suffused with fuscous; markings diffuse, dark brown finely irrorate with ochreous, sprinkled and weakly edged with blackish brown; sub-basal fascia very weak or obsolete near dorsum, outer margin diffuse, produced medially and often connected to the comparatively narrow median fascia; pre-apical spot diffuse, often connected to median fascia, usually extending to termen. Hindwing greyish brown, darker distally.

Female (fig. 8). Similar to male.

Variation. This species shows considerable minor variation in the ground colour and markings of the forewing; occasionally specimens occur with heavier and relatively distinct markings on the normal whitish grey ground colour, but more often the latter is darkened by a heavy greyish suffusion and the markings are less distinct. Melanistic forms occur in which the markings are very indistinct and mostly represented by black specks (fig. 9); in extreme forms the wing may be almost unicolorous blackish brown.

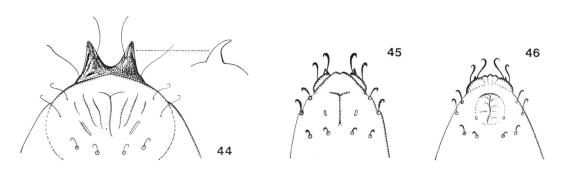

Pupal cremasters of *Cnephasia* species
Fig. 44, *C. stephensiana* (Doubleday), ♀. Fig. 45, *C. interjectana* (Haworth), ♀. Fig. 46, *C. incertana* (Treitschke), ♂

COMMENTS

Distinguished from *C. pasiuana* by the usually stronger and more variegated forewing pattern with a relatively narrow median fascia.

BIOLOGY

Ovum. Deposited singly or in batches of 2–3 eggs on the foodplant during June–August.

Larva. Head light or dark yellowish brown, with a dark brown or black ocellar area and a streak extending anteriorly from the postgenal juncture, posterior margin usually dark brown; prothoracic plate light yellowish brown or dark brown, with a yellowish medial sulcus, variably marked with black, sometimes heavily laterally or with a transverse bar, sometimes unmarked and edged with whitish laterally; abdomen finely shagreened, varying from bluish white or grey to brownish cream, or greyish green, darker dorsally; pinacula distinct, shining brown or black; anal plate yellowish brown marked with black; anal comb well developed, about 6 long dark brown or blackish prongs; thoracic legs light yellowish brown or dark brown; anal prolegs marked with black laterally. Chaetotaxy described by MacKay (1962: 25). September to June, constructing a silken hibernaculum and overwintering in an early instar, recommencing to feed in April. Polyphagous on herbaceous plants, including *Plantago*, *Lathyrus*, *Teucrium* and *Rumex*, flowers of *Ranunculus* and *Chrysanthemum*; also on cultivated strawberry (Vernon, 1971: 74) and raspberry, and in some years a pest on garden peas and beans during June. Hering (1957) records more than 130 foodplants for this species. In the first instar the larva mines the leaf, and in later instars feeds externally, drawing the leaves of a fresh shoot together or folding them to provide a shelter and feeding within on the young leaves or on the flowerhead. When feeding on the flowers of *Chrysanthemum* (pl. 15, figs. 4, 5) the larva spins the leaves and a flower bud together, causing the bud to bend over at an angle of 90 degrees and eating into it from beneath.

Pupa (text-fig. 45). Light brown; caudal bristles noticeably thickened away from their point of origin, bent into stout hooks terminally; a distinct transverse ridge ventrad of the short cremaster spines (Hering, 1951: 321, fig. 176b). June, in the folded edge of a leaf or in debris on the ground.

Imago. June to August; a common species in gardens, farmland, downland and open woodland, resting amongst vegetation during the day, often on tree trunks and branches or in exposed positions on fences, or hidden in thick hawthorn hedges; readily disturbed by beating, when it will fly quickly to another resting place. From dusk onwards both sexes fly freely and will come to light.

DISTRIBUTION

Widely distributed and generally common in lowland districts of Britain as far north as Aberdeenshire, Morayshire and Sutherland; locally common throughout Ireland. Europe to Siberia; Canary Islands; Newfoundland.

Cnephasia pasiuana (Hübner)

Pl. 36, figs. 10–12

[*Sciaphila pascuana*; Barrett, 1905: 266. *Sciaphila abrasana*; sensu Barrett, 1905: 271. *Cnephasia pasivana*; Pierce & Metcalfe, 1922: 11, pl. 4 (♂, ♀ genitalia). *C. pascuana*; Meyrick, 1928: 515; Ford, 1949: 55]

DESCRIPTION

♂♀ 15–20 mm. Sexual dimorphism not pronounced; male usually smaller than female, antenna very shortly ciliate.

Male (fig. 10). Forewing ground colour greyish brown finely irrorate with fuscous, the irroration usually not very apparent; markings obscurely indicated and sometimes almost indiscernible; when discernible in well-marked specimens, sub-basal fascia obtusely angled outwards at middle and obsolescent towards the dorsum; median fascia moderately broad throughout, its inner margin most strongly defined and usually slightly sinuous; cilia concolorous. Hindwing grey; cilia paler.

Female (fig. 11). Similar to male but forewing markings frequently stronger.

Variation. Both sexes show similar variation in the forewing ground colour, but in the female the normally drab ground colour is more often lightened by an admixture of whitish, resulting in the markings becoming more conspicuous (fig. 12). Almost unicolorous albinistic forms occur but are extremely rare. Barrett (1905: 266) mentions a unicolorous yellowish brown form found commonly on the salt marshes in the Thames estuary.

COMMENTS

The obscure forewing pattern and relatively broad median fascia are characteristic of this species.

BIOLOGY

Larva. Head light yellowish brown, marked with black laterally; prothoracic plate light yellowish brown, spotted with black; abdomen variable, yellowish brown or light greenish grey, translucent; pinacula black; anal plate yellowish brown, sometimes paler than head and indistinct; anal comb black; thoracic legs yellowish brown, terminal segments black. The larva may be found in May and June feeding on the flowers of *Chrysanthemum* (pl. 15, fig. 3), *Aster tripolium*, *Hypochoeris*, *Anthemis*, *Achillea* and other Compositae; also on those of *Ranunculus*, drawing the petals or florets together and feeding in the lower portion of the flowerhead. On the Continent it is alleged to be polyphagous on Compositae, including *Artemisia campestris*, *Centaurea scabiosa*, *Cirsium* and *Erigeron acer* (Hannemann, 1961: 36); also on *Echium*, *Thalictrum* and *Colchicum* (Razowski, 1959: 233).

Pupa. Dark shining blackish brown. June, in the folded edge of a leaf or in debris on the ground.

Imago. June and July; frequenting rough fields and marshes, resting amongst the coarse herbage during the day, often hiding in thick hawthorn bushes, but readily beaten out. Both sexes fly freely in the evening.

DISTRIBUTION

Locally common in the south of England northwards to Yorkshire. In Wales it is known from Pembrokeshire and Glamorgan, but has probably been overlooked in other parts.

In Glamorgan at Taff's Well, near Cardiff, larvae of this species have been found commonly on *Anaphalis margaritacea*, a perennial composite native to North America. In Ireland it is known from Co.'s Dublin, Cork, Kerry and Limerick.

Europe and palaearctic Asia.

Cnephasia genitalana Pierce & Metcalfe
Pl. 36, figs. 13–15

[*Cnephasia genitalana*; Pierce & Metcalfe, 1922: 12, pl. 5 (♂, ♀ genitalia); Meyrick, 1928: 515; Ford, 1949: 56]

DESCRIPTION

♂♀ 15–22 mm. Sexual dimorphism not pronounced; male usually smaller than female, forewing more angulate apically, antenna very shortly ciliate.

Male (fig. 13). Forewing ground colour chalk-white, finely irrorate with pale olive-brown; markings dark greyish brown, usually obsolescent; when discernible in well-marked specimens, sub-basal fascia narrow, sharply angulate at middle, obsolescent on dorsum; inner edge of median fascia indented above and below middle, outer margin diffuse; pre-apical spot diffuse; cilia concolorous. Hindwing greyish brown, paler basally.

Female (fig. 14). Similar to male.

Variation. Almost unicolorous specimens with a variable amount of black speckling representing the markings are frequent, while strongly marked specimens are uncommon (fig. 15); melanic forms are unknown.

COMMENTS

The whitish slate-grey general coloration and usually obsolescent markings are characteristic of this species.

BIOLOGY

Larva. Head yellowish brown, ocellar region and posterior margin dark brown; prothoracic plate varying from light to dark brown; abdomen finely shagreened, varying from greenish grey to greenish yellow, sometimes tinged with reddish or bluish; pinacula dark brown or black; anal plate weakly sclerotized, light yellowish brown; anal comb present; thoracic legs light yellowish brown. June, in spun flowers and leaves of *Senecio*, *Chrysanthemum* and *Ranunculus* (pl. 15, fig. 6). On the Continent, *Hieracium* and *Teucrium* have also been recorded as foodplants (Hannemann, 1961: 36; Razowski, 1959: 235).

Pupa. Dark brown. June and early July, in a flimsy cocoon spun amongst debris.

Imago. July and August; frequenting chalk downs and rough fields on the coast and inland.

DISTRIBUTION

In the British Isles this species is apparently local and uncommon and is known only from south-east England and South Wales. It has been recorded from Kent (Dover,

Folkestone, St. Margaret's Bay, Kingsdown), Essex (Mucking), Suffolk (Lakenheath), Norfolk (Derby Fen) and Glamorgan (Taff's Well) (Cox, 1940b: 236).

North and central Europe.

Cnephasia incertana (Treitschke)

Pl. 36, figs. 16, 17

[*Sciaphila subjectana*; Barrett, 1905: 270. *Cnephasia incertana*; Pierce & Metcalfe, 1922: 10, pl. 3 (♂, ♀ genitalia); Meyrick, 1928: 516; Ford, 1949: 56]

DESCRIPTION

♂ ♀ 14–18 mm. Sexual dimorphism moderately pronounced; male antenna weakly dentate towards middle, shortly ciliate; female usually larger, forewing less angulate apically, markings generally less conspicuous, tip of extensile ovipositor usually visible.

Male (fig. 16). Forewing ground colour greyish white, finely irrorate and suffused with greyish brown; markings greyish brown, coarsely irrorate with blackish; sub-basal fascia angulate at middle, abruptly terminated above dorsum; median fascia narrowly indented at and below middle, outer margin diffuse, irregular, usually connected medially to pre-apical spot by a weak greyish brown suffusion; cilia grey, irrorate with whitish. Hindwing grey, suffused with brownish distally; cilia paler.

Female (fig. 17). Forewing pattern similar to that of male but ground colour generally more strongly suffused and markings less conspicuous.

Variation. The forewing coloration and markings are relatively constant but show considerable minor variation in depth and intensity, varying from well-marked specimens with light ground coloration to obscurely marked specimens strongly suffused with greyish or greyish brown. Although the markings may be inconspicuous in specimens with a dark ground colour, unicolorous forms are unknown.

COMMENTS

Distinguished from *C. interjectana* by the weaker irroration in the forewing. In the female the slender extensile ovipositor is often exserted and in dry specimens may be seen without dissection, differing markedly from the floricomous ovipositor found in other British species of the genus.

BIOLOGY

Ovum. Greenish or olive-brown; deposited during June and July singly or in small batches in folds and crevices in the stems and leaves of herbaceous plants and in the bark of trees, or under lichen.

Larva. Head light brown or yellowish brown, edged with black postero-laterally, ocellar area blackish; prothoracic plate black, edged with whitish anteriorly; abdomen dull dark green; pinacula distinct, black, sometimes brownish but with black bases to setae; anal plate large, black. September to May and June, the young larva spinning a hibernaculum and overwintering, recommencing to feed in the spring. Polyphagous, feeding on *Chrysanthemum*, *Plantago*, *Rumex*, *Lotus*, *Ranunculus*, *Saxifraga* and other herbaceous and garden plants, including cultivated strawberry, beans and peas, trefoils, clover and lucerne; also on *Vitis*, *Malus* and young conifers. On the Continent it has been recorded on more than 200 plants (Hering, 1957). In the first instar the larva mines the leaves, forming short, irregular, blotch-like mines (Hering, 1957: pl. 21, fig. 185), but in later instars it lives externally, feeding in spun leaves and often twisting those of tender shoots.

Pupa (text-fig. 46). Dull black; spines of the cremaster not bent into hooks, short, their length less than the distance to a ridge lying ventrad; caudal bristles only slightly thickened away from their point of origin, bent distally to form hooks (Hering, 1951: 320, fig. 176c). June, spun up in a folded leaf or amongst debris.

Imago. June and July; frequenting hedgerows, roadside verges and woods. During the day the moth may often be disturbed in numbers by beating and sometimes is found resting on tree trunks and fences.

DISTRIBUTION

Common and generally distributed throughout the British Isles as far north as East Lothian in the east and Argyll in the west of Scotland; records from Perthshire and Morayshire are unconfirmed.

Europe, North Africa, Asia Minor and Transcaucasia.

Cnephasia abrasana (Duponchel)

This species has been included in the British list by most authors, but all specimens examined under this name proved to have been erroneously determined.

The name *abrasana* appears to have been first introduced into the British literature in the Third Supplement to Doubleday's List (1873: [3]). Dale (1886: 42; 1891: 44) records the species from Portland (W. H. Grigg) and Glanville Wooton (Dorset). Nine specimens standing under this name in the Dale collection in the University Museum, Oxford, have on genitalic examination proved to be males and females of *C. pasiuana*. A further specimen without locality data, ex collection C. G. Barrett, in the E. R. Bankes collection (British Museum (Natural History)) also proved on dissection to be a male of *pasiuana*.

Pierce & Metcalfe (1915: 101) were unable to find British examples of this species and in their work on the genitalia of the British Tortricidae (1922: 10, pl. 3) described and figured the male genitalia of a continental specimen.

The species has also been recorded in England from Durham (Darlington) (Sang, 1885: 192), Bristol, Gloucestershire, Wiltshire, Herefordshire, Lancashire and Cumberland (Barrett, 1884: 238; 1905: 272), and in Scotland from East Lothian (Balfour, 1930: 179). The Wiltshire record is based on a specimen taken by Meyrick in August, 1876 (Barrett, 1884: 238), and although Meyrick (1895: 540) included this species in his *Handbook* it is significant that he excluded it in the revised edition published in 1928.

Tortricodes alternella (Denis & Schiffermüller)

Pl. 37, figs. 1-4

[*Cheimatophila hyemana*; Barrett, 1905: 273. *Tortricodes tortricella*; Pierce & Metcalfe, 1922: 14, pl. 6 (♂, ♀ genitalia); Meyrick, 1928: 511; Ford, 1949: 54]

DESCRIPTION

♂ ♀ 19–23 mm. Sexes moderately dimorphic; antenna of male fasciculate-ciliate, costa of forewing almost straight, markings usually contrasting with light ground colour; antenna of female sparsely ciliate, costa of forewing more strongly curved, markings less distinct.

Male (fig. 1). Forewing ground colour ochreous-white, sparsely irrorate and strigulate with brown mixed with fuscous, radial veins shaded with fuscous; basal area suffused with ochreous-brown or chestnut-brown; markings somewhat diffuse; sub-basal fascia black-brown, narrow, angulate at middle, obsolescent on dorsum; median fascia broad, rather oblique, inner margin well defined, sinuous; pre-apical spot wedge-shaped; cilia grey. Hindwing greyish brown, apex and margins infuscate; cilia paler.

Female (fig. 2). Forewing ground colour extensively suffused with fuscous; markings as in male but less distinct. Hindwing fuscous-grey; cilia paler.

Variation. Strongly marked forms of the male (fig. 3) with the ground colour of the forewing varying from cream-white to silver-grey occur commonly; obscurely marked greyish forms are found in both sexes but are most frequent in the female (fig. 4).

COMMENTS

One of the earliest species to emerge and unlikely to be confused with hibernating Tortricids which may be on the wing in early spring.

BIOLOGY

Larva. Head dull brown, sometimes with yellowish mottling; prothoracic plate greenish or yellowish brown, edged with whitish anteriorly, a whitish medial sulcus, lateral margin marked with black or dark brown; abdomen dull reddish brown dorsally, yellowish white ventrally, a distinct narrow white dorsal line and a less distinct subdorsal line; pinacula large, white or yellowish white; peritreme of spiracles dark brown; anal plate yellowish; anal comb yellowish; thoracic legs blackish brown. May and June, feeding on *Quercus* and *Carpinus*, also on *Betula*, *Corylus*, *Crataegus*, *Prunus spinosa* and *Tilia*, living in spun leaves or under the turned-down edge of a leaf.

Pupa. Dark red-brown. July to February and March, in a tough cocoon of earth and dark brown silk in the ground or amongst debris on the ground (Richardson, 1885: 253).

Imago. February and March in the south, usually emerging later in the north and occurring until April; frequenting open woodland. The male flies briskly in the sunshine; the female is more sluggish but may sometimes be found at rest on tree trunks, usually dropping to the ground when disturbed. During dull weather the moth rests on twigs and branches and is not easily detected (pl. 16, fig. 1). At night both sexes may be found by searching tree trunks and branches with the aid of a lamp.

DISTRIBUTION

Common in woods, especially mixed oak, hornbeam and birch, throughout England, Wales and the Lowlands of Scotland as far north as Easter Ross. Apparently uncommon in Ireland; recorded from Co.'s Carlow, Dublin and Wicklow (Beirne, 1941: 83). Europe except the south-east.

Exapate congelatella (Clerck)

Pl. 37, figs. 5, 6

[*Enyphantes congelatella*; Pierce & Metcalfe, 1922: 14, pl. 6 (♂, ♀ genitalia). *Exapate congelatella*; Meyrick, 1928: 511; Ford, 1949: 54]

DESCRIPTION

♂ 20–23 mm, ♀ 10–11 mm (brachypterous). Sexual dimorphism pronounced; male with wings developed normally, those of female vestigial; antenna of male ciliate, length of cilia less than width of flagellum; antenna of female sparsely hirsute, with bristle-like hairs two to three times width of flagellum basally, slightly shorter apically.

Male (fig. 5). Forewing ground colour slate-grey with whitish admixture, median area of wing from base to end of cell and in distal area strongly overlaid with white, producing a silver-grey effect; markings blackish fuscous; basal fasciae obsolete; median fascia reduced to form a large orbicular spot below costa; a smaller spot at end of cell connected to dorsum by a fuscous suffusion; pre-apical spot represented by several chevron-like inwardly oblique striae. Hindwing greyish fuscous; cilia concolorous, with a distinct fuscous basal line.

Female (fig. 6). Forewing whitish grey, margins spotted with diffuse fuscous markings; median fascia indicated on costa; hindwing reduced, pad-like; wings, body and appendages clad with moderately dense erect brownish hairs interspersed among the scales.

Variation. The forewing coloration and markings of the male are relatively constant except for slight variation in the intensity of the white scaling; specimens in which this is weak usually have a greyer appearance and the markings are less distinct. Variation in the female is most apparent in the irregular distribution of the fuscous spotting on the forewing.

COMMENTS

One of the latest species to emerge in the year; the male is readily recognized by the elongate forewings with distinctly greyish coloration; the female is brachypterous.

BIOLOGY

Ovum. October to April, deposited singly or in small batches of three or four eggs on the stems and twigs of the foodplant; hatching in late spring.

Larva. Head yellowish green, spotted with black laterally; prothoracic plate yellowish green, flecked with black posteriorly; abdomen light green, a paler subdorsal line, integument shagreened. May to early July, in spun shoots of *Ligustrum*, *Rhamnus*, *Crataegus*, *Erica*, *Vaccinium*, *Calluna*, *Potentilla* and *Rubus*. On the Continent it is recorded also on *Prunus spinosa*, *Ribes*, *Pyrus*, *Syringa*, *Ulmus*, *Salix*, *Berberis*, *Anthriscus* and other shrubs.

Pupa. Dark brown. June to October, spun up in debris on the ground.

Imago. October to December; frequenting rough hedgerows and borders of woods, particularly where elm and blackthorn grow, the male being easily beaten out and caught during the day; also occurring on open moorland, the male often resting during the day on walls, posts and heather stems. The male flies freely an hour before dusk and may often be found assembling to a freshly emerged female; later it comes to light. The brachypterous female may be found after dusk at rest on the foodplants, often on the tips of the shoots.

DISTRIBUTION

Locally common along hedgerows and in wooded country in the southern counties of England, apparently very scarce or absent in the Midlands, reappearing in Staffordshire, Yorkshire and Lancashire, being locally abundant on moorland in the north of England and plentiful on the mosses in the north-west. In Scotland this species occurs plentifully on the moors in the south but is much scarcer in the north; recorded from Inverness-shire (Newtonmore), Perthshire, Kincardineshire and Sutherland. The Sutherland record is based on eight specimens taken by A. E. Griffith and erroneously recorded as *E. duratella* (Ford, 1949: 54). Re-examination of these specimens, which are in the National Museum of Wales, Cardiff, showed them to be *congelatella* (Bradley & Martin, 1956: 153). Little is known of its distribution in Wales, but it is plentiful on the moors in the northern counties. A similar situation exists in Ireland, and Beirne (1941: 84) gives records for Co.'s Galway (Connemara), Dublin, Sligo and Down, but considers it to be evidently rare and that confirmation of its occurrence in Ireland is desirable. North and central Europe; west China.

Exapate duratella Heyden

As discussed under *E. congelatella*, the determination of the specimens from Sutherland (Ford, 1949: 54), on which the record of *duratella* was based, proved to be erroneous (Bradley & Martin, 1956: 153).

Neosphaleroptera nubilana (Hübner)

Pl. 37, figs. 7–9

[*Sciaphila nubilana*; Barrett, 1905: 272. *Nephodesme nubilana*; Pierce & Metcalfe, 1922: 14, pl. 6 (♂, ♀ genitalia). *Cnephasia nubilana*; Meyrick, 1928: 516; Ford, 1949: 56]

DESCRIPTION

♂ ♀ 12–15 mm. Sexual dimorphism usually pronounced; antenna simple in both sexes; antenna of male stouter, fuscous, weakly annulate with white; female with antennal annulations more distinct, and normally with forewing ground coloration lighter and the wing pattern more distinct.

Male (fig. 7). Forewing ground colour fuscous-brown, sparsely irrorate with whitish; markings obscure, dark fuscous-brown; cilia dark grey. Hindwing fuscous-brown; cilia greyish.

Female (fig. 8). Forewing ground colour as in male, heavily irrorate with whitish, sparsely strigulate with fuscous; markings comparatively distinct, dark fuscous-brown;

sub-basal fascia obliterate towards dorsum; inner margin of median fascia shallowly sinuous; a thick irregular stria emitted from pre-apical spot or suffusion and extending to tornal area. Hindwing as in male.

Variation. Very little variation is found in the forewing markings, but in the female the normally strong whitish irroration may be reduced or absent (fig. 9).

COMMENTS

The small size and dark fuscous-brown coloration of both fore- and hindwings are characteristic.

BIOLOGY

Larva. Head yellowish brown; prothoracic plate brownish green, sometimes dotted with black; abdomen translucent, light green; pinacula concolorous with abdomen; anal plate green; thoracic legs and prolegs whitish green. September to early June, in spun shoots of *Crataegus*, *Prunus* and occasionally *Malus*, drawing the leaves together and feeding in the heart, overwintering in an early instar and recommencing to feed in the spring. On the Continent, *Pyrus* and *Betula* are also recorded as foodplants (Swatschek, 1958: 63), and the larva has been recorded as injurious to apricot in southern Moravia (Povolný, 1951). In plum orchards in Kent, larvae are often found in grease bands in which they have become trapped whilst ascending the trunks, after having been shaken from their feeding places during strong winds.

Pupa. Black. June; in the larval habitation.

Imago. July; most often found in hawthorn and blackthorn hedges, even in those closely clipped, in gardens and around farmland, quickly flying back into the bushes when disturbed. The male flies around the foodplants in sunshine by day and during the evening, and is joined by the female towards dusk.

DISTRIBUTION

Locally plentiful in the south of England and occurring as far north as Yorkshire; one specimen was taken some years ago by A. E. Wright at Witherslack, Westmorland. In Wales it has been recorded from Montgomeryshire (Michaelis, 1954a: 63). Apparently unknown in Scotland and Ireland.

Europe, except the south-west, to Asia Minor.

Eana argentana (Clerck)

Pl. 37, fig. 10

[*Ablabia argentana*; Barrett, 1905: 253. *Nephodesme argentana*; Pierce & Metcalfe, 1922: 14, pl. 6 (♂, ♀ genitalia). *Cnephasia argentana*; Meyrick, 1928: 513; Ford, 1949: 54]

DESCRIPTION

♂ ♀ 22–27 mm. Sexual dimorphism not pronounced; male usually larger than female, antenna very shortly ciliate; antenna of female simple, forewing comparatively narrow.

Male (fig. 10). Forewing silver-white, costa narrowly suffused with whitish yellow, anterior edge dark brown basally; cilia white, shaded with yellowish basally. Hindwing white, suffused with grey distally and along cubital vein.

Female. Similar to male.

Variation. The silver-white coloration is remarkably constant and no marked variation has been found in this species.

COMMENTS

The distinctive silver-white coloration of the forewing and the absence of markings readily distinguish this species from other Tortricids. It may, however, be confused with the Pyralid *Crambus perlellus* (Scopoli) which it closely resembles superficially and which occurs in similar grassland habitats. *E. argentana* can be distinguished by the much shorter labial palpus, the length of which is equal to about twice the diameter of the eye compared with approximately four times in *perlellus*, and by the hindwing venation having vein 8 (Sc + R$_1$) free and not stalked with vein 7 (R$_s$) as in *perlellus*.

BIOLOGY

Ovum. About 150 eggs are deposited in several batches on the foodplants, each batch containing 20–90 imbricate eggs, irregularly arranged (Heddergott, 1957: 335, figs. 2, 3). The eggs are embedded in a transparent secretion, often covered with scales from the female. Hatching occurs after about 10 days.

Larva. Head blackish brown, or light brown marked with blackish posteriorly; prothoracic plate blackish brown, paler anteriorly; abdomen reddish brown varying to dark brown, integument shagreened; pinacula blackish brown; anal plate dark brown; thoracic legs blackish brown exteriorly (Swatschek, 1958: 67). Nothing is apparently known about the larva in the British Isles, although Meyrick (1928: 513) suggests that it probably feeds amongst the roots of grasses. On the Continent, Heddergott (1957: 337) has described the biology in detail and states that on hatching the larva crawls away from the egg mass and constructs a silken hibernaculum about 3 mm long, in which it remains until spring before commencing to feed. Gramineae, *Salix*, various herbaceous dicotyledons and moss are the usual foodplants. In Germany (Westphalia), during an outbreak in 1955, the larva attacked the new growth of young spruce (Heddergott, 1957: figs. 5, 6). Usually the larva lives in a silken tube below ground level, but moves above ground when population density is high.

Pupa. Dark red-brown, wings lighter. June, in a tough silken cocoon in the larval habitation.

Imago. End of June and July. Frequenting grassy places, usually in mountainous regions; on the wing at dusk, with a feeble flight like that of *Crambus perlellus* which is found in similar habitats.

DISTRIBUTION

The only authentic records of this species from the British Isles are from Scotland (Perthshire), where it was first discovered in 1875 on a mountain in the Forest of Atholl by Buchanan White (1875: 85). Subsequently it was recorded from Glen Tilt (Barrett,

1905: 254), and was still plentiful in grassy areas in the glen on the 19th July, 1910, when it was taken by G. H. Conquest (1911: 156). There are unconfirmed records from Berwick (Northumberland) (Bolam, 1929: 122), and from Dilham, between Norwich and Cromer (Norfolk) (Cottam, 1903: 226). Since 1910 *argentana* has apparently not been seen in its known haunts and its present status in the British Isles is uncertain.

Europe, north-west Africa, Asia Minor to central Asia, Siberia and Japan; Kashmir, India; North America.

Eana osseana (Scopoli)

Pl. 37, figs. 11–13

[*Ablabia osseana*; Barrett, 1905: 252. *Nephodesme osseana*; Pierce & Metcalfe, 1922: 13, pl. 6 (♂, ♀ genitalia). *Cnephasia osseana*; Meyrick, 1928: 513; Ford, 1949: 54]

DESCRIPTION

♂ ♀ 16–23 mm. Sexual dimorphism not pronounced; antenna of male very shortly ciliate (pubescent-like); female usually smaller.

Male (fig. 11). Forewing ground colour pale ochreous suffused with ochreous-brown, coarsely irrorate with fuscous; markings obsolescent and reduced to diffuse fuscous-brown spots in cell area; cilia whitish ochreous. Hindwing brownish grey; cilia whitish.

Female (fig. 12). Forewing more pointed, somewhat greyer in colour and with the markings more reduced.

Variation. The ochreous-brown suffusion of the forewing varies in shade and intensity. Heavily suffused specimens often have a distinct ferruginous tinge (fig. 13), while those lightly suffused may be whitish ochreous and in extreme forms the wing appears whitish with scarcely any markings; almost unicolorous forms are found in both sexes but less frequently in the male.

COMMENTS

Distinguished from other British *Eana* species by the general coloration of the forewing. It may, however, be confused with *Cnephasia longana*, which it resembles in size and general coloration, but may be separated by the comparatively narrow forewing with the apex more produced, and the characteristic obscure markings.

BIOLOGY

Larva. Head yellowish brown, marked with black or blackish brown, ocellar area weakly and postgenal juncture strongly pigmented; prothoracic plate brown or black, a distinct medial sulcus; abdomen grey; pinacula black; anal plate black; anal comb absent; thoracic legs brown. May and June; polyphagous, living in a silken tube low down amongst the rootstocks and fibrous roots of grasses, moss and herbaceous plants, and under stones.

Pupa. June and July, spun up in or near the larval habitation.

Imago. July and August, and in northern localities in September; frequenting downland, rough meadows, pastureland and coastal and mountainous districts.

DISTRIBUTION

A locally common grassland and moorland species throughout the British Isles, including St. Kilda.

North and central Europe to Siberia; Faroes; Iceland; North America.

Eana incanana (Stephens)

Pl. 37, figs. 14, 15

[*Sciaphila incanana*; Barrett, 1905: 267. *Nephodesme incanana*; Pierce & Metcalfe, 1922: 13, pl. 5 (♂, ♀ genitalia). *Cnephasia incanana*; Meyrick, 1928: 514; Ford, 1949: 55]

DESCRIPTION

♂♀ 17–23 mm. Sexual dimorphism not pronounced; male usually smaller and forewing markings more distinct, antenna very shortly ciliate (pubescent-like).

Male (fig. 14). Forewing ground colour white suffused with grey, suffusion heavier distally, sparsely strigulate with fuscous; markings dark grey variably mixed and edged with black, an admixture of tawny or ochreous, mainly in the median area; sub-basal fascia conspicuous, curving from costa and abruptly terminated before dorsum, margins well defined, inner margin sharply inflected beyond middle in the submedian fold; median fascia oblique, inner margin variable, usually indented twice in dorsal half; pre-apical spot obscure, weakly strigulate with black, emitting a thin chain of black mixed with tawny dots extending to tornus, a series of similar dots in terminal margin; cilia greyish white. Hindwing light grey, infuscate distally; cilia paler.

Female (fig. 15). Similar to male but usually larger with forewing ground colour more greyish and markings less contrasting.

Variation. In both sexes the strength of the forewing markings varies according to the extent of the black scaling; the ground coloration is fairly constant but may be lighter or darker according to the intensity of the grey suffusion.

COMMENTS

The rounded forewing and conspicuous curved sub-basal fascia are characteristic.

BIOLOGY

Larva. Head light brown; prothoracic plate lighter brown, varying to dark brown or black, margined with black laterally; abdomen translucent, grey, tinged with greenish dorsally; pinacula dark grey, setae blackish; peritreme of spiracles black; anal plate varying from light brown to blackish brown, mottled with black; anal comb absent; thoracic legs black. The larva may be found in April and May, feeding on the flowers and developing ovaries of *Endymion non-scriptus* (pl. 16, figs. 2, 3) and *Chrysanthemum leucanthemum*, spinning a silken web over the blossom. On *Endymion* the larva is best obtained by examining the flower buds before they begin to expand, when light yellow frass exuding from between the buds indicates the presence of a larva (Ford, 1936a: 61). On the Continent it is recorded also on *Ornithogalum nutans* (Hannemann, 1961: 46) and *Vaccinium* (Razowski, 1959: 292).

Pupa. Brown, darker dorsally. June, spun up in debris on the ground.

Imago. July; frequenting woodland glades, seldom seen by day, resting amongst the foliage of trees and bushes and occasionally on trunks of trees or on the seedheads of *Endymion*. Both sexes fly at night and come to light. The moth is best obtained by collecting and rearing the larvae, or by searching with a lamp and net at night in woods where bluebells grow.

DISTRIBUTION

Local and seldom common, occurring in England north to Lancashire and Yorkshire. Barrett (1905: 268) records it from all the English counties south of London, and from Essex, Suffolk, Somerset, Herefordshire and Leicestershire, but it seems to have become scarcer with the disappearance of much of the woodland. In Wales it has been recorded from Flintshire and Caernarvonshire (Michaelis, 1954a: 63). In Scotland it is found in birch woods in Aberdeenshire, Inverness-shire, Argyll and Perthshire, but is uncommon and probably has a foodplant other than *Endymion* and *Chrysanthemum*. There are no records from the south of Scotland; apparently unknown in Ireland.

North and central Europe; Ukraine.

Eana penziana (Thunberg & Becklin)
Pl. 38, figs. 1–11

[*Sciaphila bellana*; Barrett, 1905: 257. *S. colquhounana*; Barrett, 1905: 258. *Nephodesme bellana*; Pierce & Metcalfe, 1922: 13, pl. 5 (♂, ♀ genitalia). *N. colquhounana*; Pierce & Metcalfe, 1922: 13, pl. 5 (♂, ♀ genitalia). *Cnephasia bellana*; Meyrick, 1928: 514; Ford, 1949: 54. *C. colquhounana*; Meyrick, 1928: 514; Ford, 1949: 55]

In the British Isles this species is represented by two geographically and ecologically distinct forms, one occurring inland in hill and mountain districts, the other in coastal areas and the low-lying hinterland. The hill form, f. *bellana* Curtis, is provisionally referred to the nominate subspecies, *E. penziana penziana*; the coastal form, originally described from Dublin and the Isle of Man, is known as ssp. *colquhounana* (Barrett).

DESCRIPTION

♂ ♀ 20–27 mm. Sexes similar; antenna of male very shortly ciliate.

E. penziana penziana (Thunberg & Becklin) (inland hill form, f. *bellana* Curtis).

Male (fig. 1), *female* (fig. 2). Forewing ground colour white, indistinctly strigulate with black; markings black, very variable; basal fascia reduced to a few black spots, most evident on costa; sub-basal fascia sharply angulate at middle, usually narrowed and abruptly terminating a little below; median fascia oblique, sometimes fractured and sprinkled with yellow medially, margins irregular and deeply notched, often broken and partially obliterated; pre-apical spot reduced to form heavy striae, confluent dorsad with a series of small strigulae, the latter sometimes coalesce to form a thick subterminal fascia; cilia greyish white, a grey basal line marked with dark neural dots sometimes extending into the cilia. Hindwing greyish white, greyer and faintly strigulate with grey apically.

Variation. Both sexes show a similar range of individual minor variation in wing expanse and markings. Although the forewing ground colour is almost consistently clear white, the markings are extremely variable; heavily marked specimens (fig. 3) with a trace of greyish suffusion on the white ground colour occur but are uncommon.

E. penziana colquhounana (Barrett) (coastal subspecies)

Differs from *penziana penziana* f. *bellana* in having the white ground colour of the forewing extensively irrorate or suffused with grey; the markings are essentially as in f. *bellana* but less conspicuous. Hindwing distinctly grey.

Variation. This coastal subspecies shows considerable variation throughout its range. Specimens from the Isle of Man (figs. 4–6) generally have a rather dull grey appearance, the markings being somewhat reduced and often with an increased admixture of yellow scaling, the generally darker grey coloration extending into the cilia and obscuring the neural dots.

Specimens from the Shetlands (figs. 7–9) usually have the white ground colour of the forewing less heavily suffused with grey, the markings showing moderate contrast with the whitish background. The Shetland population is extremely variable, showing a wide range of individual variation from lightly marked specimens (figs. 7, 8) reminiscent of the inland hill form, to strongly suffused fuscous forms (fig. 9).

On the west coast of Ireland, especially in the Burren district, Co. Clare, specimens have a distinctive silver-grey appearance, usually with subdued greyish black markings (figs. 10, 11).

COMMENTS

Although extremely variable, the large size and elongate forewings readily distinguish *penziana* from related species.

BIOLOGY

Ovum. Deposited on the foodplant, the period of oviposition varying from May to September according to locality; hatching in about 14 days.

Larva. Head dark reddish brown; prothoracic plate black; abdomen dark green, integument puckered; pinacula black; anal plate black; thoracic legs black. June or July to the following May or June, living in a silken tube about 5 cm long on the foodplant, hibernating in an early instar. The larva of *penziana penziana* f. *bellana*, which in the British Isles is found only in inland hill and mountain regions, feeds on *Festuca ovina*, living in a silken gallery at the roots and feeding on the tips of the grass at night (Gregson, 1873b: 360; Barrett, 1884: 243). That of *penziana colquhounana*, which occurs mainly in coastal areas in the British Isles, feeds in the root-crowns of *Plantago maritima* (South, 1893: 102) and on *Armeria maritima*. According to Ford (1949: 55) the larva of ssp. *colquhounana* lives in a silken tube extending to the roots of the foodplant, but Donovan (1902: 14) and Emmet (1968: 46) record that in western Ireland it inhabits the upper part of the sea pink or sea thrift, feeding on the leaves, and is most often to be found on plants which are not too compact and growing on the south-facing surface of rocks near the seashore. The larva will, however, live also amongst the roots, and the silken tube may be found under clumps of *Saxifraga* and *Lotus*, amongst which *Armeria* grows, on limestone rock on the Burren coast, Co. Clare (Bradley & Pelham-Clinton, 1967: 135).

Pupa. Dark brown. May to July, spun up in the larval habitation or in a crevice in a rock.

Imago. The inland form *bellana* is on the wing in June and July; the coastal subspecies *colquhounana* occurs from May to August and September. The moth is inactive during the day, resting on rocks or amongst herbage; both sexes fly after dusk and come readily to light.

DISTRIBUTION

In the British Isles this species is represented by two ecotypes, one apparently restricted to inland hill and mountain districts, the other to coastal areas. The inland form *bellana*, originally described from Midlothian, Scotland, and usually referred to the nominate continental subspecies *penziana penziana*, shows slight differences in the forewing pattern suggesting that it might represent a separate race. In England it occurs in Lancashire (Warton Crag, Witherslack), Yorkshire (Grassington), Cumberland (Cockermouth, Keswick), Westmorland and Northumberland (Belford), and in Scotland in Kirkcudbright, Midlothian (Arthur's Seat), Perthshire (Kinnoull Hill, Rannoch), Aberdeenshire (Braemar) and Inverness-shire (Aviemore).

The coastal subspecies *penziana colquhounana* was originally described from Dublin and the Isle of Man. It is widely distributed in Scotland and records include Berwickshire (Ayton, Eyemouth) (Bolam, 1929: 123), Kincardineshire (Muchalls), Morayshire (Forres), Shetland Isles (Unst), St. Kilda, Sutherland (Invernaver, Lochinver), Hebrides and Ayrshire (Ardrossan); in Ireland it is more generally distributed round the coasts and is particularly common in the south and west, including the Blasket Islands. The only known records from Wales are from the Great Ormes Head and Bardsey Island, Caernarvonshire; in England recorded from Highlaws, Northumberland. Europe to Asia Minor and Siberia.

TORTRICINI

Aleimma loeflingiana (Linnaeus)

Pl. 39, figs. 1–4

[*Dictyopteryx loeflingiana*; Barrett, 1905: 197. *Aleimma loeflingiana*; Pierce & Metcalfe, 1922: 19, pl. 8 (♂, ♀ genitalia). *Tortrix loeflingiana*; Meyrick, 1928: 507; Ford, 1949: 51]

DESCRIPTION

♂♀ 14–19 mm. Sexual dimorphism not pronounced; antenna of male very shortly ciliate.

Male (fig. 1). Forewing ground colour pale yellow, weakly strigulate and suffused with ochreous-brown or tawny, veins often similarly lined; markings somewhat diffuse, plumbeous with a variable ochreous-brown admixture; fasciae slightly oblique, partially obliterate, margins diffuse; basal fascia obsolescent, outer edge usually indicated by a thin line; sub-basal fascia excised costally, dilated medially and with outer margin often projected, sharply indented above submedian fold; median fascia at about three-fifths, excised on costa, dilated medially; pre-apical spot usually represented by two heavy striae on costa, emitting or followed by several strigulae in the terminal area; cilia pale yellow, grey around tornus, with a strong ochreous-brown basal line from apex to near tornus. Hindwing grey; cilia paler.

Female. Similar to male.

Variation. An extremely variable species, ranging from weakly marked individuals (fig. 2) in which the light yellow ground colour of the forewing is strigulate with ochreous-brown or tawny, with the fasciae obsolescent and weakly indicated only on the costa, to heavily marked forms (fig. 3) in which the markings are darkened with a blackish admixture, the sub-basal and median fasciae coalescing medially. Exceptionally heavily marked specimens sometimes occur in which the diffuse blackish markings extend to the terminal area, the yellow ground coloration being almost obliterated except along the costal margin and in the basal area. An uncommon form (fig. 4) occurs having the normal yellow ground coloration replaced by warm orange-ochreous or tawny, with the markings usually reduced.

COMMENTS

The angular shape of the forewing and the characteristic almost reticulate pattern distinguish this species.

BIOLOGY

Larva. Head, prothoracic and anal plates and thoracic legs blackish brown or black; abdomen pale green varying to brownish green, integument strongly shagreened; pinacula blackish brown or black. The fully grown larva is found in May feeding on *Quercus*, also on *Carpinus* and *Acer*, living in a rolled leaf or in a pocket formed from a folded leaf.

Pupa. Light reddish brown. June, usually in a dull white cocoon in the larval habitation, but sometimes in leaf litter.

Imago. Late June and July; frequenting oak woods and forest borders; often found on isolated oaks along roadsides and on farmland. During the day it may be obtained in numbers by shaking the lower branches of oaks, and occasionally is found at rest on the trunks, usually low down near the ground. Both sexes fly freely at dusk and sometimes come to light.

DISTRIBUTION

Common in woodland throughout most of Britain north to Argyll and Aberdeenshire; less common in the north of England and Scotland except the west. In the Orkneys recorded from the island of Hoy by J. J. Weir (1882: 2). Widely distributed and common amongst oak in Ireland.

Europe to Asia Minor and the Caucasus.

Tortrix viridana (Linnaeus)

Pl. 39, fig. 5

[*Heterognomon viridana*; Barrett, 1905: 193. *Tortrix viridana*; Pierce & Metcalfe, 1922: 19, pl. 8 (♂, ♀ genitalia); Meyrick, 1928: 507; Ford, 1949: 51]

DESCRIPTION

♂♀ 17–24 mm; males under 17 mm are known, the smallest being an individual from Oxshott (Surrey) with a wing span barely 13 mm, but such undersized specimens are rare. Sexual dimorphism not pronounced; antenna of male weakly dentate, densely ciliate, cilia less than width of flagellum, posterior margin of flagellum thinly clad with whitish scales; female usually with abdomen stouter and wings broader, antenna filiform, sparsely ciliate, more densely clad with whitish scales.

Male (fig. 5). Forewing light green (viridine), darker basally, delicately strigulate distally, head, thorax, patagium and tegula concolorous, vertex and patagium often tinged with yellow; labial palpus whitish, suffused with fuscous exteriorly; whole of costal margin of forewing narrowly edged with whitish yellow, base of costa sometimes suffused with fuscous; dorsal scale-tuft tinged with yellow; cilia whitish, a pale green sub-basal line. Hindwing light grey; cilia whitish, with a grey sub-basal line around apex and along termen.

Female. Similar to male.

Variation. This species shows little variation except in the depth and intensity of the green coloration of the forewing and in the strength of the strigulation. Specimens are sometimes found in which the green coloration is replaced by dull yellow or primrose yellow; these apparently represent a recurring genetical form and are not due to abrasion or discoloration.

COMMENTS

This species is unique amongst the British Tortricids in having a light green, almost unicolorous forewing.

BIOLOGY

Ovum. Pale yellow at first, becoming orange-brown later, lenticular, delicately sculptured, usually covered by green scales from the upper surface of the forewings. Deposited during June and July in small batches, usually on the bark of the branches, overwintering and hatching in April and early May. Sich (1916: 15) describes the life-cycle of this species in Britain, and a monographic study has been published by Gasow (1925).

Larva. Head shining blackish brown or black; prothoracic plate varying from light greenish brown to green or grey, sometimes mottled with brown, a narrow whitish medial sulcus, posterior margin edged with black, anterior margin sometimes with a whitish border, the coloration and markings being exceedingly variable (in early instar larvae the prothoracic plate is usually entirely black); abdomen light olive-green, greyish green in early instars, integument strongly shagreened; pinacula dark brown or black; anal plate dark brown or green; anal comb usually with 8 prongs; thoracic legs shining black. Late April to early June, feeding on *Quercus*, at first living in the buds, later rolling a leaf or folding the edges (pl. 16, figs. 5, 6). Local populations of this species fluctuate annually; in some years the species is very abundant and the larvae cause extensive defoliation in oak woods, sometimes descending or dropping to the ground and feeding on the undergrowth. Although *Quercus* is the preferred foodplant, the larva will feed on *Fagus*, *Acer*, *Populus*, *Salix*, *Carpinus*, *Vaccinium*, *Urtica* and other plants.

Pupa. Varying from brown to black. June, in a folded leaf of the foodplant (pl. 16, figs. 4, 7) or of the foliage below.

Imago. June and July, sometimes extending into August; frequenting oak woods and open woodland, in some years occurring in great abundance; also occasionally found sparingly on open moorland amongst *Salix* and *Vaccinium*. The moth is easily disturbed from rest amongst the foliage of the foodplants, especially oak, during the day, and is occasionally taken at light.

DISTRIBUTION

Occurring wherever there are oak woods in Britain north to Argyll and Aberdeenshire; sometimes very common in the west and south-west of Scotland but rarely so abundant as in southern England. Generally distributed and common amongst oak in Ireland.

Europe, North Africa, Asia Minor, Transcaucasia and Iran.

Spatalistis bifasciana (Hübner)

Pl. 39, fig. 6

[*Dictyopteryx bifasciana*; Barrett, 1905: 200. *Chrosis bifasciana*; Pierce & Metcalfe, 1922: 24, pl. 9 (♂, ♀ genitalia). *Spatalistis bifasciana*; Meyrick, 1928: 518; Ford, 1949: 57]

DESCRIPTION

♂ ♀ 12–14 mm. Sexual dimorphism not pronounced; antenna of male very shortly ciliate; female usually larger, antenna simple.

Male, female (fig. 6). Forewing ground colour ochreous-yellow, almost entirely overlaid with dark fuscous except in terminal area, roughened with sparse groups of raised metallic plumbeous scales; fasciae indefinite, approximately indicated by irregular lustrous plumbeous or bluish grey streaks; an ochreous-yellow spot on dorsum contiguous with basal scale-tuft, a similar smaller spot on costa at or near middle, usually with a variable admixture of ferruginous-ochreous or ochreous-yellow mesad and two or three yellow costal striae apicad; terminal area broadly clear ochreous-yellow apically, narrowing and terminating above tornus; cilia concolorous. Hindwing brownish grey; cilia paler.

Variation. General coloration and markings are very constant, the ground colour varying from pale ochreous-yellow to orange-yellow, and the overlying fuscous coloration may be blackish brown in strongly marked specimens. The extent of the ochreous-yellow coloration varies in the median area below the costal spot and, although never very extensive in British specimens examined, is stated by Razowski (1966: 126) to extend across the wing in lightly coloured continental specimens, forming a clear ochreous-yellow fascia.

COMMENTS

The small size of this species and the dark coloration and ochreous-yellow apical area of the forewing are characteristic.

BIOLOGY

Larva. Head, prothoracic plate and thoracic legs brown, prothoracic plate edged with dark brown; abdomen yellow, integument strongly shagreened; pinacula large, violet; anal plate with a brownish violet transverse stripe between anterior edge and first setae; anal comb bearing 8 prongs (Swatschek, 1958: 70). July to May, in berries of *Rhamnus catharticus*, *Frangula alnus* and *Thelycrania sanguinea*. On the Continent the larva has been reared on the fruits of *Vaccinium myrtillus* (Swatschek, 1958: 71) and *Vaccinium uliginosum* (Razowski, 1966: 127).

Imago. May and June; frequenting the margins of woods, occasionally to be found resting in sunshine on the upper surface of leaves of *Frangula*, *Quercus*, *Ilex* and *Syringa*, and rarely on fences. The moth is on the wing in the evening a little before dusk, flying rather high.

DISTRIBUTION

A rare species in the British Isles, occurring only in the south of England, locally in Essex (Plaistow), Kent (Bexley, Chattenden, Sittingbourne), Surrey (Box Hill, Haslemere, Reigate), Sussex, Hampshire (Winchester, New Forest), Isle of Wight, Dorset (Bloxworth, Corfe Castle), Somerset, Gloucestershire (Flaxley), Herefordshire (Symond's Yat), Buckinghamshire, Hertfordshire and Huntingdonshire (Warboys Wood).

Europe to eastern Siberia; Japan.

Croesia bergmanniana (Linnaeus)
Pl. 39, figs. 7–9

[*Dictyopteryx bergmanniana*; Barrett, 1905: 199. *Argyrotoxa bergmanniana*; Pierce & Metcalfe, 1922: 16, pl. 6 (♂, ♀ genitalia); Meyrick, 1928: 518; Ford, 1949: 324]

DESCRIPTION

♂ ♀ 12–15 mm. Sexual dimorphism not pronounced; antenna of male very shortly ciliate.

Male, female (fig. 7). Head and thorax bright yellow; ground colour of forewing bright yellow, coarsely reticulate and partly suffused with ferruginous-orange, more heavily in distal half; markings reduced, lustrous bluish grey, slightly refractive, ferruginous-edged; basal and median fasciae strongest on costa; median fascia narrow, extending to tornus from costa at about three-quarters; pre-apical spot developed into a linear marking extending around the apex and along subterminal margin to tornal area; a usually conspicuous small black scale-tuft in submedian fold above middle of dorsum; cilia concolorous, grey around tornus. Hindwing brownish grey; cilia paler.

Variation. Strongly marked specimens with intensified ferruginous-orange coloration and higher refraction occasionally occur; however, there is a greater tendency towards reduction of the markings and a clearer yellow ground colour (fig. 8). Very pale lightly marked forms are known from the Dundrum sandhills, Co. Down, where almost unicolorous specimens are found (fig. 9).

COMMENTS

The bright yellow ground colour of the forewing and concolorous head and thorax are distinctive of this species.

BIOLOGY

Ovum. Deposited on the stem of the foodplant during June and July; overwintering until early April.

Larva. Head, prothoracic plate and thoracic legs shining blackish brown or black, plate narrow; abdomen greenish grey or yellowish white, sometimes bright yellow in the full-grown larva; pinacula concolorous with abdomen; anal plate dark brown or black, sometimes yellow marked with brown in the full-grown larva; anal comb present. April and May, feeding in spun leaves and shoots of *Rosa* (pl. 17, fig. 1), when young folding a leaf in the middle and living within, later spinning the terminal leaves together or to a flower bud; in small roses such as *Rosa pimpinellifolia*, the leaves may be drawn together and twisted over like a hood. *Rhamnus catharticus* and *Frangula alnus* are also recorded as foodplants on the Continent.

Pupa. Light brown or brownish yellow, head and thoracic region dark brown. May and June, in the folded edge of a leaf or in the larval habitation.

Imago. June and July; frequenting especially hedgerows and gardens, resting during the day on rose bushes and easily disturbed. Both sexes fly of their own accord from late afternoon onwards; occasionally coming to light.

DISTRIBUTION

Very common in Britain, occurring northwards to Inverness-shire and Aberdeenshire, but in Scotland common only in the south. Widely distributed and common amongst rose in Ireland, particularly on coastal sandhills and in the limestone area of the Burren, Co. Clare.

North and central Europe to Siberia; North America.

Croesia forsskaleana (Linnaeus)
Pl. 39, figs. 10–12

[*Dictyopteryx forskaleana*; Barrett, 1905: 203. *Tortrix forskaleana*; Pierce & Metcalfe, 1922: 20, pl. 8 (♂, ♀ genitalia). *Argyrotoxa forskaleana*; Meyrick, 1928: 517; Ford, 1949: 56]

DESCRIPTION

♂ ♀ 12–16 mm. Sexual dimorphism not pronounced; antenna of male very shortly ciliate; female usually larger.

Male (fig. 10). Forewing ground colour primrose-yellow, conspicuously reticulate with ferruginous-ochreous or ferruginous-orange, the reticulation partly formed by the radial veins which are similarly lined; markings dark grey mixed with ferruginous; basal fasciae only weakly indicated by strigulae; median fascia linear from costa, strongly angulate before middle, thence broadening to form a greyish suffusion enclosing two small black scale-tufts above middle of dorsum; a thick lustrous blackish grey mixed with ferruginous terminal line from before apex to near tornus; cilia yellow along termen, grey at tornus. Hindwing light ochreous-yellow, paler at base; cilia pale yellow.

Female. Similar to male.

Variation. Weakly marked specimens lacking the greyish suffusion and pronounced ferruginous-orange reticulation (fig. 11) are common, as also are specimens with the markings and reticulation intensified (fig. 12).

COMMENTS

The pale yellow ground colour and strongly reticulate pattern of the forewing distinguish this species.

BIOLOGY

Larva. Head yellowish green, shining and unmarked, ocellar region brownish; prothoracic plate greenish; abdomen whitish yellow, translucent; prolegs not sclerotized laterally; pinacula minute, slightly lighter in colour than ground colour of abdomen; anal plate yellowish green; anal comb present; thoracic legs light green. When full-grown the larva becomes bright yellow-green prior to pupation. September to May and June, hibernating in an early instar and recommencing to feed in the spring in spun leaves or flowers, later in a longitudinally rolled leaf, on *Acer campestre* (pl. 17, fig. 2) and *A. pseudoplatanus*.

Pupa. Bright yellow, changing to yellowish brown. June and July, in a white silken cocoon spun in a folded leaf of the foodplant.

Imago. July and August; frequenting gardens and mixed woodland. During the day the moth rests on the leaves of maple and sycamore and is rather lethargic and easily boxed.

DISTRIBUTION

A locally common woodland species in England and Wales; apparently rare in Scotland, although Bolam (1929: 118) found it generally common and often abundant in Berwickshire and Roxburghshire; one specimen was taken in Stirlingshire in 1962 by E. C. Pelham-Clinton. Records of this species in Ireland from Co.'s Sligo and Kildare are unconfirmed and Beirne (1941: 84) considers its occurrence there to be doubtful. Europe to north Caucasus; North America.

Croesia holmiana (Linnaeus)

Pl. 39, fig. 13

[*Dictyopteryx holmiana*; Barrett, 1905: 202. *Peronea holmiana*; Pierce & Metcalfe, 1922: 22, pl. 9 (♂, ♀ genitalia); Meyrick, 1928: 521; Ford, 1949: 57]

DESCRIPTION

♂ ♀ 12–15 mm. Sexes similar; antenna of male very shortly ciliate; female usually larger.

Male (fig. 13). Forewing ground colour orange-ochreous, infused with ferruginous except at apex and basal area above dorsum, a sprinkling of black scales in tornal and distal areas; a conspicuous triangular, often black-edged, white costal marking beyond middle, markings otherwise poorly defined, pale violet; outer margin of basal fascia slightly oblique, often interrupted at middle; sub-basal fascia from before middle of costa, narrow, weak on dorsum and sometimes with wedge of ground colour; lower half of median fascia extending from white costal marking, sometimes weak internally; pre-apical spot obsolescent, represented by strigulae in terminal area; cilia pale yellow, fuscous around tornus. Hindwing light grey, infuscate terminally.

Female. Similar to male.

Variation. Constant except for minor variation in the extent of the ferruginous coloration which, when strong, gives the wing a rich orange appearance.

COMMENTS

Immediately recognized by the conspicuous white costal marking on the forewing.

BIOLOGY

Larva. Head yellowish varying to light brown or red-brown, ocellar region and posterior margin blackish; prothoracic plate black, or dark brown with black posterior and lateral margins; abdomen varying from bright yellow to pale green, integument weakly shagreened, pinacula concolorous; anal comb present; thoracic legs brown. May and June, feeding on *Crataegus*, *Rosa*, *Rubus*, *Malus*, *Pyrus*, *Prunus* and *Cydonia*, spinning together two leaves at their margins and living within.

Pupa. Dark yellow or orange-yellow. June and July, in the larval habitation or a folded leaf.

Imago. July and August; frequenting thick hedges of hawthorn and blackthorn, and also orchards, gardens and margins of woods. Readily disturbed from its resting place during the day, usually on a leaf close to the ground, but quickly returning to cover. The moth flies freely after dusk, sometimes coming to light or sugar.

DISTRIBUTION

Widely distributed in Britain north to Dunbartonshire; common in many places in the south but becoming progressively scarcer northwards. Generally distributed but local in Ireland.

North and central Europe; Asia Minor.

Acleris latifasciana (Haworth)

Pl. 39, figs. 14–17

[*Peronea schalleriana*; sensu Barrett, 1905: 239. *Argyrotoxa comariana*; sensu Pierce & Metcalfe, 1922:

pl. 6 (♂, ♀ genitalia). *A. schalleriana*; sensu Pierce & Metcalfe, 1922: 17, pl. 7 (♂, ♀ genitalia). *Peronea schalleriana*; sensu Meyrick, 1928: 523. *P. latifasciana*; Ford, 1949: 58]

DESCRIPTION

♂ ♀ 15–20 mm. Polymorphic, without pronounced sexual dimorphism; four recurring forms are recognized from the British Isles (Sheldon, 1931: 30).

Male, female (fig. 14). Forewing ground colour silver-white, diffusely strigulate with mixed grey and black scales; markings brownish red; outer margin of basal fascia diffuse; sub-basal fascia developed on dorsum, diffusely confluent with basal fascia; costal blotch extending to tornus, inner edge well defined, outer edge diffuse; terminal area sprinkled with black; cilia grey. Hindwing brownish grey.

Variation. The nominate form is the least common and shows slight variation in the intensity of the greyish suffusion which, in some specimens, may be sufficiently heavy almost to obliterate the silver-white ground colour. The most distinctive and frequent form (fig. 15, f. *labeculana* Freyer) has the ground colour suffused with grey, giving the wing a cinereous appearance, the basal fascia obsolescent, only a vestige of the sub-basal fascia, and the costal blotch lunate, rich red-brown or chestnut-brown. A similarly patterned form (fig. 16, f. *comparana* Hübner) has the ground colour pale yellow-brown varying to ochreous, with the costal blotch and sub-basal fascia dull bluish black. A common but more variable form (fig. 17, f. *perplexana* Barrett) has the ground colour grey-brown, varying considerably in shade, the costal blotch relatively inconspicuous, its inner margin extending across the wing in some specimens.

COMMENTS

Certain forms of this species resemble forms of *A. comariana* and are differentiated under that species. The brownish f. *perplexana* may be confused with *A. schalleriana* f. *castaneana*, but may be distinguished by the comparatively smooth costa of the forewing.

BIOLOGY

Ovum. The greenish grey eggs are deposited singly at the base of a bud of the foodplant in late August and September, overwintering and hatching the following spring. Sheldon (1919: 272) describes the life-cycle of this species.

Larva. Head yellowish brown, ocellar region and postero-lateral margin dark brown; prothoracic plate yellowish brown or brownish green, a well-defined darker subdorsal spot on the posterior margin; abdomen whitish green, pinacula concolorous. May and June, feeding between spun leaves and flowers of *Crataegus*, *Prunus*, *Rosa*, *Rubus*, *Sorbus*, *Salix* and *Vaccinium*. On the Continent recorded also on *Symphytum officinale*, *Filipendula ulmaria* and *Populus*; records of this species on *Azalea* and *Rhododendron* probably refer to *comariana* (Obraztsov, 1963: 218).

Pupa. Reddish brown. July and August, in a silken cocoon spun up in a folded leaf or in the larval habitation; about 21 days duration.

Imago. August and September; frequenting lanes, open woodland and margins of woods, hiding during the day in bramble bushes and other vegetation from which it may be beaten out, flying quickly to another bush when disturbed. Both sexes fly freely at dusk and will come to light.

DISTRIBUTION

Widely distributed and common throughout the British Isles north to Wester Ross. North and central Europe to eastern Siberia, China and Japan; North America.

Acleris comariana (Lienig & Zeller)
Pl. 40, figs. 1–5

[*Peronea comariana*; Barrett, 1905: 243. *Argyrotoxa comariana*; Pierce & Metcalfe, 1922: 17. *A. comparana*; sensu Pierce & Metcalfe, 1922: pl. 7 (♂, ♀ genitalia). *Peronea comariana*; Meyrick, 1928: 523; Ford, 1949: 58]

DESCRIPTION

♂♀ 13–18 mm. Polymorphic, without pronounced sexual dimorphism; seven recurring forms are recognized from the British Isles, excluding the nominate form which is known only from the Continent.

Male, female (fig. 1, f. *fuscana* Sheldon). Forewing ground colour dark grey or fuscous-brown, varying in shade, with weak yellowish brown strigulation; markings obsolescent, sometimes partially indicated by a slightly darker shade of ground colour and small tufts of raised blackish scales; cilia concolorous with ground colour, with a dark sub-basal line. Hindwing grey, paler basally, terminal margin often infuscate; cilia concolorous, with a dark sub-basal line.

This form occurs commonly in Lancashire and is the only obscurely marked form of *comariana* known in the British Isles. It closely resembles the continental nominate form in which the forewing is thickly overlaid with shining grey strigulae through which the brownish yellow ground colour penetrates as ripples and streaks.

Variation. The major forms of this species were enumerated by Sheldon (1925: 281), and the seven found in the British Isles were illustrated in colour by Fryer (1928: 157, pl. 2) in a genetical study of the polymorphism in this species. More recently, Turner (1968: 489) discusses the major forms in a study of the ecological genetics.

With the exception of f. *fuscana*, named forms of *comariana* occurring in the British Isles have the costal blotch relatively well developed and conspicuous. These forms fall into two groups according to the colour of the costal blotch; in one group the blotch varies from fuscous to deep blue-black, in the other from light brown to reddish brown. In the former group three distinct forms are recognized, the ground colour and markings differing in each form, as follows: ground colour light beige, weakly strigulate with darker shade, sub-basal fascia vestigial (fig. 2, f. *potentillana* Gregson); ground colour yellowish brown with weak strigulation (fig. 3, f. *comparana* Sheldon); ground colour whitish grey coarsely irrorate with fuscous, suffusion from costal blotch to dorsum reddish brown (f. *fasciana* Sheldon), this form differing from f. *latifasciana* Sheldon (fig. 5) only in the colour of the costal blotch. In the second group, in which the costal blotch is reddish brown, one form has the ground colour light grey with weak strigulation (fig. 4, f. *proteana* Herrich-Schäffer); another (fig. 5, f. *latifasciana* Sheldon) has the ground colour whitish grey with variable fuscous irroration and a reddish brown suffusion from the costal blotch to the dorsum; a third form (f. *brunneana* Sheldon) resembles f. *proteana* but differs in the weakly reticulate reddish brown ground colour, with the sub-basal fascia obsolescent.

COMMENTS

In *comariana* the forms f. *latifasciana*, f. *proteana* and f. *potentillana* are analagous to the nominate form of *A. latifasciana*, *latifasciana* f. *labeculana* and *latifasciana* f. *comparana*, respectively.

A. comariana may be distinguished from *latifasciana* by its almost invariably smaller size. Differences in the biology also help in distinguishing the two species; *comariana* is bivoltine, moths being found on the wing in June and July and from late August to October, while *latifasciana* is univoltine, the moths occurring in August and September. There is thus only a certain amount of overlapping, as worn specimens of *latifasciana* may be found in September when the second brood of *comariana* is in full emergence.

BIOLOGY

Ovum. Chorion sculptured, silver-white or whitish green, the general coloration varying according to the season; eggs which overwinter become brick red, those from the early summer generation change very little in colour. The eggs are deposited singly low down on the foodplant, usually on the stipules or on the petioles. Eggs laid by the first generation of moths (June and July) hatch in about ten days; those of the second generation (late August to October) overwinter and hatch the following spring. The biology of this species has been described by Balachowsky (1966: 605) and Petherbridge (1920: 6).

Larva. Head yellow or yellowish brown, ocellar region and postero-lateral margin marked with blackish; prothoracic plate yellowish brown, posterior margin dark brown or black; abdomen varying in colour from whitish green to yellowish green, yellowish brown or dull green, an indistinct darker subdorsal line; pinacula brown; thoracic legs yellowish brown. In early instar larvae the head and prothoracic plate are entirely black, the abdomen is paler in colour and the thoracic legs are blackish brown. The larva may be distinguished from that of *latifasciana* by the blackish brown posterior margin of the prothoracic plate which in *latifasciana* is marked with a well-defined dark subdorsal spot on either side, and by the brown pinacula which in *latifasciana* are concolorous with the integument. May and June, and again in July and August; feeding in a folded leaf or spun leaves of *Potentilla palustris*, *Fragaria vesca*, cultivated strawberry and *Geum rivale*. When feeding on cultivated strawberry the larva also attacks the flowers, distorting or preventing the fruit from developing. It has also been recorded on glasshouse azaleas (Fryer, 1931: 199). As previously mentioned, records of *latifasciana* on *Azalea* and *Rhododendron* on the Continent are probably referable to *comariana*.

Pupa. Reddish brown, wings yellowish brown. June and July, and those producing the second generation, in August and September; in a silken cocoon in the larval habitation.

Imago. Bivoltine, the first generation of moths occurring from mid-June to July, the second generation from late August to October or early November. Frequenting wet heathland, bogs and marshes; occasionally found on dry, well-drained localities such as at Portland, Dorset. In certain areas it is a recurrent pest on cultivated strawberry and serious infestations have been reported from strawberry fields in the Wisbech area, Cambridgeshire; King's Lynn area, Norfolk; the Vale of Evesham, Worcestershire; south Lancashire; and the Dee Valley, Cheshire (Turner, 1968: 491). Although widespread in south-east England it is not known to attack strawberries in Kent and Essex. The moth hides during the day amongst the foodplants and other herbage, flying during the afternoon and again at dusk.

DISTRIBUTION

Widely distributed and locally common in lowland areas in Britain north to Lanarkshire and West Lothian. Widely distributed but local in Ireland.

Europe to east Russia; China and Japan; North America.

Acleris caledoniana (Stephens)
Pl. 40, figs. 6–8
[*Peronea caledoniana*; Barrett, 1905: 242. *Argyrotoxa caledoniana*; Pierce & Metcalfe, 1922: 18, pl. 7 (♂, ♀ genitalia). *Peronea caledoniana*; Meyrick, 1928: 522; Ford, 1949: 58]

DESCRIPTION

♂ ♀ 12–17 mm. Sexes similar; female usually larger.

Male (fig. 6). Forewing rather narrow, tending to be somewhat acute apically; ground colour varying from yellowish brown to deep ferruginous, often with a greyish tinge, usually with at least a trace of obscure greyish fuscous reticulation; markings indistinct; sub-basal fascia sometimes indicated by scattered black scales; costal blotch usually a darker shade of the ground colour; cilia concolorous with ground colour. Hindwing grey, infuscate terminally, more coarsely on inner margin; cilia paler.

Female (fig. 7). Similar to male.

Variation. Although very variable within the limited range described above, distinctive colour forms are unknown. Specimens with the ground coloration reddish brown tend to have stronger markings, and occasionally individuals are found, especially in the Scottish Highlands, with a comparatively well-defined reddish brown costal blotch (fig. 8, f. *rufimaculana* Sheldon).

COMMENTS

This species is closely allied to the variable *A. comariana* and although showing little difference in genitalic characters may be readily distinguished by the shape of the forewing, which is relatively narrow and apically more acute. Differences in the bionomy help in distinguishing these two species: *caledoniana* is univoltine, moths being found on the wing from July to September, usually frequenting high moorland, whilst *comariana* is bivoltine, moths occurring in late June and July and from late August to October, frequenting lowland areas.

BIOLOGY

Larva. Head yellowish brown, a black marking on postero-lateral margin; prothoracic plate green; abdomen green, darker dorsally (Peers, 1864: 63). June and July, on *Vaccinium myrtillus* (Michaelis, 1956: 235), spinning the leaves together or to a stem on the upper shoots and living within; also on *Vaccinium vitis-idaea, Myrica gale, Alchemilla alpina, Rubus chamaemorus, Potentilla* and *Pinus contorta*.

Pupa. Reddish brown, wings paler. Late June and July, in the larval habitation or between spun leaves.

Imago. July to September; a typically high moorland species frequenting mountain mosses, boggy moors and steep hillside slopes, occurring most frequently from 500–2,000 ft, but also found in bogs at lower elevations, such as at Borth, Cardiganshire (Michaelis, 1954a: 65). The moths are often to be seen in numbers flying over the foodplants on sunny afternoons, and will also fly in the afternoon in dull weather if the temperature is high.

DISTRIBUTION

Widely distributed in the northern parts of Britain, occurring commonly on hills throughout the mainland of Scotland. Recorded from the Orkneys (Weir, 1882: 2). In northern England and southwards to Herefordshire and Somerset (Exmoor) it is apparently confined to high moorland, occurring also on the Isle of Man above 1,300 ft (Hayward, 1929b: 50); other records from the south of England (Kent, Dorset and Devon) have not been confirmed and are possibly referable to *comariana*. Widely distributed in North Wales, occurring on the Glyders (Caernarvonshire) up to 2,000 ft. Locally common on hills in Ireland (Beirne, 1941: 85).

Abroad, apparently recognized only from Poland (Silesia) (Razowski, 1966: 210).

Acleris sparsana (Denis & Schiffermüller)

Pl. 40, figs. 9–11

[*Peronea sponsana*; Barrett, 1905: 236. *Eclectis sponsana*; Pierce & Metcalfe, 1922: 18, pl. 7 (♂,♀ genitalia). *Peronea sponsana*; Meyrick, 1928: 525. *P. sparsana*; Ford, 1949: 59]

DESCRIPTION

♂♀ 18–22 mm. Sexual dimorphism not pronounced; female usually larger. Polymorphic, the nominate and three colour forms being known from the British Isles (Sheldon, 1930: 243).

Male (fig. 9). Forewing ground colour grey finely irrorate with whitish; markings obsolescent, indicated by ferruginous and pale ochreous striae, most apparent costally; a few scattered raised black scales, including slight tufts near base of median fold; cilia light grey. Hindwing grey, paler basally.

Female. Similar to male.

Variation. In the obscurely marked nominate form a wide range of minor variation is found in the ferruginous striae which, when strong, may produce a slightly marbled effect. A very rare form occurs which lacks the ferruginous striae and has the forewing ground colour whitish grey, coarsely irrorate and striate along the veins with fuscous.

Two major colour forms occur not infrequently; in one the forewing ground colour is light grey, the large triangular costal blotch and the basal fascia are reddish brown, well developed and conspicuous, the sub-basal fascia only weakly developed on the costa (fig. 10, f. *favillaceana* Hübner); in the other the forewing ground colour is whitish grey, the costal blotch greyish fuscous, well defined and conspicuous, and the basal and sub-basal fasciae are well developed on the costa (fig. 11, f. *haworthana* Sheldon).

COMMENTS

The relatively broad forewing and predominately grey coloration are characteristic.

BIOLOGY

Larva. Head brownish green or light green, ocellar region blackish; prothoracic plate light green, with a black lateral marking; abdomen light green, darker dorsally; pinacula shining, concolorous with integument; anal plate light green; anal comb present; thoracic legs light green. June to early August, feeding on the leaves of *Fagus* (pl. 18, fig. 1) and *Acer pseudoplatanus* (pl. 18, fig. 2), occasionally attacking the seeds of the latter. In early instars the larva lives in a slight web on the underside of a leaf, later spinning two leaves together, one above the other, and living in a fold of the upper leaf, leaving its retreat to feed on the edges of the leaves. *Carpinus*, *Sorbus*, *Betula*, *Quercus*, *Populus* and *Rubus* are recorded as additional foodplants on the Continent.

Pupa. Light brown. July and August, in the larval habitation or spun up in a slight cocoon amongst debris.

Imago. August to October, hibernating until the following spring (Williams, 1915: 220), specimens having been taken as late as the 25th May (Meldola, 1915: 198). Frequenting woodland and suburban parks and gardens, especially where beech and sycamore grow; resting amongst the foliage during the day, fluttering to the ground when disturbed. At dusk the moth flies freely and is attracted to ivy bloom and sugar, and frequently comes to light.

DISTRIBUTION

Widely distributed and generally common in woodland containing beech and sycamore, occurring throughout the British Isles north to the Inner and Outer Hebrides and Orkney, but rather uncommon in Scotland north of Argyll.

Europe to the Ukraine.

Acleris rhombana (Denis & Schiffermüller)

Pl. 40, figs. 12–16

[*Dictyopteryx contaminana*; Barrett, 1905: 204. *Peronea reticulata*; Pierce & Metcalfe, 1922: 23, pl. 9 (♂, ♀ genitalia). *P. contaminana*; Meyrick, 1928: 521; Ford, 1949: 57]

DESCRIPTION

♂ ♀ 13–19 mm. Sexual dimorphism not pronounced; apex of forewing subfalcate in both sexes; more prominent in female, with costa shallowly concave beyond middle, hindwing usually greyer. Polymorphic, five forms being recognized by Sheldon (1931: 79).

Male (fig. 12). Forewing ground colour dark ferruginous-brown, variably suffused with fuscous, indistinctly strigulate with ferruginous, veins lined with fuscous and combining with strigulation to form a reticulate pattern; markings reduced, dark fuscous-brown; basal and sub-basal fasciae diffuse, often discernible only on costa; median fascia narrow, angulate outwards above and inwards below middle, narrowing and often obsolescent towards dorsum; discocellular scale-tuft moderately developed when present; pre-apical spot small, diffuse, usually shallowly excised on costa; cilia on costa before apex ferruginous, grey along termen except apices of those in upper half which are white, a rather conspicuous black sub-basal line, often edged with plumbeous inwardly. Hindwing whitish grey, indistinctly strigulate with grey; cilia concolorous, with a dark grey sub-basal line around apex, obsolescent towards tornus.

Female. Forewing coloration and pattern similar to that of male, hindwing greyer.

Variation. Nearly all the named forms of this very variable species occur frequently and show considerable minor variation. A common form occurs (fig. 13, f. *contaminana* Hübner) in which the forewing markings are relatively conspicuous, the ground colour being light ochreous-yellow; in this form the median fascia is produced medially and joins the pre-apical spot. In another common form (fig. 14, f. *obscurana* Donovan) the markings are conspicuous, the ground colour is dark orange and strongly reticulate, but the median fascia is separate from the pre-apical spot which is usually small and in some specimens may be obsolete; the median fascia is often dilated and strong on the dorsum but weak on the costa. In f. *ciliana* Hübner (fig. 15), which is the commonest form of this species, the forewing ground colour varies from light to dark ochreous-brown, and the median fascia is usually well developed only in the costal half. Occasionally specimens are found with the forewing markings as in f. *ciliana* but with the ground colour light ochreous-yellow and strongly reticulate (f. *reticulata* Ström). Least common is a dark form (fig. 16, f. *dimidiana* Frölich) in which the forewing ground colour is dark fuscous-brown with the markings obsolescent.

COMMENTS

The subfalcate apex of the forewing together with the reticulate pattern and comparatively conspicuous median fascia distinguish this species.

BIOLOGY

Ovum. The yellow-green eggs are deposited singly or in small batches on the bark of the branches or trunk of the foodplant during late summer and early autumn, over-wintering and hatching the following spring (Balachowsky, 1966: 610).

Larva. Head and prothoracic plate black, towards maturity changing to brown or yellow-brown, or prothoracic plate sometimes changing to green; abdomen light green or yellowish green; pinacula paler than integument; anal plate green; anal comb present, yellowish; thoracic legs black. April to June, feeding on *Crataegus* (pl. 17, fig. 3), *Prunus*, *Malus*, *Pyrus* and *Rosa*; also occasionally on *Prunus laurocerasus*, *Corylus*, *Sorbus* and *Quercus*. The young larva eats into the buds (Sich, 1918: 70), later feeding externally and spinning the leaves, usually those of the terminal shoots, and flowerheads.

Pupa. Reddish brown or dark brown, wings paler. June and July, in a silken cocoon in the larval habitation or between spun leaves of the foodplant. This species is remarkable in the genus *Acleris* for its comparatively long pupal phase.

Imago. August to October; frequenting especially hedgerows in which hawthorn and blackthorn grow, often occurring in numbers, also on scrubby downland and in orchards. The moth may be beaten out during the day and easily captured; it flies gently about the foodplants at dusk and occasionally comes to light. In the autumn the moth may be obtained at night by searching ivy blossom and hawthorn hedgerows with a lamp.

DISTRIBUTION

Widely distributed and very common in England and Wales. Common in the Lowland districts of Scotland, scarcer northwards but recorded from the north-east to Caithness and the Orkneys where the larva is common on *Sorbus*; apparently absent in the north-west and the Outer Hebrides. Common and widely distributed in Ireland.

North and central Europe to Asia Minor; North America.

Acleris aspersana (Hübner)

Pl. 40, figs. 17, 18

[*Peronea aspersana*; Barrett, 1905: 249; Pierce & Metcalfe, 1922: 22, pl. 9 (♂, ♀ genitalia); Meyrick, 1928: 522; Ford, 1949: 58]

DESCRIPTION

♂ 13–16 mm, ♀ 11–14 mm. Sexual dimorphism moderately pronounced; female usually smaller, forewing relatively narrow and with costa shallowly concave, hindwing retuse.

Male (fig. 17). Forewing ground colour ochreous-yellow, faintly strigulate or suffused with orange-yellow, veins weakly emphasized with orange-brown; markings dark plumbeous mixed with deeper ochreous and ferruginous; basal fascia developed on costa, sub-basal on dorsum; costal blotch extending to apex, inner margin almost straight, strongly edged with ferruginous, outer margin irregular, weakly edged with ferruginous; a trace of black on costa before apex; termen finely edged with ferruginous; cilia pale yellow, suffused with reddish ochreous around apex, grey on dorsum before tornus, with a pale ferruginous-brown sub-basal line. Hindwing pale grey, infuscate distally.

Female (fig. 18). Forewing ground colour ochreous-yellow as in the male but heavily suffused with ferruginous; markings obscure, costal blotch mixed with thick plumbeous striae; cilia paler. Hindwing grey, infuscate distally; cilia paler.

Variation. This species shows considerable minor variation but extreme forms appear to be unknown. In the male the strigulation on the forewing may be increased and the veins more heavily lined, producing a strong reticulate pattern; very rarely the forewing is strongly suffused with ferruginous and the plumbeous striae are more extensive and extend dorsad from the costal blotch. In pale specimens of the female the costal blotch is usually better defined, though never as clearly as in the male, and the basal area may be less heavily suffused and lighter ochreous-yellow; in darker specimens the blackish content of the costal blotch is usually increased.

COMMENTS

The small size of this species and the bright orange-yellow general coloration of the forewing are distinguishing features.

BIOLOGY

Larva. Head light brown, ocellar region and postero-lateral margin darker; prothoracic plate light brown or shining green, posterior margin sometimes dark brown or blackish; abdomen translucent, whitish green varying to bluish green or greenish brown, darker dorsally; pinacula concolorous with integument; anal comb bearing six prongs; thoracic legs dark brown. May and June, on *Spiraea*, *Potentilla* (pl. 17, fig. 4), *Fragaria vesca*, also *Filipendula ulmaria*, *Sanguisorba minor* (pl. 17, fig. 5), *Geum*, *Alchemilla vulgaris* agg., *Rubus*, *Dryas octopetala*, *Helianthemum chamaecistus* and *Salix aurita*, living in a folded leaf or within a spinning of two or three leaves drawn together.

Pupa. Yellowish brown. June and July, in the larval habitation or a folded leaf.

Imago. July and August, but apparently varying in time of appearance according to locality; in western Ireland the moth has been taken as early as June and as late as September. Most plentiful in open country away from woodland, and occurring in a variety of habitats from the chalk downs in the south of England to the mountains of the west and north, and the rough meadows and low-lying fenland of East Anglia. In Scotland it is found in grassy places on hills and mountains, and in marshes at lower elevations. The moth flies freely in the afternoon and evening, and at night frequently comes to light.

DISTRIBUTION

Locally common throughout the British Isles, including St. Kilda.

North and central Europe; Iceland; south-east Siberia.

Acleris tripunctana (Hübner)

Pl. 41, figs. 1–6

[*Peronea ferrugana*; sensu Barrett, 1905: 246 (partim); sensu Pierce & Metcalfe, 1922: 21, pl. 8 (♂, ♀ genitalia); sensu Meyrick, 1928: 522; sensu Ford, 1949: 58]

DESCRIPTION

♂♀ 15–19 mm. Sexual dimorphism not pronounced. Polymorphic; Sheldon (1931: 60) recognizes five forms but, because of confusion with *A. ferrugana*, the infrasubspecific nomenclature is confused.

Male, female (fig. 1). Forewing ground colour pale ochreous varying to reddish ochreous, sometimes with a trace of delicate darker strigulation and a few scattered black scales;

markings obsolescent, reduced to an atrophied, tripunctate, black mixed with ferruginous costal blotch; diminutive black scale-tufts usually present, a small one dorsad of submedian fold at about one-third, a similar tuft vertically above, and a larger black mixed with white tuft in disc at apex of costal blotch; cilia concolorous with ground colour, a darker sub-basal line. Hindwing grey, infuscate distally; cilia pale yellow, with a greyish sub-basal line.

Variation. The majority of the forms of this species are analogous to those of *ferrugana*. In the British Isles the nominate form appears to be uncommon. The forewing ground colour is usually uniform, but in some specimens the strigulation is stronger; frequently the spots of the costal blotch are confluent and form a triangular reddish brown or blackish grey marking with a light centre; in some specimens the ground colour is whitish cream, often irrorate with black (fig. 2). The remaining forms are more frequent and include that analagous to the nominate form of *ferrugana*; in this form (figs. 3, 4) the forewing ground colour shows the usual variation, with faint strigulation and suffusion, and diffuse weakly developed ochreous-brown markings. The most heavily marked form of this species, and probably the most frequent, is f. *bifidana* Haworth (fig. 5) in which the forewing ground colour shows a wide range of variation as in the nominate form but is most frequently reddish brown, and the costal blotch is enlarged and produced dorsad to form an angulate median fascia extending to the tornus. A not uncommon form (fig. 6, f. *notana* Donovan) occurs with the forewing coarsely irrorate with black dots.

COMMENTS

A. tripunctana can usually be distinguished from *ferrugana*, which it most closely resembles superficially, by the more diminutive black scale-tuft dorsad of the submedian fold on the forewing at about one-third. This tuft often consists of only two or three black scales and sometimes is absent altogether. In *ferrugana* it is almost invariably present and much more prominent, but the scales become dislodged easily and some may be lost in worn specimens or during setting. A slight difference between the two species is found in the shape of the forewing which in *tripunctana* has the apex more strongly produced and the termen shallowly concave near the middle; in *ferrugana* the termen is almost straight. This difference in wing shape is best seen by damping the wing with toluene or a similar wetting agent and examining it from the underside.

The two species may be more reliably separated by examination of the genitalia, which show very clear-cut diagnostic characters. In *tripunctana*, the male genitalia (text-fig. 47) have the valva comparatively broad, with the ventral margin of the sacculus almost straight. The aedeagus (text-fig. 48) has a small external sclerite laterally beyond the middle on the right, from which extends a short projection; the vesica contains two comparatively stout, unequal cornuti. The female genitalia (text-fig. 51) have comparatively long antero-lateral projections on the sterigma, and the antrum is weakly sclerotized, cylindrical and narrow.

BIOLOGY

Larva. Head and prothoracic plate blackish brown or black, plate with a narrow medial sulcus, anterior margin sometimes narrowly paler; abdomen light green, pinacula concolorous or paler, well defined; anal comb usually with four short prongs; thoracic legs black. In the south, May to early June, and a second generation in August and September; in the north, June and July; feeding in spun leaves of *Betula*, rather bunched together (pl. 19, fig. 1), *Alnus* (Sheldon, 1931: 63) and *Myrica*. On the Continent, *Quercus*, *Fagus*, *Populus tremula*, *Prunus cerasus*, *Pyrus communis* and *Rubus idaeus* are recorded also as foodplants (Swatschek, 1958: 74), but the confusion of this species with *ferrugana* makes it doubtful whether these records are entirely correct. In the British Isles the species, so far as is known, is attached primarily to *Betula*.

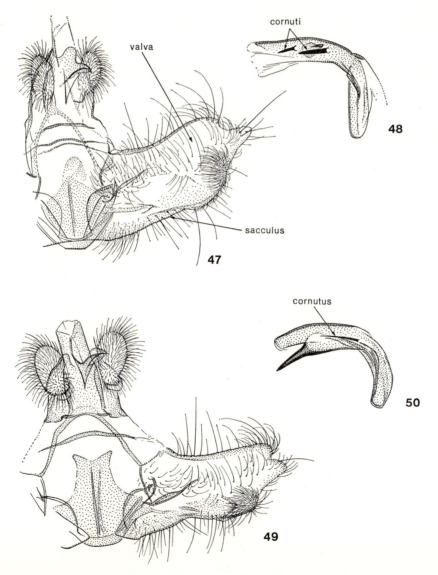

Male genitalia of *Acleris* species

Figs. 47, 48, *A. tripunctana* (Haworth). Fig. 47, genital armature. Fig. 48, aedeagus
Figs. 49, 50, *A. ferrugana* (Denis & Schiffermüller). Fig. 49, genital armature. Fig. 50, aedeagus

Pupa. Light brown. In the south, May and June, and those producing the second generation of moths in September and October; in the north, July; usually in the larval habitation, but sometimes spun up amongst debris.

Imago. Bivoltine in the south, the first generation emerging in July, the second in October and overwintering until April; in Scotland univoltine, emerging in August and overwintering until May. Frequenting birch woods, resting amongst the foliage and readily beaten out during the day in the summer and late autumn. The overwintering generation is on the wing until very late in the autumn, or even well into winter if it is mild, before going into hibernation, seeking shelter amongst dense masses of dried leaves, etc. The moths of this generation sometimes break hibernation when the weather is mild during the winter (Jacobs, 1931: 41), and may then be found with the aid of a lamp at night sitting on the bare twigs and branches in woods and hedgerows.

DISTRIBUTION

Generally distributed and common in birch woods throughout the British Isles north to Sutherland and the Outer Hebrides.

Europe; Faroes; Iceland; North America.

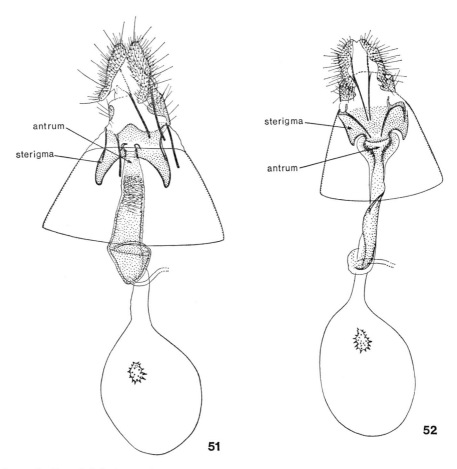

Female genitalia of *Acleris* species
Fig. 51, *A. tripunctana* (Haworth). Fig. 52, *A. ferrugana* (Denis & Schiffermüller)

Acleris ferrugana (Denis & Schiffermüller)

Pl. 41, figs. 7–10

[*Peronea ferrugana*; Barrett, 1905: 246 (partim). *P. fissurana*; Pierce & Metcalfe, 1922: 23, pl. 9 (♂, ♀ genitalia); Meyrick, 1928: 523; Ford, 1949: 58]

DESCRIPTION

♂ ♀ 14–18 mm. Sexual dimorphism not pronounced. Polymorphic; four forms are recognized by Sheldon (1931: 63).

Male (fig. 7). Forewing ground colour pale ochreous-brown varying to ferruginous-brown, faintly strigulate with a darker shade, a sprinkling of black scales, especially in distal half of wing and along termen; a usually prominent black scale-tuft dorsad of submedian fold at about one-third, a tuft of black mixed with whitish scales in disc, contiguous with apex of costal blotch; markings obsolescent, costal blotch obscure, brown with plumbeous admixture; cilia concolorous with ground colour, darker around apex, a dark sub-basal line. Hindwing grey, infuscate distally, sometimes faintly strigulate; cilia pale yellow, with a greyish sub-basal line.

Female. Similar to male.

Variation. Although generally less variable this species has forms analagous to those found in *A. tripunctana*. A very common form (fig. 8, f. *trimaculana* Pierce & Metcalfe) has the forewing ground colour paler than in the nominate form, the coloration of the costal blotch is intensified but the blotch is atrophied and broken into two and sometimes three separate spots which, in extreme forms with dark reddish brown ground colour, resemble the nominate form of *tripunctana*. Specimens with coarse black irroration (fig. 9, f. *multipunctana* Pierce & Metcalfe) are uncommon. Rarely, irrorated specimens occur in which the ground colour is cream-white and the costal blotch fulvous or reddish brown (fig. 10).

COMMENTS

The superficial similarity and parallel variation of *ferrugana* and *tripunctana* make differentiation of the two species difficult. A useful character in fresh specimens of *ferrugana* is the presence of the small but comparatively prominent black scale-tuft on the forewing dorsad of the submedian fold; in *tripunctana* this tuft is very small or absent. A slight difference is also apparent in the shape of the forewing, as described under *tripunctana*.

The structure of the genitalia of these two species is quite distinct in both sexes; in the males the external aedeagal differences in the form of the lateral prong can often be seen without dissection.

In *ferrugana*, the male genitalia (text-fig. 49) have the valva gently tapered, with the ventral margin of the sacculus concave. The aedeagus (text-fig. 50) has a very long and prong-like lateral projection on the right, and the vesica contains a single needle-like cornutus. The female genitalia (text-fig. 52) have comparatively short antero-lateral projections on the sterigma, and the antrum is strongly sclerotized, dilated laterally and with nodular encrustations.

BIOLOGY

Larva. Similar to that of *tripunctana* but more translucent. Head brown or black; prothoracic plate brownish green, edged with brown posteriorly; abdomen green, thoracic legs dark brown (Benander, 1965: 1). May and early June, and those producing a second generation of moths in August, feeding on *Quercus*, living between two leaves spun one above the other (pl. 19, fig. 2); also on *Salix caprea*. According to Sheldon (1931: 63), larvae which he reared from ova on oak refused to eat the leaves of any other tree. On the Continent, *Fagus* has also been recorded as a foodplant (Van Deurs, 1956: 34; Benander, 1965: 1).

Pupa. Similar to that of *tripunctana*. Late June, those producing the second generation of moths in early September.

Imago. Bivoltine, the first generation emerging in July, the second in September and October and overwintering until the following spring. Frequenting oak woods; the moth rests during the day high up amongst the foliage of overhanging boughs of large oaks (Huggins, 1928: 17).

DISTRIBUTION

Generally distributed and common in oak woods throughout the British Isles. Europe.

Acleris shepherdana (Stephens)

Pl. 41, figs. 11, 12

[*Peronea shepherdana*; Barrett, 1905: 247; Pierce & Metcalfe, 1922: 22, pl. 9 (♂, ♀ genitalia); Meyrick, 1928: 523; Ford, 1949: 58]

DESCRIPTION

♂ ♀ 13–16 mm. Sexual dimorphism not pronounced.

Male (fig. 11). Forewing ground colour cinnamon-brown, strigulate with darker brown, veins similarly lined, producing a reticulate pattern, especially in the distal area; markings reduced, dark ferruginous-brown; basal fascia obsolete; sub-basal fascia represented by a fine irregular stria, developing into a blackish grey suffusion on dorsum contiguous with a rather prominent black scale-tuft dorsad of submedian fold; costal blotch well developed, outer margin broadly suffused with blackish plumbeous, inner margin emphasized by scattered raised black scales; a usually prominent yellowish mixed with black discocellular scale-tuft; usually several striae of raised black scales in distal area and an incomplete black line along termen; cilia whitish yellow, grey at apex and tornus, with broad sub-apical and sub-basal lines. Hindwing whitish grey, infuscate and faintly strigulate with grey distally, terminal margin narrowly edged with dark grey; cilia whitish grey, with a greyish sub-basal line along termen.

Female (fig. 12). Similar to male.

Variation. This species shows only minor variation in the depth of coloration and in the markings of the forewing.

COMMENTS

The distinctive cinnamon-brown coloration together with the reticulate pattern and the blackish plumbeous outer margin of the costal blotch of the forewing is characteristic.

BIOLOGY

Larva. Head yellowish or greyish brown, epistomal region paler, ocellar region and postero-lateral margin marked with dark brown; prothoracic plate whitish grey or green, posterior margin edged with blackish brown; abdomen green varying to light greyish or yellowish green, paler laterally; pinacula inconspicuous, concolorous with integument; anal plate green, anterior margin sometimes marked with blackish brown; thoracic legs brown. May and June, feeding on *Filipendula ulmaria*, spinning together the leaves of the young shoots and living within (pl. 19, fig. 3), often gnawing a midrib and causing the tip of the leaf to shrivel. On the Continent, *Sanguisorba officinalis* and *Alchemilla* are also recorded as foodplants.

Pupa. Light brown. June and July, in a slight cocoon in spun shoots or the folded edge of a leaf, or in debris on the ground.

Imago. August and September; frequenting fens and marshy places, hiding during the day low down amongst dense vegetation and herbage. The moth is not easily disturbed but may be found by parting the herbage down to ground level and searching; in the evening it is more readily flushed from rest. Both sexes fly at night and come readily to light.

DISTRIBUTION

Local, occurring most plentifully in the fens of East Anglia; also found in marshy places in Essex (Rowney Wood, Debden), Kent (Brook near Wye), Hampshire, Isle of Wight, Dorset, Wiltshire, Berkshire, Hertfordshire and Lancashire. A single record from East Lothian, Scotland (Evans, 1897: 102), is unconfirmed and is almost certainly referable to *A. aspersana*. Unknown in Ireland; Beirne (1941: 86) states that the record from Co. Sligo is erroneous and refers to *aspersana*.

North and central Europe; south-east Siberia.

Acleris schalleriana (Linnaeus)
Pl. 41, figs. 13–16

[*Peronea logiana*; sensu Barrett, 1905: 228. *Eclectis logiana*; sensu Pierce & Metcalfe, 1922: 18, pl. 7 (♂ genitalia). *E. hastiana*; sensu Pierce & Metcalfe, 1922: 18, pl. 7 (♂ genitalia). *Peronea lipsiana*; sensu Pierce & Metcalfe, 1922: 22, pl. 9 (♀ genitalia). *P. logiana*; sensu Meyrick, 1928: 524. *P. schalleriana*; Ford, 1949: 59]

DESCRIPTION

♂ ♀ 15–19 mm. Sexual dimorphism not pronounced. Polymorphic; four forms are recognized by Sheldon (1930: 195).

Male (fig. 13). Forewing ground colour whitish grey, sparsely sprinkled with violet-brown scales, especially along termen, sometimes mixed with ochreous in distal area, a raised ochreous mixed with violet-brown discocellular scale-tuft near apex of costal blotch and contiguous with its inner margin, a small violet-brown mixed with white scale-tuft dorsad of submedian fold at about one-third; markings reduced, plumbeous with violet-brown and black admixture; basal and sub-basal fasciae developed on costa, outer margin of sub-basal indicated by sparse violet-brown scales; costal blotch well developed, triangular, extending to near apex; an incomplete line of raised scales along inner margin of costal blotch; cilia grey, apices paler, with a dark grey basal line. Hindwing grey, infuscate distally; cilia paler, with a dark sub-basal line.

Female. Similar to male.

Variation. In the nominate form the forewing ground colour varies from whitish grey to cream-white. A form (fig. 14, f. *falsana* Hübner) occurs with the costal blotch fractured, the sub-apical spot sometimes being completely separated, the whitish ground colour showing through. A further form (fig. 15, f. *plumbosana* Haworth) occurs with rather distinct yellowish brown or light reddish brown ground colour and a moderately strong costal blotch; in f. *castaneana* Haworth the blotch is obscure and the wing is almost unicolorous reddish brown. In some populations, especially in the Burren, Co. Clare, the ground colour is predominantly greyish brown or distinctly grey, with the markings variably developed but never strong and well defined (fig. 16).

COMMENTS

The moderately rough-scaled anterior margin of the forewing and the enlarged costal blotch, which extends to near the apex, distinguish this species from *A. comariana* and *A. latifasciana*.

BIOLOGY

Larva. Head brownish yellow or brownish green; prothoracic plate similar to head, or shining green, marked with blackish brown postero-laterally; abdomen light green or yellowish green, sometimes tinged with grey, darker dorsally; pinacula concolorous with integument; anal plate green, marked with brown; anal comb yellowish; thoracic legs light greenish brown. June to August, feeding on *Viburnum lantana* and *V. opulus*, twisting the leaves and living in a small pocket within.

Pupa. Pale brown. August and September, in the larval habitation (pl. 20, fig. 1).

Imago. August to October, overwintering until the following May. Frequenting scrub, woods and hedges on calcareous soils where *V. lantana* grows, and on damp soils and boggy areas where *V. opulus* occurs. By day the moth rests amongst the foodplants and other bushes and is readily disturbed by beating, flying quickly to another hiding place. It is more active in the spring after hibernation.

DISTRIBUTION

Locally common in the south of England, especially on the North Downs and the South Downs, also at Yarmouth (Isle of Wight) and in the New Forest (Hampshire); occurring north to Yorkshire but apparently not found in Cheshire and very rare in north Lancashire, being absent in the south of that county. In Wales it is found locally in the south but is apparently absent in the north. In Scotland the species has been recorded from Argyll and Roxburghshire, but these records are unconfirmed. In Ireland there are records from north Kerry (Caternane) and Co. Antrim (Belfast) (Beirne, 1941: 86), and it has been taken plentifully in the Burren, Co. Clare, where larvae were found feeding on *V. opulus* in July (Bradley & Pelham-Clinton, 1967: 135).

Holarctic in distribution, occurring throughout Europe to central Asia except southern Italy and Greece; also in North America where it is represented by a distinct subspecies (Obraztsov, 1963: 233).

Acleris variegana (Denis & Schiffermüller)

Pl. 42, figs. 1–5

[*Peronea variegana*; Barrett, 1905: 232; Pierce & Metcalfe, 1922: 23, pl. 9 (♂, ♀ genitalia); Meyrick, 1928: 524; Ford, 1949: 59]

DESCRIPTION

♂ ♀ 14–18 mm. Sexual dimorphism not pronounced. Polymorphic; Sheldon (1931: 2) recognizes eight forms in the British Isles.

Male (fig. 1). Forewing ground colour white, basal half variably strigulate with grey, a concentration of shining grey or plumbeous striae on dorsum before middle, distal half overlaid with dark ochreous-brown except for scattered white scales and several small groups of raised scales in apical and tornal areas; markings poorly defined, basal fascia indicated by grey suffusion on costa; sub-basal fascia represented on dorsum by an outward-oblique black mixed with ferruginous patch containing a large erect scale-tuft in the submedian fold; median fascia varying from ochreous-brown to ferruginous-brown with plumbeous admixture and scattered black mixed with white raised scales forming small tufts along the inner margin; pre-apical spot obscure, similarly coloured, containing plumbeous admixture; cilia ochreous-brown, with a dark sub-basal line. Hindwing grey, paler basally; cilia concolorous, with a dark sub-basal line.

Female. Similar to male.

Variation. The variation within the recurring major forms of this species is considerable, but mostly the forms are distinct and do not intergrade. Although a common and widespread species the genetics of the various forms have apparently not yet been studied. Consequently there is some uncertainty concerning the validity of the infrasubspecific nomenclature. Sheldon (1931: 2) distinguishes eight forms which are defined below, but both Obraztsov (1956: 135) and Razowski (1966: 366) recognize only six forms and differ from Sheldon in the application of the names.

The nominate form is common throughout the range of the species; almost as common is the distinctive form (fig. 2, f. *asperana* Fabricius) in which the forewing is sharply demarcated at the middle, the whole of the basal half being white except for a fuscous suffusion at the base of the costal margin and an indication of the sub-basal fascia on the dorsum, usually in the form of two short vertical black bars, the large scale-tuft below the submedian fold being white, while in contrast the distal half of the wing is blackish fuscous intermixed with ferruginous and shining plumbeous scaling. A very similar form (f. *nyctemerana* Hübner) differs only in having the basal half of the forewing distinctly tinged with pale ochreous. Another form (fig. 3, f. *cirrana* Curtis), less common but usually found wherever the species occurs, has the forewing ground colour light ochreous, the sub-basal fascia developed on the dorsum, and the median fascia obsolescent dorsally but well developed on the costa and confluent with the pre-apical spot, forming a dark fuscous or blackish fuscous triangular costal blotch. An apparently uncommon form (f. *brunneana* Sheldon) resembles the nominate form except that the forewing ground colour is ochreous-brown; this form of *variegana* is often found with *A. permutana*, which it resembles, and was commonly known to early collectors as the "*permutana* form". An uncommon form (fig. 4, f. *argentana* Sheldon), apparently most frequent in the north of England, has the markings obsolescent except for the fuscous costal blotch, and the white ground colour is sprinkled with blackish scales, especially in the distal area. An extreme and very rare form (f. *albana* Humphreys & Westwood) has the forewing almost unicolorous white with only sparse fuscous scaling in the apical area, especially along the costa and termen. A melanic form (fig. 5, f. *caeruleoatrana* Strand) is not uncommon and has the forewing blackish plumbeous with somewhat darker irregular transverse marbling.

COMMENTS

The form f. *brunneana* resembles the closely related species *permutana* in general appearance but may be distinguished by the erect scale-tuft in the submedian fold.

BIOLOGY

Ovum. 100–150 eggs are usually deposited singly or in small batches on the leaves of the foodplant, usually along the mid-rib on either surface.

Larva. Head and prothoracic plate yellowish or greenish brown, head with blackish brown ocellar and postero-lateral markings; abdomen yellowish green or light green, pinacula concolorous or paler; anal plate green; thoracic legs yellowish brown. The larva may be found from May to early July, feeding on Rosaceae, especially species of *Rosa*, including garden varieties, and *Rubus*, *Sanguisorba minor*, *Crataegus*, *Prunus*, especially *P. spinosa*, *Malus* and *Pyrus*; living in the folded edge of a leaf or between loosely spun leaves, at first skeletonizing and in later instars eating the leaves from the edges. Also recorded on various other plants, including *Ulmus*, *Vaccinium*, *Corylus* and *Berberis*.

Pupa. Light brown. July, in the larval habitation or spun up amongst dead leaves on the ground.

Imago. Apparently univoltine in the British Isles, with an emergence period from July to September; often common in gardens and orchards and along roadside hedges, borders of woods and rough downland. The moth rests amongst the foliage during the day and is easily disturbed by beating; at dusk both sexes are on the wing and later come to light and occasionally to sugar. In North America this species has adapted to a double-brooded life cycle, having been introduced from Europe (Powell, 1964: 78).

DISTRIBUTION

Common and widely distributed, occurring in suitable localities throughout the British

Isles as far north as the Orkneys; generally common in the south of Scotland, becoming less frequent northwards. In Scotland the larva feeds on *Rosa pimpinellifolia* on the Sutherland machair, and is also found on this plant in the limestone area of the Burren, Co.'s Clare and Galway, in western Ireland.

Europe to Asia Minor, central Asia, China and Japan; north-west Africa; North America.

Acleris permutana (Duponchel)
Pl. 42, fig. 6
[*Peronea permutana*; Barrett, 1905: 231; Pierce & Metcalfe, 1922: 21, pl. 8 (♂, ♀ genitalia); Meyrick, 1928: 524; Ford, 1949: 59]

DESCRIPTION
♂ ♀ 15–20 mm. Sexual dimorphism not pronounced.

Male (fig. 6). Forewing ground colour whitish ochreous or cream, basal half weakly strigulate with yellowish ochreous and sprinkled with black intermixed with grey scales, forming small tufts; a strong suffusion of shining plumbeous on dorsum before middle, extending across sub-basal fascia, distal half of wing suffused with reddish ochreous, weakly strigulate; markings poorly developed, deep ferruginous-brown; basal and sub-basal fasciae weakly indicated on costa, sub-basal fascia represented on dorsum by an outward-oblique marking extending to middle of wing, overlaid with plumbeous dorsally, with several black striae along dorsal margin, at its apex a heavy circular patch of black scales straddling the submedian fold and from the centre of which a prominent tuft of semi-erect ferruginous-brown scales arises; inner margin of median fascia well defined, roughened with small scale-tufts, often mixed with black along inner margin and on dorsum; pre-apical spot diffuse, mixed with roughened black scales, produced towards lower half of termen as a narrow band; cilia ferruginous-brown, grey around tornus, with a darker sub-basal line. Hindwing whitish grey, suffused with brown and weakly strigulate distally; cilia concolorous, with a grey sub-basal line.

Female. Similar to male.

Variation. One of the most constant species of the genus, distinct forms apparently being unknown.

COMMENTS
Distinguished from *A. variegana* by the comparatively rich pale ochreous and reddish ochreous coloration, the generally less variegated wing pattern in which the costal blotch is not developed as in some forms of *variegana*, and the large recumbent scale-tuft in the submedian fold, which in *variegana* is erect.

BIOLOGY
Ovum. The eggs are deposited in small batches on the leaves and stems of the foodplant in August and September (Gregson, 1880a: 45).

Larva. Head light brown or yellowish brown, ocellar region and postero-lateral margin marked with blackish brown; prothoracic plate yellow-brown, with a large and a small dark brown spot on each side; abdomen yellow or yellowish green; thoracic pinacula and legs brown; abdominal pinacula concolorous with integument. The larva may be found from June to early August, feeding in spun leaves and shoots of *Rosa pimpinellifolia*. On the Continent it is known to feed on other species of *Rosa*, and has also been recorded on *Prunus spinosa* (Swatschek, 1958: 76).

Pupa. Dark brown. August, in a white silken cocoon spun amongst the leaves of the foodplant or dead leaves below.

Imago. August to September, apparently confined to sandhills and other coastal habitats where burnet rose occurs. The moth is very retiring and during the day is best flushed from rest amongst its foodplant by means of a bee-smoker. Both sexes fly at dusk and later come to light.

DISTRIBUTION

Local and scarce, the discontinuous distribution in the British Isles indicating that this species is possibly overlooked in some coastal areas where burnet rose is plentiful. In England it is known from Dungeness (Kent) (Wakely, 1966: 101), Beachy Head, Falmer and Patcham (Sussex) (South, 1897: 220), east Cornwall, near Derby (Derbyshire) (Hulme, 1962: 83), and Wallasey and New Brighton sandhills (Cheshire), where it was first discovered in Britain in 1848 (Mansbridge, 1905: 115) and where it can still be taken. Its only inland locality was Barnes Common (Surrey) (Gregson, 1880a: 46) and, although Tutt (1897: 209) considered the records from this locality to be doubtful, there are specimens with authentic data (J. A. Clark collection) in the British Museum (Natural History). In Wales it was first taken in 1876 at Penmaenbach (Caernarvonshire) by Gregson (1880a: 46), and has since been found on the Hiraethog Mountains (Denbighshire) and at Rhuddlan (Flintshire) (Michaelis, 1954a: 65), and in South Wales at Porthcawl (Glamorgan). In Ireland it has been recorded from the Dublin and Galway coasts (Beirne, 1941: 86); more recently it has been taken in the Burren, Co. Clare (Bradley & Pelham-Clinton, 1967: 136).

West, central and southern Europe to Asia Minor and central Asia.

Acleris boscana (Fabricius)

Pl. 42, figs. 7–9

[*Leptogramma boscana*; Barrett, 1905: 217. *Peronea boscana*; Pierce & Metcalfe, 1922: 21. *P. scabrana*; sensu Pierce & Metcalfe, 1922: 21 pl. 8 (♂, ♀ genitalia). *P. boscana*; Meyrick, 1928: 527; Ford, 1949: 60]

DESCRIPTION

♂♀ 15–18 mm. Sexual dimorphism not pronounced. Polymorphic; Sheldon (1931: 124) recognizes two forms in the British Isles.

Male (fig. 7). Forewing ground colour cream-white, a few scattered pale ochreous dots or striae most evident in distal area, these often roughened with black scales; markings reduced, plumbeous-brown variably mixed and edged with black; basal and sub-basal fasciae weakly indicated on costa by striae, outer margin of sub-basal indicated by small black scale-tufts, one above and one below submedian fold; median fascia narrow, well developed but excised on costa, extending to middle of wing, obsolescent dorsally, inner margin indicated by raised pale ochreous intermixed with black scales; pre-apical spot lunate, weakly excised, usually somewhat darker than median fascia, the two often partially confluent and forming a tripunctate costal blotch; cilia concolorous with ground colour. Hindwing light grey, infuscate distally; cilia cream-white, with darker sub-basal line.

Female. Similar to male.

Variation. The nominate form is represented by the first or summer generation and shows only minor individual variation. The second or autumn generation represents a distinct form (fig. 8, f. *ulmana* Duponchel) in which the forewing appears farinaceous, the ground colour varying from grey to silver-grey (fig. 9), the surface roughened with raised concolorous scales often intermixed with fulvous and black; the median fascia and pre-apical spot vary from greyish fuscous mixed with fuscous in dark specimens to deep purplish brown in light specimens, usually with a distinct cluster of raised black scales in the discocellular area, these sometimes forming an elongate black streak.

COMMENTS

Specimens of the autumn generation of this species resemble *A. logiana*, but are usually greyer in general coloration. However, in the British Isles these two species are apparently allopatric and their distributions do not overlap, that of *boscana* not extending as far north as Scotland where *logiana* occurs.

BIOLOGY

Ovum. Greenish in colour, deposited in small batches on the leaves of the foodplant.

Larva. Head and prothoracic plate dark brown or black, plate sometimes darker pigmented laterally; abdomen green, pinacula paler; anal plate dark brown or black; thoracic legs dark brown or black. May and June, those producing the second generation of moths in August and early September. Feeding on *Ulmus*, living at first in a folded or curled edge of a leaf, later between two leaves spun one above the other (pl. 20, fig. 2), gnawing the upper and lower surfaces from within; in the last instar the larva occasionally comes out to feed on the edges of the leaves (West, 1877: 304).

Pupa. Light reddish brown. June and early July, and again in September; in a silken cocoon between two spun leaves, occasionally in a folded leaf.

Imago. Bivoltine; the first generation emerges at the end of June and in July, the second in September and October, overwintering until the following March. During the day the moth rests amongst the foliage of elm, and in warm weather may be disturbed by beating; those of the summer generation fly actively when dislodged, while those of the autumn generation are lethargic and flutter or drop to the ground. During the winter, the moths of the second generation rest on elm twigs with the wings wrapped around the bark and in this situation are protected by their cryptic grey coloration (West, 1906: 232).

DISTRIBUTION

Locally common amongst elms in the southern counties of England, occurring northwards to Yorkshire but apparently rare. A record from North Wales (Caernarvonshire) is unconfirmed (Michaelis, 1954a: 66); the species is apparently unknown in Scotland and Ireland.

Europe to east Siberia.

Acleris logiana (Clerck)

Pl. 42, figs. 10–12

[*Leptogramma niveana*; Barrett, 1905: 219. *Peronea niveana*; Pierce & Metcalfe, 1922: 20, pl. 8 (♂, ♀ genitalia); Meyrick, 1928: 527. *P. logiana*; Ford, 1949: 60]

DESCRIPTION

♂ ♀ 18–22 mm. Sexual dimorphism not pronounced. Polymorphic; Sheldon (1931: 101) recognizes three forms in the British Isles.

Male (fig. 10). Forewing moderately rough-scaled, ground colour white suffused and weakly strigulate with grey, strigulae often irrorate with black; markings obsolescent, greyish fuscous, reduced mainly to small patches and strigulae on costa; median fascia sometimes indicated by weak pale ochreous suffusion at middle and on dorsum, its inner margin demarcated by an almost continuous ridge of raised scales; an elongate fuscous-black dash in discal area; pre-apical spot indicated by grey mixed with ochreous suffusion; termen edged with fuscous-black; cilia concolorous with ground colour, shaded with pale ochreous basally. Hindwing light grey, infuscate distally, with darker strigulation apically and along termen; cilia whitish grey, with a grey sub-basal line along termen.

Female. Similar to male.

Variation. The three forms are well differentiated but show considerable minor variation. In the common nominate form the greyish suffusion and strigulation of the forewing ground colour varies in extent but is never heavy; in some specimens the median fascia may be more evident as a greyish suffusion in the discal area; more variable is the development of the discocellular spot which ranges from a conspicuous sharply defined linear dash to an indistinct fuscous suffusion.

The two named forms occur not infrequently, the more common (fig. 11, f. *tripunctana* Sheldon) has a distinct tripunctate fuscous-black costal blotch; the second, a less common form (fig. 12, f. *scotana* Stephens), is essentially similar to the nominate form in general coloration and markings but has diffuse longitudinal orange-brown or fulvous streaks or patches; these orange-brown or fulvous markings are variable and may be reduced to a sub-basal streak or patch on the dorsum.

COMMENTS

The nominate form and f. *tripunctana* of this species resemble specimens of the autumn generation of *A. boscana*, but the two species are separated by their allopatric distribution in the British Isles.

BIOLOGY

Ovum. Bright red; April, deposited singly or in small batches on the twigs of the foodplant.

Larva. Head dark brown or reddish brown; prothoracic plate brown, black laterally; abdomen light green or whitish green, integument weakly shagreened, pinacula concolorous; anal comb with six prongs; thoracic legs blackish brown. June and July, feeding in spun leaves of *Betula*.

Pupa. Light brown. Late July and August, between spun leaves of the foodplant.

Imago. September to April; frequenting birch woods. In September and October, and later if the weather is mild, the moth may be found at rest during the day on the branches or trunks of birch trees; in the spring, after hibernation, it is more often to be found at rest on the trunks.

DISTRIBUTION

In the British Isles this species is known only from Scotland, where it occurs in old birch woods in the Highlands northwards from Perthshire to east Sutherland. Known localities include Aberdeenshire, Inverness-shire (Aviemore), Morayshire (Forres), Perthshire (Black Wood, Glen Fender, Glen Tilt, Moncreiffe Hill) and Sutherland (Invershin). Apart from a record from East Lothian (Balfour, 1930: 179) the species is apparently unknown from southern Scotland.

North and central Europe to east Siberia and Japan; North America.

Acleris umbrana (Hübner)

Pl. 42, figs. 13, 14

[*Peronea umbrana*; Barrett, 1905: 223; Pierce & Metcalfe, 1922: 20, pl. 8 (♂, ♀ genitalia); Meyrick, 1928: 525; Ford, 1949: 59]

DESCRIPTION

♂♀ 18–22 mm. Sexual dimorphism not pronounced; forewing of male somewhat more pointed, costa less strongly arched at base than in female. Polymorphic; Sheldon (1930: 223) recognizes three forms in the British Isles.

Male (fig. 13). Forewing ground colour fuscous-brown with a weak chestnut-brown admixture, suffused with violaceous-plumbeous; fasciae obsolescent, wing traversed longitudinally by a velvet-black streak extending from base to apex, partially edged with chestnut-brown, obliquely interrupted at middle by a prominent ridge-like tuft of pale ochreous scales, apical half of streak narrower, often indented or constricted in places with violet-plumbeous scales, bordered by several small tufts of dark fuscous scales; a large conspicuous scale-tuft dorsad of submedian fold, a small tuft above dorsum towards tornus, and a smaller pre-tornal tuft; termen thinly edged with chestnut-brown; cilia concolorous with ground colour, grey around tornus. Hindwing light grey, weakly strigulate with fuscous; cilia concolorous.

Female. Similar to male.

Variation. The three forms of this species show considerable variation and tend to intergrade; in the nominate form the forewing ground colour varies from fuscous-brown to ochreous-brown, the longitudinal black streak may be broader throughout its length and have the margins diffuse and with heavier purplish suffusion, or narrower and more sharply demarcated, sometimes with the apical half foreshortened and not reaching the apex; in specimens with the streak narrower the two small tufts in the tornal area are usually darker and may be black. Rarely the black streak is obsolescent and variants without any strong indication of it are referable to f. *brunneana* Sheldon. More frequently the black streak is multifractured into thin radiating lines spreading across the wing, the ground colour being pale ochreous and the plumbeous suffusion often less evident (fig. 14, f. *lamprana* Sheldon).

COMMENTS

This species is similar to certain forms of *A. hastiana*; it may be distinguished by differences in the shape of the costa of the forewing, which in both sexes of *hastiana* appears shallowly concave beyond the middle; also, the two small but well-developed scale-tufts present in the tornal area of the forewing in *umbrana* are not found in *hastiana*.

BIOLOGY

Larva. Head and prothoracic plate black; abdomen light green; thoracic legs black. June and July, feeding on *Sorbus*, *Prunus padus*, *Thelycrania sanguinea*, *Crataegus*, *Salix*, *Carpinus* and *Alnus*.

Imago. August to April, hibernating during the winter; frequenting woodland, resting rather high up amongst the foliage of trees during the day and rarely seen, or amongst blackthorn, from which it may occasionally be beaten out. The moth flies at night; it is attracted to ivy bloom in the autumn and to sallow blossom in the spring.

DISTRIBUTION

A local species, apparently never very common and sometimes not observed in its known haunts for periods of several years. Recorded only in England, occurring in forest and woodland localities in Essex, Sussex, Hampshire (New Forest, Selborne), Dorset (Bloxworth, Somerset border (Hayward, 1929a: 19)), Somerset (Hazelbury Plucknett, Misterton (Hayward, 1927: 44)), Oxfordshire (Nettlebed) and Northamptonshire (Whittlebury Forest) northwards to Westmorland and Northumberland. Until the turn of the century this species was plentiful in the New Forest and also in Whittlebury Forest (Barrett, 1905: 224). Hodgkinson (1869: 224) recorded it from Westmorland (Witherslack) and according to H. N. Michaelis it occurred in Lancashire (Windermere) prior to 1890.

North, central and eastern Europe.

Acleris hastiana (Linnaeus)
Pl. 43, figs. 1–22
[*Peronea hastiana*; Barrett, 1905: 224. *Eclectis hastiana*; Pierce & Metcalfe, 1922: 18, pl. 7 (♀ genitalia).

Peronea maccana; sensu Pierce & Metcalfe, 1922: 21, pl. 8 (♂ genitalia). *P. hastiana*; Meyrick, 1928: 526; Ford, 1949: 60]

DESCRIPTION

♂♀ 17–22 mm. Sexual dimorphism not pronounced. Polymorphic; Sheldon (1923: 76; 1930: 148) recognizes about eighty forms found in the British Isles.

Male (fig. 1). Forewing with costa strongly curved basally, straight from before middle to apex, coarsely fringed with scales which are slightly expanded on the curvature to near middle, causing apical half of costa sometimes to appear shallowly concave; ground colour whitish grey sparsely irrorate with purplish brown; markings purplish brown, suffused with plumbeous and plumbeous-black especially at base and along costal and dorsal margins; basal and sub-basal fasciae moderately well defined, confluent towards costa; inner margin of median fascia well defined, indicated by roughened scales, a moderate scale-tuft containing an admixture of pale ochreous near middle, fascia extended distad across most of wing except for small areas of whitish ground colour; pre-apical spot not differentiated; veins shaded with deep ferruginous-brown distally; a small blackish scale-tuft in sub-basal fascia immediately below submedian fold, a similar small tuft a little above fold and one or two (not always apparent) minute tufts in fold before white tornal area; cilia grey, sometimes mixed with ferruginous, with a whitish grey sub-basal line. Hindwing brownish grey, infuscate and often strigulate distally; cilia grey, paler basally.

Female. Similar to male.

Variation. *A. hastiana* and the following species, *A. cristana*, are the most polymorphic of the British species of Tortricidae, each with a large number of widely differing forms. In *hastiana* many intermediate forms occur and intergrade to such an extent that it is sometimes hardly possible to separate them; Sheldon (1923: 100) stressed this difficulty in his monograph on this species.

The forewing pattern consists of two basic phenotypes, one with typical transverse tortricoid markings, the other with longitudinal markings. The transverse markings are variously represented by the basal, sub-basal and median fasciae, and the pre-apical spot. Forms with longitudinal markings may have a broad median streak from the middle of the base to the apex, often emphasized by lighter coloration in the costal part of the wing, or a broad streak along the dorsum, or dark radiating striations. A number of recurrent forms also occur which have a combination of the transverse and longitudinal markings, the latter being superimposed on the former which are usually reduced or obsolescent. Although Sheldon (1923; 1930) made a comprehensive study of the variation of *hastiana*, and Obraztsov (1956: 137) tabulated over 90 separately recognized named forms of this species, apparently no genetical studies have been recorded and any arrangement of the forms is therefore arbitrary.

As the range of variation in *hastiana* is too extensive to be comprehensively covered in the present work, a selection of forms is illustrated to show as wide a range as possible. Of the group with transverse markings the following forms are illustrated: f. *subfasciana* Sheldon (fig. 2), f. *rufifasciana* Sheldon (fig. 3), f. *griseana* Sheldon (fig. 4), f. *plumbeana* Sheldon (fig. 5). These forms show a gradual transition to those in which the markings are obsolescent as in f. *albisparsana* Sheldon (fig. 6) and f. *flavicapitana* Sheldon (fig. 7); in the latter the scale-tuft in the median fascia is distinctive.

A number of forms occur in which the transverse markings are obsolescent towards the dorsum but are well developed costally, the median fascia and pre-apical spot forming a distinctive subtrapezoidal (fig. 8, f. *albimaculana* Sheldon) or triangular costal blotch (fig. 9, f. *autumnana* Hübner).

In the group with longitudinal markings, the colour of the median streak varies from ferruginous (fig. 10, f. *plumbeostriana* Sheldon) or pale yellow (fig. 11, f. *ramostriana* Stephens) to pure white or black, or a combination of pale yellow and white (fig. 12, f. *mayrana* Hübner). The variously coloured streaks of these forms may be combined with different ground colours varying from white to grey and pale yellow to ochreous-brown; a form occurs with whitish ground colour and a black median streak (f. *sagittana* Sheldon). Occasionally the costal half of the wing is paler, especially that adjacent to the median streak (fig. 13, f. *divisana* Hübner). In some forms the streak is overlaid with darker coloration (fig. 14, f. *leucophaeana* Humphreys & Westwood) varying from deep reddish brown to black.

In forms with a pronounced dorsal streak the wing usually appears purplish brown or black, the streak varying from ochreous tinged with ferruginous (fig. 15, f. *scoticana* Sheldon) to white (f. *griseovittana* Sheldon). One of the most prominent of the striated forms is f. *radiana* Hübner (fig. 16).

An almost infinite number of combinations of the forms with longitudinal and transverse markings are found and include f. *combustana* Hübner (fig. 17), f. *sheldonana* Metcalfe (fig. 18), f. *brunneostriana* Sheldon (fig. 19) and f. *mixtana* Sheldon (fig. 20).

Two unicolorous forms are illustrated, f. *aquilana* Hübner (fig. 21) and f. *albana* Sheldon (fig. 22).

COMMENTS

A number of forms of this highly polymorphic species resemble certain forms of *cristana*, but *hastiana* can be distinguished by the relatively straight costa of the forewing and the presence of numerous small scale-tufts. Forms resembling *A. abietana* and *A. umbrana* may be differentiated by referring to the comments on those species.

BIOLOGY

Ovum. Whitish green, changing later to deep reddish brown. In early spring the eggs are deposited singly or in batches of 2 or 3 in crevices in the bark of twigs and branches of the foodplant (Sheldon, 1923: 79); in the summer the eggs are deposited singly on the terminal shoots (Huggins, 1924: 282). An account of the life history of *hastiana* in Britain is given by Sheldon (1923: 75).

Larva. Head brownish green or light green, ocellar region and postero-lateral margin marked with blackish brown; prothoracic plate brownish green or light green; abdomen light green, pinacula concolorous; in earlier instars the head and plate are black and the abdomen yellowish green or bluish green. Late April to early June, those producing a second generation of moths from July to September and October, feeding on various species of *Salix*, showing a preference for small-leaved varieties such as *Salix repens*;

not apparently on *S. caprea*. In west Galway the main foodplant is *S. aurita*, but it also occurs on *S. repens* and *S. fragilis*. The larva lives between spun leaves, drawing together the leaves or folding them, or spinning together the leaves of a young shoot, often fastening them to the stem (pl. 21, fig. 1). Michaelis (1958: 123) records *Populus alba* as a foodplant in Lancashire, and Emmet (1971: 9) found a larva feeding on *Prunus* in west Galway. In most of Scotland and northern Ireland, where this species is univoltine, the larvae feed in July and early August.

Pupa. Brown or blackish brown. June and July, and again in September and October; in Scotland and northern Ireland, the pupa is found from the end of July to early September. Concealed in a flimsy silken cocoon, usually in spun leaves or a folded leaf away from the larval habitation, or amongst debris on the ground. In wet areas pupation occurs more frequently in the larval habitation.

Imago. Bivoltine in England and Wales and northwards to Wigtownshire in south-west Scotland, and in west and south-west Ireland. The first generation emerges in June, the second at the end of August, somewhat later in Ireland, remaining on the wing until the end of September or early October before hibernating until the early spring. North of Argyll, over the greater part of Scotland, this species is univoltine and the larva is full-grown in early August, the moth appearing from the beginning of September; imagines of the univoltine populations are usually larger than those of the bivoltine populations. The Scottish univoltine populations extend at least as far south as Berwickshire in the east. In the north-west the univoltine population is abundant at sea-level, although in Argyll it is rare at that elevation despite the abundance of suitable sallows but becomes increasingly common up to at least 900 ft above sea-level. There are very few records from south Argyll and the Clyde area and it is not known whether the species there is univoltine or bivoltine. In Ireland the species is known to be bivoltine in Co.'s Cork, Kerry and Galway, but univoltine in Co. Tyrone and possibly Co. Mayo. The available evidence therefore suggests that it is bivoltine in the south and south-west and univoltine in the north.

The moth is rarely seen on the wing during the day, resting low down at the base of sallow bushes, where it may be found if the surrounding herbage is pulled back. Occasionally moths may be netted flying over sallow bushes at dusk, but the main flight period is apparently after midnight, when they are occasionally taken at light.

DISTRIBUTION

Widely distributed throughout the British Isles north to the Orkneys, but apparently unknown from St. Kilda and the Shetlands. In southern England it is extremely local, but occurs commonly in certain localities in Kent (Sandwich, Faversham, Sittingbourne, Dungeness), and coastal areas of the Isle of Wight, Devon and Cornwall. The species is more widespread in the eastern counties and is especially abundant in the fens of East Anglia. Noted localities in Wales and north-west England are the coastal sandhills in Glamorgan (Porthcawl) and from Cheshire (Wallasey) northwards to Cumberland. In Scotland it is more plentiful in the west and north-west, being especially common in coastal areas and outlying islands. Locally common throughout Ireland.

Europe to east Siberia and Japan; North Africa; North America.

Acleris cristana (Denis & Schiffermüller)

Pl. 44, figs. 1–17

[*Peronea cristana*; Barrett, 1905: 220; Pierce & Metcalfe, 1922: 20, pl. 8 (♂, ♀ genitalia); Meyrick, 1928: 526; Ford, 1949: 60]

DESCRIPTION

♂♀ 18–22 mm. Sexual dimorphism not pronounced. Polymorphic; Sheldon (1917: 245) discusses the variation in this species. More than 100 forms are known from the British Isles and are enumerated by W. B. L. Manley (1973) in a monographic study.

Male (fig. 1). Forewing with costa shallowly emarginate beyond middle, apex slightly retuse, ground colour dark brown, usually darker basally, sometimes with weak strigulation; transverse markings obsolete, sometimes a weak indication of a large dark triangular costal blotch; a conspicuous longitudinal white streak along dorsum, the upper edge contiguous with submedian fold from near base to tornus, deeply emarginate at base; a large conspicuous erect wedge-like tuft or "button" of coarse scales near middle of wing before end of discal cell, contiguous with a diffuse black or black-brown dash reaching to end of cell; scattered comparatively minute tufts of brown or black mixed with white scales, several in tornal area, two more or less evenly spaced between end of cell and apex, one near base above submedian fold and one pure white tuft immediately below fold in angle formed by emargination of white dorsal streak; cilia more or less concolorous with wing, with a dark grey sub-basal line finely edged basally with pale ochreous and finely irrorate outwardly with whitish. Hindwing greyish fuscous, paler basally; cilia grey, with a dark sub-basal line.

Female. Similar to male.

Variation. The comments on the range of variation in *A. hastiana* also apply in principle to *cristana*. In the latter species the forewing ground colour varies from white to ochreous, ochreous-brown, dark brown, purplish brown, and black. The usually prominent discocellular scale-tuft, characteristic of this species, varies in colour from white or cream to brown, orange, and black; in some forms the tuft may be obsolescent and reduced to a few scales.

In *cristana* the typical transverse tortricoid markings are usually obsolete but when present are very weak and represented by the basal and sub-basal fasciae and the costal blotch. Forms with longitudinal markings, which are represented by subcostal, median, and subdorsal streaks, are more frequent. The dorsal streak varies from white to cream, yellow and various shades of orange; in some forms the streak is fractured and incomplete. Less frequently a somewhat diffuse longitudinal subcostal streak is present, reaching from base to costa before apex, varying in colour from orange-brown to dark chestnut and purplish black. The costal area of the wing may be concolorous either with this streak or with the ground colour. A short median streak, reaching from the base to end of cell and confluent with the subcostal streak, is present in many forms, and varies from yellowish orange to reddish orange and deep chestnut. Almost unicolorous light brown forms without markings also occur.

The numerous recurring forms of *cristana* are illustrated in the monograph by

W. B. L. Manley (1973); in the present work a limited number of forms only has been selected to show the range of variation.

A not uncommon form occurs in which the forewing ground colour is similar to that in the nominate form but the dorsal streak is yellow and the scale-tufts are cream (fig. 2, f. *ochreapunctana* Clark). Less common is a form with the ground colour charcoal, the dorsal streak white and the scale-tufts whitish grey (fig. 3, f. *albonigrana* Clark). Another not uncommon form, also with a white dorsal streak and scale-tufts, has typical dark brown ground colour but a chestnut-brown subcostal streak concolorous with the costal area (fig. 4, f. *chantana* Curtis). A very common form (fig. 5, f. *spadiceana* Haworth) with a similar but lighter costal area has the white dorsal streak fractured and incomplete, the ground colour purplish brown and the scale-tufts blackish.

Two forms with a conspicuous orange median streak are illustrated; in one the white dorsal streak is present (fig. 6, f. *albovittana* Stephens), but in the other the dorsal streak is not developed (fig. 7, f. *desfontainana* Fabricius). A form in which the median streak is chestnut-brown and almost concolorous with the dark subcostal streak has the conspicuous white dorsal streak but the scale-tufts are vestigial (fig. 8, f. *alboruficostana* Clark); a similar form occurs but lacks the dorsal streak, while the median and subcostal streaks are concolorous (fig. 9).

Two common almost unicolorous forms are illustrated (fig. 10, f. *merlana* Clark; fig. 11, f. *profanana* Fabricius).

Four not uncommon forms which show combinations of the dorsal streak and transverse markings are f. *subcapucina* Desvignes (fig. 12); f. *cristalana* Donovan (fig. 13); f. *subfulvovittana* Clark (fig. 14); f. *ustulana* Sheldon (fig. 15). Except for the absence of the dorsal streak, f. *semiustana* Curtis (fig. 16) resembles f. *ustulana*. The last form illustrated has the forewing ground colour blue-black with reduced black transverse markings and a contrasting bright orange dorsal streak (fig. 17, f. *atrana* Clark).

COMMENTS

The very large coarse-scaled tuft situated towards the end of the discal cell on the forewing is peculiar to *cristana* and is present in most forms, readily distinguishing this species from others in the genus *Acleris*.

BIOLOGY

Ovum. Light grey and opalescent when first laid, later changing to reddish brown and similar in colour to the bark of a twig; the eggs are deposited singly or in batches of 3 or 4 on the twigs of the foodplant during the latter half of March and early April following pairing in early spring (Sheldon, 1917: 220; Whittingham, 1931: 238; 1934: 90).

Larva. Head yellowish brown, epistomal region darker; prothorax greenish brown, anterior and posterior margins with darker shading, pinacula black; abdomen grey-green, paler ventrally, a weak narrow dark green dorsal line, pinacula distinct, lighter than integument; spiracles very inconspicuous; anal plate grey-green, with several minute black dots distributed over its surface (Sheldon, 1917: 221). End of April to early July, feeding on *Prunus spinosa*, *Crataegus*, *Malus* and *Pyrus aria*, living in spun leaves and occasionally on the flowers and fruit. Sheldon (1917: 246) found that in captivity the larva would also eat cultivated *Prunus* and *Fagus*, *Quercus*, *Carpinus*, *Betula*, *Ulmus*, *Tilia* and *Corylus*.

Pupa. Brown, wings darker. July and August, in a flimsy cocoon spun in a folded leaf or amongst debris on the ground.

Imago. August to November, then hibernating, concealing itself in old yew trees and similar dense cover, and on the wing again from March until May. The moth frequents forests and wooded areas and is most often to be found in dense thickets of blackthorn, hawthorn and other trees and shrubs, resting with the wings partly wrapped round the twig. During the day the moths are best obtained by penetrating the dense thickets and beating, the moths dropping to the ground when dislodged. Sheldon (1917: 248) describes in detail the collecting of this species and its behaviour. The normal flight period apparently begins at dusk and the moth occasionally comes to light.

DISTRIBUTION

Occurring in wooded areas in the southern counties of England north to Herefordshire, Worcestershire and Norfolk, but local and generally uncommon except in favoured localities such as the New Forest (Hampshire) and west Surrey. Barrett (1905: 223) includes Yorkshire in the distribution and refers to a single record from Scotland (Perthshire) by F. B. White. Apparently unknown in Wales, and rare in Ireland, occurring in Co.'s Kerry, Cork and Antrim (Beirne, 1941: 87). This species is subject to considerable fluctuation in its populations and in some years is much more abundant than in others; after a wet summer and autumn it is often very plentiful in its favoured haunts.

Central Europe to east Siberia and Japan.

Acleris hyemana (Haworth)

Pl. 45, figs. 1-4

[*Peronea mixtana*; Barrett, 1905: 229; Pierce & Metcalfe, 1922: 21, pl. 8 (♂, ♀ genitalia); Meyrick, 1928: 526. *P. hyemana*; Ford, 1949: 60]

DESCRIPTION

♂ ♀ 12-19 mm. Sexual dimorphism moderately pronounced; female usually smaller, forewing relatively narrow throughout and with apex more pronounced, markings often more contrasting than in male. Polymorphic; Sheldon (1930: 242) recognizes three forms in the British Isles.

Male (fig. 1). Forewing ground colour silver-grey; markings reddish brown varying to purplish brown, extensive and diffuse on dorsum, reducing ground colour to small patches, irregularly sprinkled with blackish brown; basal fascia indeterminate; sub-basal fascia usually indicated by a blackish brown line on dorsum; median fascia usually narrow and well developed costally, often darkened with black medially, dilated and diffuse dorsally; pre-apical spot rather large; cilia yellowish brown, overlaid basally with broad scales concolorous with ground colour of wing, ochreous-brown at apex, grey at tornus. Hindwing light grey, infuscate and weakly strigulate distally; cilia concolorous.

Female (fig. 2). Markings denser and coloration more intense than in male.

Variation. The nominate form is the most common and shows considerable minor variation. An uncommon form occurs with reduced fuscous markings (fig. 3, f. *griseana* Sheldon). A unicolorous form (fig. 4, f. *brunneana* Sheldon) with the coloration varying from dark reddish brown to fuscous-brown occurs throughout the range of this species in the British Isles.

COMMENTS

The narrow forewings with prominent apex, especially in the female, and the purplish brown and silver-white coloration are characteristic.

BIOLOGY

Ovum. Deposited on the foodplant in April and May.

Larva. Head yellowish brown or greenish brown, ocellar region dark brown; prothoracic plate concolorous with abdomen or brighter green; abdomen light green or yellowish green, with slightly darker medio-dorsal line; integument weakly shagreened; pinacula yellowish brown; anal comb present; thoracic legs greenish brown. June to August, feeding on *Calluna* and *Erica* (pl. 21, fig. 3) in a slight web on the top shoots, with a preference for the flowers; also found on *Pinus contorta* in Scotland.

Pupa. Dark brown. August and September, in an earthen cocoon amongst moss or surface debris; sometimes in a slight cocoon in the larval habitation.

Imago. September and October, overwintering until the following April; frequenting heaths and moors. In the autumn the moth is usually rather lethargic, hiding during the day amongst heather and ling, or amongst leaf litter during dry weather. It is more active in the spring, the male flying in the afternoon sunshine, the female sitting on the tops of the foodplants. The cryptic coloration, resembling that of a dried shoot of the previous year's bloom, makes the moth difficult to detect when at rest.

DISTRIBUTION

Common on heaths and moors throughout the British Isles, including the Outer Hebrides, but apparently not recorded from the Orkneys and Shetlands. Very common on moors in Scotland and parts of Wales, and in Ireland generally distributed and common on mountains and bogs.

North and central Europe.

Acleris lipsiana (Denis & Schiffermüller)

Pl. 45, figs. 5–7

[*Peronea lipsiana*; Barrett, 1905: 234; Pierce & Metcalfe, 1922: 22, pl. 9 (♂ genitalia); Meyrick, 1928: 525; Ford, 1949: 59]

DESCRIPTION

♂ ♀ 17–23 mm. Sexual dimorphism not pronounced; female usually smaller. Polymorphic; Sheldon (1930: 273) recognizes four forms.

Male (fig. 5). Forewing ground colour reddish brown varying to chestnut-brown; markings obscure and diffuse, forming a greyish pruinose suffusion, especially in costal half; wing sparsely irrorate with black, mainly in distal half; two small adjacent discocellular whitish yellow scale-tufts; cilia grey. Hindwing grey, sometimes finely strigulate with dark grey distally; cilia paler.

Female. Similar to male.

Variation. The nominate form, which has two whitish yellow discocellular scale-tufts on the forewing, is rare in the British Isles; these scale-tufts are lacking in the most common form (fig. 6, f. *sudoriana* Hübner). Very rarely specimens occur with the greyish pruinose suffusion extending over the whole of the wing and completely obliterating the reddish brown ground colour (f. *griseana* Sheldon). Another uncommon form (fig. 7, f. *costimaculana* Sheldon) has the grey suffusion reduced, the sub-basal fascia weakly indicated on the dorsum, and the costal part of the median fascia and the pre-apical spot fused and forming a rather diffuse triangular costal blotch.

COMMENTS

The common form of this species, f. *sudoriana*, which lacks the whitish yellow discocellular scale-tufts, closely resembles *A. rufana* f. *purpurana*, but may be distinguished by the comparatively broader forewing and less prominent apex, and the more uniform pruinose coloration.

BIOLOGY

Ovum. Yellowish green, deposited in small batches on the leaves of the foodplant in the spring. The early stages of this species have been described by Sheldon (1919: 254).

Larva. Head light chestnut-brown, epistomal region darker; prothoracic plate greenish brown, postero-lateral margin marked with black; abdomen brownish green, pinacula concolorous; thoracic legs brown. June and July, on *Myrica gale*, *Vaccinium myrtillus* and *V. vitis-idaea*, spinning the leaves together to form a tube and living within. The tube is similar to that constructed by the larva of *A. maccana*, but is considerably longer and sometimes is as much as 5–7 cm in length. *Betula* and *Malus sylvestris* are also recorded foodplants.

Pupa. Blackish brown. July and August, in the larval habitation or spun up in surface debris.

Imago. Late August to October, overwintering until the following April or May; frequenting hill bogs, mountains and moorlands.

DISTRIBUTION

Local and apparently restricted to the hills and mountains of the north, occurring in north Lancashire and Westmorland (Foulshaw Moss, Witherslack) in England, and in Scotland from Renfrewshire and Arran to Aberdeenshire, Morayshire, Inverness-shire and Sutherland. In Scotland it is much less common than other *Acleris* species associated with *Myrica* and *Vaccinium* in the Highlands; records from outside the central Highlands require confirmation because of possible confusion with *A. rufana*.
Europe to Siberia.

Acleris rufana (Denis & Schiffermüller)

Pl. 45, figs. 8–12

[*Peronea rufana*; Barrett, 1905: 235; Pierce & Metcalfe, 1922: 22, pl. 9 (♂, ♀ genitalia); Meyrick, 1928: 524; Ford, 1949: 59]

DESCRIPTION

♂♀ 17–21 mm. Sexual dimorphism not pronounced. Polymorphic; Sheldon (1930: 275) recognizes six forms.

Male (fig. 8). Forewing ground colour light yellow-ochreous, sparsely flecked with black mixed with reddish brown; a concentration of slightly raised reddish brown scales in disc; markings ash-grey, suffused with plumbeous and plumbeous-black admixture, reduced to a semi-elliptical sub-basal spot on dorsum, often with a reddish brown admixture, and a large triangular costal blotch in which the plumbeous-black suffusion forms a somewhat radiate pattern; cilia whitish yellow, overlaid at base with broad scales concolorous with ground colour of wing, a reddish brown sub-basal line, apices often tinged with reddish brown. Hindwing light grey, weakly strigulate distally; cilia cream-white.

Female. Similar to male.

Variation. The forewing ground colour and pattern show considerable variation. The nominate form is the least variable, the ground colour being relatively constant and the markings either weak and more diffuse or intensified. The other forms vary from almost complete loss of markings to the development of conspicuous longitudinal subcostal and dorsal streaks. A common form (fig. 9, f. *ochreana* Sheldon) occurs in which the forewing is almost unicolorous ochreous, except for the minute black flecks found in most forms of this species, and sometimes an indication of the sub-basal spot. A rather uncommon form (fig. 10, f. *albicostana* Sheldon), found throughout the range of this species in the British Isles, has the forewing ground colour pale ochreous and the costa broadly suffused with white basally, the suffusion often extending to the apex. A form with the forewing deep purplish brown densely irrorate with grey and the costa narrowly edged with whitish grey suffusion occurs not uncommonly (fig. 11, f. *purpurana* Sheldon). Two forms with the forewing conspicuously marked with longitudinal streaks occur commonly. In one of these (fig. 12, f. *apiciana* Hübner) the ground colour is densely suffused or overlaid with whitish grey, a conspicuous purplish brown streak extends from near the base to near the apex and a similar but more diffuse streak extends along the dorsum. The other form (f. *ochreostriana* Sheldon) is similar except that the ground colour is not suffused with whitish grey, and the streaks are purplish black or black.

COMMENTS

The not uncommon almost unicolorous purplish brown form f. *purpurana* closely resembles *A. lipsiana* f. *sudoriana*, but differs in the comparatively narrow forewing with more prominent apex, the usually pronounced whitish suffusion on the costal margin, and the distinct reddish brown terminal line.

BIOLOGY

Larva. Head black, blackish brown or brown; prothoracic plate shining jet black;

abdomen greyish green or blackish green, greyish in early instars; pinacula indistinct, paler than integument; anal plate yellowish green, sometimes mottled with blackish; anal comb with six prongs; thoracic legs black (Sheldon, 1919: 271). June to early August, on *Myrica gale*, drawing the leaves neatly together towards the stem and feeding within (pl. 21, fig. 2); also on *Salix caprea*, *Populus alba* and *Filipendula ulmaria*. On the Continent, *Rubus idaeus* is an additional foodplant.

Pupa. Dark brown, wings smooth. August, in a silken cocoon amongst dead leaves or in the earth.

Imago. Late August to October, overwintering until the following April; frequenting moorlands and poorly drained woodland. During the day the moth hides low down amongst the foodplants or nearby herbage until the afternoon when it comes up and sits on the tops of the plants, flying gently about as the evening advances and later coming to light. In the spring, after hibernation, it may frequently be found at night sitting on the tips of *Myrica* stems.

DISTRIBUTION

Widely distributed and locally common in suitable localities throughout Britain as far north as Sutherland. In Ireland it is recorded from Co.'s Cork, Dublin, Mayo and Donegal, and is evidently rare (Beirne, 1941: 86).

North and central Europe; Siberia; Japan.

Acleris lorquiniana (Duponchel)

Pl. 45, figs. 13–15

[*Peronea lorquiniana*; Barrett, 1905: 244; Pierce & Metcalfe, 1922: 23, pl. 9 (♂, ♀ genitalia); Meyrick, 1928: 522; Ford, 1949: 57]

DESCRIPTION

♂ ♀ 15–21 mm. Sexual dimorphism not pronounced. Polymorphic; Sheldon (1931: 99) recognizes three forms in the British Isles.

Male, female (fig. 13, f. *flavana* Sheldon). Forewing ground colour light brownish ochreous, coarsely irrorate with scattered black scales, venation paler and rather prominent, sometimes a trace of ferruginous-brown between veins; scale-tufts obsolete, but usually an indication of a blackish discocellular spot; markings obsolete; cilia concolorous with ground colour, apices paler. Hindwing whitish grey, slightly darker and coarsely strigulate with fuscous distally; cilia yellowish white.

Variation. The nominate form of this species does not occur in the British Isles and is apparently rare on the Continent; it differs only slightly from f. *flavana* in having the basal area and costa of the forewing suffused with white, but according to Razowski (1966: 271) intermediate forms are known on the Continent. In the British Isles f. *flavana* is the most common form of *lorquiniana* and shows considerable minor variation: the venation may be less prominent and the wing almost unicolorous except for the black irroration, the discocellular spot may be conspicuous with a brownish admixture, and the black irroration may be heavier and more extensive, often then accompanied by heavier strigulation in the hindwing. Two strikingly marked forms occur, neither of them commonly. In one (fig. 14, f. *uliginosana* Humphreys & Westwood) the blackish discocellular spot is greatly enlarged. In the other (fig. 15, f. *striana* Sheldon), a rather rare form, the forewing is traversed longitudinally by a prominent ferruginous-brown streak arising from the mesothorax, broadening medially and reaching to the apex, extending into the cilia; usually this streak is darkened internally by a heavy fuscous suffusion, leaving only the margins clear ferruginous-brown. Combinations of these two forms are known to occur in which the characteristic markings of both are reduced in intensity and comparatively inconspicuous.

COMMENTS

A distinctive species, readily recognizable by the narrow forewing with its apex acutely angled, and the brownish ochreous general coloration.

BIOLOGY

Larva. Head yellowish brown; prothoracic and anal plates shining green; abdomen pale green or yellowish green, with a darker subdorsal line, pinacula concolorous. May and June, and those producing the second generation of moths in late July and August; feeding on *Lythrum salicaria*. Larvae of the first generation feed in the young shoots; those of the second generation feed in the flower spikes, eating the flowers and seeds.

Pupa. Pale brown. Late May and June, and again in August and September; spun up in debris on the ground.

Imago. Bivoltine, the first generation on the wing in June and July, the second in September and October. During the day the moth is lethargic and is seldom seen, hiding amongst herbage in its fenland haunts, but after dark comes readily to light.

DISTRIBUTION

Locally abundant in the fens and broads of Norfolk (Barton, Horning, Potter Heigham, Ranworth, Stalham) and Cambridgeshire (Chippenham, Wicken); also taken at Gosport (Hampshire) (Wakely, 1966: 100) and Freshwater (Isle of Wight).

Europe to the Ukraine.

Acleris abietana (Hübner)

Pl. 46, figs. 1–3

[*Acleris abietana*; Pelham-Clinton, 1967: 151, pl. 8 (♂, ♀ genitalia)]

DESCRIPTION

♂ ♀ 19–23 mm. Sexual dimorphism not pronounced; female usually larger.

Male (fig. 1). Costa of forewing fringed with cilia-like scales; ground colour of wing ochreous-brown, varying in shade, coarsely irrorate and strigulate with fuscous; markings blackish brown, diffuse; basal fascia poorly developed, strongest towards costa; sub-basal fascia demarcated distally by an almost continuous concave ridge of raised scales, produced medially as a short spur; inner margin of median fascia similarly indicated, with a prominent pale ochreous scale-tuft at middle; pre-apical spot ill-defined, a small pale scale-tuft near its apex; a moderately large scale-tuft above dorsum near base of submedian fold, followed by a smaller tuft at middle of fold and one towards tornus; cilia concolorous with wing. Hindwing grey, weakly infuscate distally; cilia paler.

Female (figs. 2, 3). Similar to male.

Variation. On the Continent seven forms of this species are recognized by Obraztsov (1956: 149). With the exception of f. *confixana* Hübner, in which the forewing is more uniformly pale brown with the markings reduced, the various forms have certain areas of the ground colour much paler. In f. *costialba* Obraztsov the costal area between the sub-basal and median fascia is whitish, forming a conspicuous blotch; in f. *dorsialba* Obraztsov the middle of the dorsum to the tornal area is whitish; in f. *lutiplaga* Rebel the dorsal margin is whitish yellow; in f. *mitterbergeriana* Hauder the basal fascia is ochreous-yellow, the outer margin well defined and strongly convex; and in f. *opacana* Hübner the forewing is almost uniformly blackish brown.

COMMENTS

Superficially similar to *A. hastiana* but distinguished by the more numerous and prominent scale-tufts of the forewing and the roughened scales of the costa, which in *abietana* are longer and more pronounced than in any other of the British *Acleris* species.

BIOLOGY

The immature stages and life history of this species in Britain are not yet known. On the Continent, the larva is found in June and July, feeding on *Pinus*, *Abies* and *Picea*, living in a slight web between the needles. Patočka (1960: 125) gives the following descriptions.

Larva. Head light reddish or yellowish brown, with black postero-lateral markings, ocellar region indistinct; prothoracic plate light brown, marked with blackish brown postero-laterally; abdomen greyish green or greyish brown; pinacula dark brown; anal plate light brown, with dark mottling; anal comb present; thoracic legs blackish.

Pupa. Reddish brown; late July and August, in the larval habitation.

Imago. Late October, probably overwintering.

DISTRIBUTION

This species is a comparatively recent addition to the British list. D. L. Coates took a female on 11th October, 1965 and a male on 14th August, 1966 in a mercury vapour light-trap at Aberfoyle, Perthshire (Pelham-Clinton, 1967: 151).

North, central and eastern Europe.

Acleris maccana (Treitschke)
Pl. 46, figs. 4–6
[*Peronea maccana*; Barrett, 1905: 226; Pierce & Metcalfe, 1922: 21, pl. 8 (♀ genitalia); Pierce & Metcalfe, 1935: 114, pl. 67 (♂ genitalia); Meyrick, 1928: 525; Ford, 1949: 59]

DESCRIPTION

♂♀ 17–22 mm. Sexual dimorphism not pronounced; female usually smaller. Polymorphic; Sheldon (1930: 222) recognizes three forms in the British Isles.

Male (fig. 4). Forewing ground colour whitish grey, irrorate and strigulate with grey mixed with brown, often with a purplish admixture; markings moderately well developed, rather diffuse, plumbeous-grey mixed with purplish or reddish brown; basal and sub-basal fasciae confluent, weak on dorsum, outer margin of sub-basal fascia strengthened medially with dark purplish brown mixed with blackish scales, often roughened and forming a raised scale-tuft in submedian fold; median fascia edged along inner margin with dark purplish brown mixed with roughened black scales, diffuse distally and confluent costally with pre-apical spot the apex of which extends beyond middle of wing; cilia more or less concolorous with wing basally, apices grey. Hindwing light grey, darker distally, terminal margin sometimes weakly strigulate; cilia paler, sometimes with dark grey neural dots.

Female. Similar to male.

Variation. A rare but sometimes locally common form (fig. 5, f. *canescana* Sheldon) occurs in which the forewing ground colour is white weakly irrorate with grey or fuscous, with the markings developed in only the costal half of the wing. A common form (fig. 6, f. *suffusana* Sheldon) occurring throughout the range of this species has the forewing ground colour heavily overlaid with dark grey, leaving only a few scattered white scales, most pronounced along the margins of the markings which are obscure and indicated mainly by dark purplish brown lines and striae intermixed with roughened black scales.

COMMENTS

This species differs from *A. hastiana* in the shape of the forewing, the costa of which is relatively smooth-scaled and evenly arched, while in *hastiana* the wing is broader basally, with the costa rough-scaled, strongly arched basally and shallowly concave beyond the middle.

BIOLOGY

Larva. Head yellowish brown or brownish green, ocellar region black, a distinct black marking in the postero-lateral margin; prothoracic plate concolorous with head; abdomen light bluish green, pinacula prominent, concolorous; anal plate weakly developed (Sheldon, 1919: 253; Benander, 1929: 136). June and July, feeding on *Vaccinium* and *Myrica gale*, spinning the leaves of a shoot together to form a tube and living within; on *Myrica* the spinning is more twisted and untidy than that of *A. rufana*.

Pupa. Reddish brown, head, thorax and wings lighter. July and August, in the larval habitation.

Imago. August to October, hibernating until the following spring. Frequenting woodland and open moorland on hills and mountains. During the day the moth rests amongst the foodplants, but flies freely at night and comes to light and sugar. In the central

Highlands of Scotland the most usual foodplants appear to be *Vaccinium myrtillus* and *V. vitis-idaea*, while on the west coast of Argyll the foodplant is *Myrica*.

DISTRIBUTION

The distribution of this species in the British Isles is uncertain. It appears to be restricted to the mountain districts of Scotland, occurring in Perthshire, Argyll, Inverness-shire, Aberdeenshire, Morayshire and Ross and Cromarty. Records from south-eastern Scotland (East Lothian, Roxburghshire and Midlothian) are unconfirmed; a specimen determined as *A. maccana* in the Balfour collection and examined by E. C. Pelham-Clinton proved to be *A. latifasciana*. In England the species has been recorded from Essex (Epping Forest) (Barrett, 1905: 227), Lancashire (Manchester) (Stainton, 1859: 223) and Northumberland (Hexham) (Maling, 1877: 280), but these records are unconfirmed and very dubious. In Ireland it has been recorded from Co.'s Mayo and Donegal, but Beirne (1941: 86) considers the records unreliable.

Central, north and north-east Europe; Iceland; North America.

Acleris literana (Linnaeus)

Pl. 47, figs. 1–20

[*Leptogramma literana*; Barrett, 1905: 216. *Oxigrapha literana*; Pierce & Metcalfe, 1922: 24, pl. 9 (♂, ♀ genitalia). *Peronea literana*; Meyrick, 1928: 526; Ford, 1949: 60]

DESCRIPTION

♂♀ 18–22 mm. Sexual dimorphism not pronounced. Polymorphic; Sheldon (1921b: 129) and Bradley (1962: 117) discuss the named forms of this species, of which about 30 are known from the British Isles.

Male, female (fig. 1). Forewing ground colour light green, roughened with scattered groups of raised scales which are sometimes tipped with whitish and appear paler; markings black, reduced and fractured into sharply defined hieroglyphic-like shapes; basal fascia obsolescent, variably indicated by a few black scales at extreme base and one or two strigulae on costa, or as a triangular sub-dorsal marking as shown in f. *griseana* (fig. 2); sub-basal fascia narrow, slightly outward-oblique from costa to submedian fold, sharply angulate and produced distad as a short spur in fold, with a prominent black mixed with green scale-tuft below; inner margin of median fascia indicated by black costal strigulae followed by an almost continuous unbroken transverse ridge of raised green scales, these mixed with black towards dorsum, a heavy elongate discocellular dash, bifurcate distally, the upper spur confluent with a rather diminutive plumbeous patch which, together with a small circular or angular spot in disc, is derived from the atrophied pre-apical spot; usually several strigulae scattered along costa; cilia greyish white, a pale green sub-basal line, black neural dots at base along termen and around tornus. Hindwing brownish grey, infuscate distally; cilia paler, with a dark sub-basal line.

Variation. The recurring forms of this species fall into two fairly well-defined groups according to the structure of the scaling of the forewing. In the group that includes the nominate form the forewing is relatively smooth-scaled except for scattered groups of raised scales and the several distinct scale-tufts common to most species in the genus *Acleris*, notably that situated in the disc (the discocellular tuft), that dorsad of the submedian fold, and the two small tufts in the tornal area; in the second group the whole of the wing surface is rough-scaled or squarrose and the principal tufts are less prominent. This difference in scaling is accentuated in some forms of the squarrose group by the coloration of individual scales being paler or lacking pigmentation.

As there is no genetical information available any arrangement of the colour forms is necessarily somewhat arbitrary. It is therefore convenient to separate them into two groups according to scaling with the result that certain recurring markings and patterns, notably the heavy black irroration and the bifurcate discocellular dash, are found in both groups, but generally are weaker in the squarrose group.

There is considerable minor variation within some of the major forms, especially in the markings of the nominate form; also, in this and many of the other forms there is sometimes a weak suffusion of pale ochreous on the ground coloration. Mostly the forms are distinct and do not intergrade, but not infrequently combinations occur, the markings of one form being superimposed on the other, and while some may be distinctive, for example, the combination of typical *literana* and f. *irrorana*, apparent in f. *irroroliterana*, others may be difficult to place.

Group 1. Forewing mostly smooth-scaled, scale-tufts distinct:

f. *griseana* Sheldon (fig. 2): rare; forewing markings as in nominate form but ground colour greenish grey or light slate-grey;

f. *irroroliterana* Bradley (fig. 3): rare; similar to nominate form but with coarse black irroration superimposed over the whole wing;

f. *nigroliterana* Bradley (fig. 4): rare; markings extensively and broadly edged with plumbeous-black suffusion;

f. *brunneana* Sheldon (fig. 5): uncommon; markings extensively developed, diffuse, with a strong admixture of brown or fulvous, the dorsal margin heavily suffused with blackish;

f. *fulvodorsana* Bradley (fig. 6): rare; ground colour and markings as in nominate form but having in addition a conspicuous semi-elliptical deep fulvous sub-basal patch on the dorsum;

f. *fulvomaculana* Sheldon (fig. 7): rare; similar to *fulvodorsana* but with the dorsal patch ferruginous and having in addition four or five elongate ferruginous patches distributed over the wing; similar ferruginous patches are found in f. *aerugana* in the squarrose group;

f. *olivariana* Bradley (fig. 8): rare; markings as in nominate form but ground colour deep olive except in apical area and margins of costa and dorsum, with a ferruginous sub-basal patch on the dorsum;

f. *squamulana* Hübner (fig. 9): rare; forewing ground colour olive-green, markings reduced, a ferruginous sub-basal patch on the dorsum, whitish suffusion in apical area, along submedian fold and on dorsum near tornus;

f. *romanana* Fabricius (fig. 10): rare; markings obsolete except discocellular dash which is prominent but reduced and lacks the lower spur;

f. *fulvoliterana* Sheldon (fig. 11): uncommon; similar to f. *brunneana* but with strong subcostal and median fulvous streaks converging towards base, a weak semi-elliptical deep fulvous dorsal patch overlaid by a short pale fulvous streak;

f. *subfulvoliterana* Bradley: uncommon; resembling a very pale f. *fulvoliterana*, the longitudinal fulvous streak being only weakly developed and incomplete with the lower fork interrupted immediately distad of the furcation, the black markings as in the nominate form and the sub-basal patch on the dorsum absent;

f. *fulvirrorana* Bradley: rare; similar to f. *fulvoliterana* but with coarse black irroration superimposed over the whole wing;

f. *subliterana* Bradley: rare; resembling a weakly marked example of the nominate form without the black basal marking, but differing in having a vestige of a longitudinal streak in the form of two elongate pale ochreous patches, one from near middle of base, the other in the discal area dorsad of discocellular dash.

Group 2. Forewing squarrose, scale-tufts comparatively indistinct:

f. *suffusana* Sheldon (fig. 12): uncommon; similar to f. *fulvoliterana* but with the markings and green ground colour partially obscured by blackish suffusion;

f. *tricolorana* Haworth: the most common of the forms with fulvous streaks; similar to f. *suffusana* but differing in having the ground colour distinctly mixed with whitish grey or white;

f. *squamana* Fabricius (fig. 13): one of the commonest forms of *literana* found in the British Isles; Sheldon (1921b: 161) records that at least half of the specimens taken in the New Forest were referable to this form. A characteristic of this and other forms in this group is the coarse texture of the ground colour scaling; the typically light green colour is usually darkened with a variable admixture of deep moss green, and the black markings are often reduced or obsolescent; very rarely a diffuse black patch is present on the dorsum near the base;

f. *dorsosquamana* Bradley: rare; similar to f. *squamana* but with a semi-elliptical ferruginous sub-basal patch on the dorsum;

f. *mixtana* Sheldon (fig. 14): a common form similar to f. *squamana* but having the ground colour extensively mixed with white, producing a variegated appearance, and the black markings somewhat stronger;

f. *suavana* Herrich-Schäffer: rare; black markings as in nominate form but with the variegated ground colour of f. *mixtana*; no British specimens have been seen which exactly fit the description but examples of f. *mixtana* in which the black markings are exceptionally heavy approach it;

f. *flavana* Sheldon (fig. 15): an extremely rare form in which the ground colour is brownish buff and the markings are greatly reduced; except for the buff ground colour this form resembles a strongly marked f. *squamana*;

f. *irrorana* Hübner (fig. 16): a common form with the ground colour variegated as in f. *mixtana* but coarsely irrorate with black dots;

f. *dorsomaculana* Bradley: not uncommon; similar to f. *mixtana* but distinguished by the semi-elliptical fulvous sub-basal patch on the dorsum as in f. *fulvodorsana*; in some examples this patch is slightly olivaceous;

f. *aerugana* Hübner: a common form similar to f. *squamana* but differing in having well-defined ferruginous blotches as in f. *fulvomaculana*;

f. *nigromaculana* Sheldon (fig. 17): not uncommon; a variable form with the ground colour as in f. *squamana* but less squarrose, with several large prominent blackish blotches; these blotches are not sharply defined as in the nominate form but are somewhat diffuse and suffused with plumbeous centrally; in some specimens a strong blackish irroration extends along the dorsum from the basal spot to the middle;

f. *fulvana* Sheldon (fig. 18): rare; similar to f. *squamana* but characterized by the extensive fulvous mixed with blackish suffusion along the dorsum from near the base, where it is broad, obliquely to the termen;

f. *fulvomixtana* Stephens: not uncommon; similar to f. *nigromaculana* in markings, with a light fulvous longitudinal streak similar to that in f. *fulvoliterana* but very much weaker and incomplete, the lower fork being interrupted just beyond the furcation;

f. *subfulvomixtana* Bradley (fig. 19): rare; similar to f. *fulvomixtana* but differing in having the black markings reduced and represented by only a few black scales;

f. *adustana* Bradley (fig. 20): rare; a combination of f. *fulvomixtana* and f. *irrorana*, having the coloration and markings of the former with the black irroration of the latter superimposed.

COMMENTS

Although an extremely variable species, *literana* is readily distinguished by the light green or olive-green ground colour and black markings of the forewing, and the presence of characteristic fulvous or ferruginous markings in some of the forms.

BIOLOGY

Ovum. Pearl-grey in colour; deposited singly or in groups of two or three on the branches and twigs of the foodplant. So far as is known the moths do not pair until after hibernation, oviposition taking place in the spring. Sheldon (1921b: 129) describes the early stages and rearing of this species. In captivity, the females should be given fresh sprigs of foodplant in April and May on which to oviposit.

Larva. Head dark brown, ocellar region and postero-lateral margin marked with blackish brown; prothoracic plate greyish green or yellowish green, with a medial sulcus and a small characteristic blackish brown marking on the outer half of the posterior margin; abdomen tapered caudally, light greyish green, integument shagreened, dorsal vessel clearly visible as a dark line; pinacula concolorous with integument; thoracic legs shining black. May and June, on *Quercus*, spinning the leaves together and skeletonizing them; it is easily dislodged from its feeding place by beating. According to Sheldon (1921b: 132) the larva is easily distinguished from other oak-feeding tortricid larvae feeding at the same period by the prothoracic markings and the greyish green coloration of the tapered abdomen. In captivity it will also eat *Malus*, *Betula*, *Tilia*, *Fagus* and *Acer campestre*.

Pupa. Light reddish brown, wings paler. Late June and July, spun up between leaves of the foodplant.

Imago. August to October, overwintering until the following April and May. The moth rests during the day on lichen-covered boughs of oak and other trees, occasionally on the trunks, and may be disturbed by beating. It is best obtained in the late summer and autumn, as it is seldom seen in the spring after hibernation.

DISTRIBUTION

Local, occurring in Britain north to Easter Ross. Most frequent in the southern counties of England, where in some years it is locally common in the New Forest (Hampshire) and other forests and in oak woods and wooded areas from Kent to Cornwall and the Wye Valley; rare in the Midlands and northern counties of England. Local and uncommon in Wales; records include Denbighshire (Llangollen), Caernarvonshire (Bangor, Conway Valley and Llandudno) and Merionethshire (Aberdovey, Arthog, Fairbourne and Maentwrog) in the north, and Glamorgan in the south. In Scotland it is recorded from Roxburghshire (Trowmill, Hawick), Berwickshire (Burnmouth), East Lothian (Biel, Whittinghame), Midlothian (Edinburgh), West Lothian, Renfrewshire (Clyde Valley and Isle of Arran), Dunbartonshire (Luss), Argyll (Port Appin), Perthshire (Kinnoul Hill, Moncreiffe Hill), and Easter Ross (Evanton), which is the most northerly record known from Scotland. In Ireland it is very local, being known from west Co. Cork (Glengarriff) (Huggins, 1953b: 252), Co. Kerry (Killarney), Co. Fermanagh (Tempo) and east Co. Tyrone (Beirne, 1941: 87).

North and central Europe, Italy, Sardinia, Russia; Asia Minor.

Acleris emargana (Fabricius)

Pl. 46, figs. 7–11

[*Teras caudana*; Barrett, 1905: 214. *Rhacodia emargana*; Pierce & Metcalfe, 1922: 23, pl. 9 (♂, ♀ genitalia). *Peronea caudana*; Meyrick, 1928: 521; Ford, 1949: 57]

DESCRIPTION

♂♀ 16–22 mm. Sexual dimorphism not pronounced. Polymorphic; Sheldon (1930:121) recognizes six forms in the British Isles.

Male (fig. 7). Forewing subfalcate, costa deeply emarginate beyond middle, fringed with long cilia-like scales basad and apicad of emargination; ground colour yellow-ochreous, strigulate with ferruginous-brown, veins lined with brown or grey-brown, producing a reticulate pattern most pronounced in basal half; markings greyish brown diffusely mixed with ferruginous; basal and sub-basal fasciae represented by striae forming concave transverse lines; inner margin of median fascia sharply defined, fascia diffuse distally, confluent with pre-apical spot which is atrophied and ill-defined, emitting long striae reaching to tornal area; cilia with a plumbeous sub-basal line from below apex, ferruginous terminally from apex to near tornus, grey at tornus. Hindwing light grey, margins narrowly infuscate, strigulate with grey distally; cilia concolorous, with a grey sub-basal line. Thorax light grey, margins narrowly infuscate.

Female. Similar to male.

Variation. The deep costal emargination characteristic of this species shows considerable variation and in some specimens may be relatively shallow (figs. 10, 11). In the nominate form the forewing ground colour varies from whitish ochreous to orange-ochreous. The most common form of this species in the British Isles (fig. 8, f. *griseana* Sheldon) has the markings obsolescent and the forewing ground colour is light ochreous coarsely overlaid with a grey suffusion except in the distal and tornal areas; an uncommon and similar almost unicolorous brown or grey-brown form occurs with a strong red-brown, fulvous or orange streak along the dorsal margin (f. *caudana* Fabricius); an equally uncommon but generally distributed form (figs. 9, 10, f. *excavana* Donovan) has the ground colour varying from light to dark ferruginous-brown without the heavy strigulation or reticulation basally. The apparently rare form f. *ochracea* Stephens has the forewing unicolorous pale ochreous. An uncommon form but found throughout the range of the species has the ground colour dark fuscous-grey heavily strigulate with ferruginous-brown mixed with roughened black scales, the markings obsolescent and reduced to striae and transverse lines, sometimes mixed with raised black scales (fig. 11, f. *fuscana* Sheldon).

COMMENTS

The pronounced emargination on the costa of the forewing is distinctive of this species.

BIOLOGY

Larva. Head greenish brown or yellowish brown; prothoracic and anal plates light green or yellowish green; abdomen bright green varying to pale green, pinacula concolorous; anal comb with six prongs; thoracic legs light green. The larva may be found in May and June, feeding on *Salix caprea* and other sallows, *Populus* and *Betula*, turning down the edge of a leaf, living within and eating the tip, or in spun shoots; also on *Corylus* and *Alnus*.

Pupa. Light brown or reddish brown, thorax shining, abdomen dull. June and July, in the larval habitation or spun up in fallen leaves or under moss on tree trunks.

Imago. July to September; frequenting woods, marshes and hedgerows, especially where sallows are well established. During the day the moth rests amongst the sallows (pl. 21, figs. 1, 2) and other vegetation, but is readily disturbed by beating, fluttering to the ground like a piece of dead leaf.

DISTRIBUTION

Generally distributed throughout the British Isles. Common in England and Wales, becoming scarcer northwards; comparatively local in Scotland, occurring as far north as Sutherland, the Outer Hebrides and the Orkneys. Common in marshes in Ireland (Beirne, 1941: 85); in the Burren, Co. Clare, the moth has been taken in October (Bradley & Pelham-Clinton, 1967: 136).

Europe to Siberia, north China and Japan; North America.

Plate 22

COCHYLIDAE (PHALONIIDAE)

Figures 1–22

The colour illustrations of moths
on Plates 22–47
are approximately × 2·4 natural size

1   *Hysterosia inopiana* (Haworth), ♂. (Corfe, Dorset)

2   *Hysterosia inopiana* (Haworth), ♀. --

3   *Hysterosia schreibersiana* (Frölich), ♂. --

4   *Hysterosia sodaliana* (Haworth), ♂. (Sanderstead, Surrey)

5   *Hysterophora maculosana* (Haworth), ♂. (Corfe, Dorset)

6   *Hysterophora maculosana* (Haworth), ♀. (Bloxworth, Dorset)

7   *Phtheochroa rugosana* (Hübner), ♂. (Bexley, Kent)

8   *Phalonidia manniana* (Fischer von Röslerstamm), ♂. (Byfleet, Surrey)

9   *Phalonidia manniana* (Fischer von Röslerstamm), ♀. (Bixley, Norfolk)

10 *Phalonidia minimana* (Caradja), ♂. (Corfe Castle, Dorset)

11 *Phalonidia minimana* (Caradja), ♀. (Corfe Castle, Dorset)

12 *Phalonidia permixtana* (Denis & Schiffermüller), ♂. --

13 *Phalonidia vectisana* (Humphreys & Westwood), ♂. (Hoo, Kent)

14 *Phalonidia vectisana* (Humphreys & Westwood), ♀. (Shoreham, Sussex)

15 *Phalonidia vectisana* (Humphreys & Westwood), ♂. (Shoreham, Sussex)

16 *Phalonidia vectisana* (Humphreys & Westwood), ♂. (Ballyconneely, Co. Galway)

17 *Phalonidia alismana* (Ragonot), ♂. (Chandler's Ford, Hants)

18 *Phalonidia luridana* (Gregson), ♂. (Cotswolds, Glos)

19 *Phalonidia affinitana* (Douglas), ♂. (Benfleet, Essex)

20 *Phalonidia affinitana* (Douglas), ♀. (Benfleet, Essex)

21 *Phalonidia gilvicomana* (Zeller), ♂. (Surrey)

22 *Phalonidia curvistrigana* (Stainton), ♂. (Darenth, Kent)

Plate 23

COCHYLIDAE (PHALONIIDAE)

Figures 1–16

COCHYLIDAE (PHALONIIDAE)

PLATE 23

1 *Stenodes alternana* (Stephens), ♂. (Deal, Kent)

2 *Stenodes alternana* (Stephens), ♀. (Deal, Kent)

3 *Stenodes straminea* (Haworth), ♂. (Quernmore, Kent)

4 *Agapeta hamana* (Linnaeus), ♂. (Gravesend, Kent)

5 *Agapeta hamana* (Linnaeus), ♂. (Gravesend, Kent)

6 *Agapeta hamana* (Linnaeus), ♂. (Corfe Castle, Dorset)

7 *Agapeta hamana* (Linnaeus), ♂. (Burren, Co. Clare)

8 *Agapeta zoegana* (Linnaeus), ♂. (Corfe Castle, Dorset)

9 *Agapeta zoegana* (Linnaeus), ♂. (Abbot's Wood, Sussex)

10 *Agapeta zoegana* (Linnaeus), ♂. (Abbot's Wood, Sussex)

11 *Agapeta zoegana* (Linnaeus), ♀. (Wicken, Cambs)

12 *Aethes tesserana* (Denis & Schiffermüller), ♂. (Otford, Kent)

13 *Aethes tesserana* (Denis & Schiffermüller), ♂. (Purbeck, Dorset)

14 *Aethes tesserana* (Denis & Schiffermüller), ♂. --

15 *Aethes tesserana* (Denis & Schiffermüller), ♂. --

16 *Aethes tesserana* (Denis & Schiffermüller), ♂. (Otford, Kent)

Plate 24

COCHYLIDAE (PHALONIIDAE)

Figures 1–16

COCHYLIDAE (PHALONIIDAE)

PLATE 24

1   *Aethes rutilana* (Hübner), ♂. (Betchworth, Surrey)

2   *Aethes piercei* Obraztsov, ♂. (Tilgate, Sussex)

3   *Aethes piercei* Obraztsov, ♂. (Corfe, Dorset)

4   *Aethes piercei* Obraztsov, ♀. (Tilgate, Sussex)

5   *Aethes hartmanniana* (Clerck), ♂. (Westcott, Surrey)

6   *Aethes hartmanniana* (Clerck), ♀. --

7   *Aethes margarotana* (Duponchel), ♂. (Essex)

8   *Aethes williana* (Brahm), ♂. (Street, Devon)

9   *Aethes cnicana* (Westwood), ♂. (Corfe, Dorset)

10   *Aethes rubigana* (Treitschke), ♂. (Corfe, Dorset)

11   *Aethes smeathmanniana* (Fabricius), ♂. (Bognor, Sussex)

12   *Aethes margaritana* (Haworth), ♂. (Brighton, Sussex)

13   *Aethes dilucidana* (Stephens), ♂. (Worthing, Sussex)

14   *Aethes francillana* (Fabricius), ♂. (Swanage, Dorset)

15   *Aethes beatricella* (Walsingham), ♂. (Folkestone, Kent)

16   *Commophila aeneana* (Hübner), ♂. (Chattenden, Kent)

1

2

3

4

5

6

7

8

9

10

11

12

13

14

15

16

Plate 24

Plate 25

COCHYLIDAE (PHALONIIDAE)

Figures 1–21

COCHYLIDAE (PHALONIIDAE)

PLATE 25

1 *Eupoecilia angustana angustana* (Hübner), ♂. --

2 *Eupoecilia angustana angustana* (Hübner)
   f. *fasciella* Donovan, ♂. (New Forest, Hants)

3 *Eupoecilia angustana thuleana* (Vaughan), ♂. (Shetlands)

4 *Eupoecilia ambiguella* (Hübner), ♂. (New Forest, Hants)

5 *Cochylidia implicitana* (Wocke), ♂. (Hoo, Kent)

6 *Cochylidia heydeniana* (Herrich-Schäffer), ♀. (Bexley, Kent)

7 *Cochylidia subroseana* (Haworth), ♂. (Chiddingfold, Surrey)

8 *Cochylidia rupicola* (Curtis), ♂. (Silverdale, Lancs)

9 *Cochylidia rupicola* (Curtis), ♀. (Silverdale, Lancs)

10 *Falseuncaria ruficiliana* (Haworth), ♂. (Purbeck, Dorset)

11 *Falseuncaria ruficiliana* (Haworth), ♂. (Aviemore, Inverness-shire)

12 *Falseuncaria degreyana* (McLachlan), ♂. (Thetford, Norfolk)

13 *Falseuncaria degreyana* (McLachlan), ♂. --

14 *Cochylis roseana* (Haworth), ♂. (Mucking, Essex)

15 *Cochylis flaviciliana* (Westwood), ♂. (Coulsdon, Surrey)

16 *Cochylis dubitana* (Hübner), ♂. (Erith, Kent)

17 *Cochylis hybridella* (Hübner), ♂. (Dymchurch, Kent)

18 *Cochylis atricapitana* (Stephens), ♂. (The Crumbles, Sussex)

19 *Cochylis atricapitana* (Stephens), ♀. (The Crumbles, Sussex)

20 *Cochylis pallidana* Zeller, ♂. (Studland, Dorset)

21 *Cochylis nana* (Haworth), ♂. (Bexley, Kent)

Plate 26

TORTRICIDAE: TORTRICINAE

Figures 1–11

1 *Pandemis corylana* (Fabricius), ♂. (Bexley, Kent)

2 *Pandemis corylana* (Fabricius), ♀. --

3 *Pandemis corylana* (Fabricius), ♂. --

4 *Pandemis corylana* (Fabricius), ♀. --

5 *Pandemis cerasana* (Hübner), ♀. (Bexley, Kent)

6 *Pandemis cerasana* (Hübner), ♂. (Oxted, Surrey)

7 *Pandemis cerasana* (Hübner), ♀. (Wicken, Cambs)

8 *Pandemis cerasana* (Hübner), ♂. (St. John's Wood, London)

9 *Pandemis cerasana* (Hübner), ♂. (Bexley, Kent)

10 *Pandemis cinnamomeana* (Treitschke), ♂. (Hindhead, Surrey)

11 *Pandemis cinnamomeana* (Treitschke), ♀. (Hindhead, Surrey)

1

2

3

4

5

6

7

8

9

10

11

Plate 27

TORTRICIDAE: TORTRICINAE

Figures 1–11

PLATE 27

1   *Pandemis heparana* (Denis & Schiffermüller), ♂.
    (Bexley, Kent)

2   *Pandemis heparana* (Denis & Schiffermüller), ♀.
    (Eastbourne, Sussex)

3   *Pandemis heparana* (Denis & Schiffermüller), ♂.
    (Wicken, Cambs)

4   *Pandemis heparana* (Denis & Schiffermüller), ♀.
    (Stoborough, Dorset)

5   *Pandemis heparana* (Denis & Schiffermüller), ♂.
    (Bexley, Kent)

6   *Pandemis dumetana* (Treitschke), ♂. --

7   *Pandemis dumetana* (Treitschke), ♀. (Wicken, Cambs)

8   *Argyrotaenia pulchellana* (Haworth), ♂. --

9   *Argyrotaenia pulchellana* (Haworth), ♀. (Ashdown
    Forest, Sussex)

10 *Archips oporana* (Linnaeus), ♂. (Oxshott, Surrey)

11 *Archips oporana* (Linnaeus), ♀. (Oxshott, Surrey)

1

2

3

4

5

6

7

8

9

10

11

Plate 28

TORTRICIDAE: TORTRICINAE

Figures 1–11

PLATE 28

1   *Archips podana* (Scopoli), ♂. --

2   *Archips podana* (Scopoli), ♀. (Lewisham, Kent)

3   *Archips podana* (Scopoli), ♂. (Buckingham Palace Garden, London)

4   *Archips podana* (Scopoli), ♀. (Birmingham, Warwick)

5   *Archips podana* (Scopoli), ♂. (Beaconsfield, Bucks)

6   *Archips betulana* (Hübner), ♂. (King's Lynn, Norfolk)

7   *Archips betulana* (Hübner), ♀. (King's Lynn, Norfolk)

8   *Archips crataegana* (Hübner), ♂. (Isle of Wight)

9   *Archips crataegana* (Hübner), ♀. (New Forest, Hants)

10   *Archips crataegana* (Hübner), ♂. (New Forest, Hants)

11   *Archips crataegana* (Hübner), ♀. (New Forest, Hants)

1

2

3

4

5

6

7

8

9

10

11

Plate 29

TORTRICIDAE: TORTRICINAE

Figures 1–11

PLATE 29

1 *Archips xylosteana* (Linnaeus), ♂. (Polegate, Sussex)

2 *Archips xylosteana* (Linnaeus), ♀. (Bexley, Kent)

3 *Archips xylosteana* (Linnaeus), ♂. (Bexley, Kent)

4 *Archips xylosteana* (Linnaeus), ♀. (Coombe Wood, Kent)

5 *Archips rosana* (Linnaeus), ♂. --

6 *Archips rosana* (Linnaeus), ♀. (Wallington, Surrey)

7 *Archips rosana* (Linnaeus), ♀. (Southport, Lancs)

8 *Choristoneura diversana* (Hübner), ♂. (Halstead, Kent)

9 *Choristoneura diversana* (Hübner), ♀. (Halstead, Kent)

10 *Choristoneura hebenstreitella* (Müller), ♂. (Eltham, Kent)

11 *Choristoneura hebenstreitella* (Müller), ♀. (Chislehurst, Kent)

1

2

3

4

5

6

7

8

9

10

11

Plate 30

TORTRICIDAE: TORTRICINAE

Figures 1–12

PLATE 30

1   *Choristoneura lafauryana* (Ragonot), ♂. (King's Lynn, Norfolk)

2   *Choristoneura lafauryana* (Ragonot), ♂. (King's Lynn, Norfolk)

3   *Choristoneura lafauryana* (Ragonot), ♀. (King's Lynn, Norfolk)

4   *Cacoecimorpha pronubana* (Hübner), ♂. (Eastbourne, Sussex)

5   *Cacoecimorpha pronubana* (Hübner), ♂. (Eastbourne, Sussex)

6   *Cacoecimorpha pronubana* (Hübner), ♀. (Eastbourne, Sussex)

7   *Syndemis musculana musculana* (Hübner), ♂. (Shirley, Surrey)

8   *Syndemis musculana musculana* (Hübner), ♀. (Shirley, Surrey)

9   *Syndemis musculana musculana* (Hübner), ♂. (Aviemore, Inverness-shire)

10 *Syndemis musculana musculana* (Hübner), ♀. (Aviemore, Inverness-shire)

11 *Syndemis musculana musculinana* (Kennel), ♂. (Unst, Shetlands)

12 *Syndemis musculana musculinana* (Kennel), ♀. (Shetlands)

1

2

3

4

5

6

7

8

9

10

11

12

Plate 31

TORTRICIDAE: TORTRICINAE

Figures 1–13

PLATE 31

1   *Ptycholomoides aeriferanus* (Herrich-Schäffer), ♂.
    (Westwell, Kent)

2   *Aphelia viburnana* (Denis & Schiffermüller), ♂.
    (Studland, Dorset)

3   *Aphelia viburnana* (Denis & Schiffermüller), ♀.
    (Studland, Dorset)

4   *Aphelia viburnana* (Denis & Schiffermüller), ♂.
    (Funton, Kent)

5   *Aphelia viburnana* (Denis & Schiffermüller), ♂.
    (Roydon, Norfolk)

6   *Aphelia viburnana* (Denis & Schiffermüller), ♀.
    (Studland, Dorset)

7   *Aphelia paleana* (Hübner), ♂. (Bare, Lancs)

8   *Aphelia paleana* (Hübner), ♀. (Derby)

9   *Aphelia paleana* (Hübner), ♂. (Southwold, Suffolk)

10 *Aphelia paleana* (Hübner), ♀. (Derby)

11 *Aphelia unitana* (Hübner), ♂. (Morpeth,
    Northumberland)

12 *Aphelia unitana* (Hübner), ♀. (Burren, Co. Clare)

13 *Aphelia unitana* (Hübner), ♂. (Morpeth,
    Northumberland)

Plate 32

TORTRICIDAE: TORTRICINAE

Figures 1–16

PLATE 32

1 *Clepsis senecionana* (Hübner), ♂. (Tilgate, Sussex)

2 *Clepsis senecionana* (Hübner), ♀. (Matley Bog,
New Forest, Hants)

3 *Clepsis senecionana* (Hübner), ♀. (Corfe, Dorset)

4 *Clepsis rurinana* (Linnaeus), ♂. (Cotswolds,
Gloucester)

5 *Clepsis spectrana* (Treitschke), ♂. (Dry Sandford,
Berks)

6 *Clepsis spectrana* (Treitschke), ♀. (Dry Sandford,
Berks)

7 *Clepsis spectrana* (Treitschke), ♂. (Derby)

8 *Clepsis spectrana* (Treitschke), ♀. (Derby)

9 *Clepsis spectrana* (Treitschke), ♂. (Burnley, Lancs)

10 *Clepsis consimilana* (Hübner), ♂. (Bexley, Kent)

11 *Clepsis consimilana* (Hübner), ♀. (Corfe, Dorset)

12 *Clepsis consimilana* (Hübner), ♂. (Bexley, Kent)

13 *Epiphyas postvittana* (Walker), ♂. (Rosewarne,
Cornwall)

14 *Epiphyas postvittana* (Walker), ♀. (Porthtowan,
Cornwall)

15 *Epiphyas postvittana* (Walker), ♂. (Brea, Cornwall)

16 *Epiphyas postvittana* (Walker), ♀. (Falmouth, Cornwall)

1

2

3

4

5

6

7

8

9

10

11

12

13

14

15

16

Plate 33

TORTRICIDAE: TORTRICINAE

Figures 1–13

PLATE 33

1 *Adoxophyes orana* (Fischer von Röslerstamm), ♂. (S.E. Kent)

2 *Adoxophyes orana* (Fischer von Röslerstamm), ♀. (S.E. Kent)

3 *Ptycholoma lecheana* (Linnaeus), ♂. (Cudham, Kent)

4 *Ptycholoma lecheana* (Linnaeus), ♀. (Essex)

5 *Ptycholoma lecheana* (Linnaeus), ♂. (Yarmouth, Isle of Wight)

6 *Lozotaeniodes formosanus* (Geyer), ♀. (Byfleet, Surrey)

7 *Lozotaenia forsterana* (Fabricius), ♂. (Bexley, Kent)

8 *Lozotaenia forsterana* (Fabricius), ♂. (Camghouran, Perthshire)

9 *Lozotaenia forsterana* (Fabricius), ♀. (Bexley, Kent)

10 *Paramesia gnomana* (Clerck), ♂. (England)

11 *Paraclepsis cinctana* (Denis & Schiffermüller), ♂. (Adisham, Kent)

12 *Epagoge grotiana* (Fabricius), ♂. (Bexley, Kent)

13 *Epagoge grotiana* (Fabricius), ♀. (England)

Plate 34

TORTRICIDAE: TORTRICINAE

Figures 1–16

1 *Capua vulgana* (Frölich), ♂. (Darenth, Kent)

2 *Capua vulgana* (Frölich), ♀. (Bexley, Kent)

3 *Philedone gerningana* (Denis & Schiffermüller), ♂. (Arnside Knot, Westmorland)

4 *Philedone gerningana* (Denis & Schiffermüller), ♀. (Farleton Fell, Westmorland)

5 *Philedonides lunana* (Thunberg), ♂. --

6 *Philedonides lunana* (Thunberg), ♂. (Goyts Moss, Derby)

7 *Philedonides lunana* (Thunberg), ♀. --

8 *Ditula angustiorana* (Haworth), ♂. (Bexley, Kent)

9 *Ditula angustiorana* (Haworth), ♀. (Bexley, Kent)

10 *Pseudargyrotoza conwagana* (Fabricius), ♂. (Corfe Castle, Dorset)

11 *Pseudargyrotoza conwagana* (Fabricius), ♂. (Oxted, Surrey)

12 *Pseudargyrotoza conwagana* (Fabricius), ♀. (Bexley, Kent)

13 *Sparganothis pilleriana* (Denis & Schiffermüller), ♂. (Ventnor, Isle of Wight)

14 *Sparganothis pilleriana* (Denis & Schiffermüller), ♀. (Yarmouth, Isle of Wight)

15 *Sparganothis pilleriana* (Denis & Schiffermüller), ♂. (Wych, Herts)

16 *Sparganothis pilleriana* (Denis & Schiffermüller), ♀. (Studland, Dorset)

1

2

3

4

5

6

7

8

9

10

11

12

13

14

15

16

Plate 35

TORTRICIDAE: TORTRICINAE

Figures 1–15

PLATE 35

1 *Olindia schumacherana* (Fabricius), ♂. (Westcott, Surrey)

2 *Olindia schumacherana* (Fabricius), ♀. (Westcott, Surrey)

3 *Isotrias rectifasciana* (Haworth), ♂. (Gravesend, Kent)

4 *Isotrias rectifasciana* (Haworth), ♀. (Darenth, Kent)

5 *Eulia ministrana* (Linnaeus), ♂. (Corfe Castle, Dorset)

6 *Eulia ministrana* (Linnaeus), ♂. (Scotland)

7 *Eulia ministrana* (Linnaeus), ♀. (Meathop, Westmorland)

8 *Cnephasia longana* (Haworth), ♂. (Bexley, Kent)

9 *Cnephasia longana* (Haworth), ♀. (Bexley, Kent)

10 *Cnephasia longana* (Haworth), ♀. (Folkestone, Kent)

11 *Cnephasia communana* (Herrich-Schäffer), ♂. (Dry Sandford, Berks)

12 *Cnephasia communana* (Herrich-Schäffer), ♀. (Dry Sandford, Berks)

13 *Cnephasia conspersana* Douglas, ♂. (Sandown, Isle of Wight)

14 *Cnephasia conspersana* Douglas, ♀. (Arnside, Westmorland)

15 *Cnephasia conspersana* Douglas, ♀. (Arnside, Westmorland)

Plate 36

TORTRICIDAE: TORTRICINAE

Figures 1–17

1 *Cnephasia stephensiana* (Doubleday), ♂. (Quernmore, Kent)

2 *Cnephasia stephensiana* (Doubleday), ♀. (Quernmore, Kent)

3 *Cnephasia stephensiana* (Doubleday), ♂. (Bexley, Kent)

4 *Cnephasia stephensiana* (Doubleday), ♂. (Kingsdown, Kent)

5 *Cnephasia stephensiana* (Doubleday) f. *octomaculana* Curtis, ♂. (Paisley, Renfrewshire)

6 *Cnephasia stephensiana* (Doubleday) f. *octomaculana* Curtis, ♀. (Paisley, Renfrewshire)

7 *Cnephasia interjectana* (Haworth), ♂. (Corfe Castle, Dorset)

8 *Cnephasia interjectana* (Haworth), ♀. (Lymington, Hants)

9 *Cnephasia interjectana* (Haworth), ♂. (Scunthorpe, Lincs)

10 *Cnephasia pasiuana* (Hübner), ♂. (Osterley, Middlesex)

11 *Cnephasia pasiuana* (Hübner), ♀. (Halstead, Kent)

12 *Cnephasia pasiuana* (Hübner), ♀. (Navestock, Essex)

13 *Cnephasia genitalana* Pierce & Metcalfe, ♂. (Dover, Kent)

14 *Cnephasia genitalana* Pierce & Metcalfe, ♀. (Dover, Kent)

15 *Cnephasia genitalana* Pierce & Metcalfe, ♂. (Kingsdown, Kent)

16 *Cnephasia incertana* (Treitschke), ♂. (Bexley, Kent)

17 *Cnephasia incertana* (Treitschke), ♀. (Halstead, Kent)

1

2

3

4

5

6

7

8

9

10

11

12

13

14

15

16

17

Plate 37

TORTRICIDAE: TORTRICINAE

Figures 1–15

PLATE 37

1 *Tortricodes alternella* (Denis & Schiffermüller), ♂. (Oxshott, Surrey)

2 *Tortricodes alternella* (Denis & Schiffermüller), ♀. --

3 *Tortricodes alternella* (Denis & Schiffermüller), ♂. (Sutherland)

4 *Tortricodes alternella* (Denis & Schiffermüller), ♀. (Seale, Surrey)

5 *Exapate congelatella* (Clerck), ♂. (East Malling, Kent)

6 *Exapate congelatella* (Clerck), ♀. (Bexley, Kent)

7 *Neosphaleroptera nubilana* (Hübner), ♂. (Cranmore, Isle of Wight)

8 *Neosphaleroptera nubilana* (Hübner), ♀. (Cranmore, Isle of Wight)

9 *Neosphaleroptera nubilana* (Hübner), ♀. (Oxted, Surrey)

10 *Eana argentana* (Clerck), ♂. (Glen Tilt, Perthshire)

11 *Eana osseana* (Scopoli), ♂. (Moorfoot Hills, Midlothian)

12 *Eana osseana* (Scopoli), ♀. --

13 *Eana osseana* (Scopoli), ♂. (Otford, Kent)

14 *Eana incanana* (Stephens), ♂. --

15 *Eana incanana* (Stephens), ♀. (Doncaster, Yorks)

1

2

3

4

5

6

7

8

9

10

11

12

13

14

15

Plate 38

TORTRICIDAE: TORTRICINAE

Figures 1–11

PLATE 38

1 *Eana penziana penziana* (Thunberg & Becklin)
f. *bellana* Curtis, ♂. (Warton Crag, Lancs)

2 *Eana penziana penziana* (Thunberg & Becklin)
f. *bellana* Curtis, ♀. (Keswick, Cumberland)

3 *Eana penziana penziana* (Thunberg & Becklin)
f. *bellana* Curtis, ♀. (Keswick, Cumberland)

4 *Eana penziana colquhounana* (Barrett), ♂.
(Isle of Man)

5 *Eana penziana colquhounana* (Barrett), ♀.
(Isle of Man)

6 *Eana penziana colquhounana* (Barrett), ♀.
(Isle of Man)

7 *Eana penziana colquhounana* (Barrett), ♂.
(Unst, Shetlands)

8 *Eana penziana colquhounana* (Barrett), ♂.
(Unst, Shetlands)

9 *Eana penziana colquhounana* (Barrett), ♀.
(Unst, Shetlands)

10 *Eana penziana colquhounana* (Barrett), ♂.
(Ballynalackan, Co. Clare)

11 *Eana penziana colquhounana* (Barrett), ♀.
(R. Caher, Co. Clare)

Plate 39

TORTRICIDAE: TORTRICINAE

Figures 1–17

PLATE 39

1 *Aleimma loeflingiana* (Linnaeus), ♂. (Bexley, Kent)

2 *Aleimma loeflingiana* (Linnaeus), ♀. (Isle of Wight)

3 *Aleimma loeflingiana* (Linnaeus), ♀.
(New Forest, Hants)

4 *Aleimma loeflingiana* (Linnaeus), ♂. --

5 *Tortrix viridana* (Linnaeus), ♂. (Corfe Castle,
Dorset)

6 *Spatalistis bifasciana* (Hübner), ♀. (Bexley, Kent)

7 *Croesia bergmanniana* (Linnaeus), ♀. (Bexley, Kent)

8 *Croesia bergmanniana* (Linnaeus), ♂. (Hoy, Orkneys)

9 *Croesia bergmanniana* (Linnaeus), ♂. (Co. Down)

10 *Croesia forsskaleana* (Linnaeus), ♂. (Bexley, Kent)

11 *Croesia forsskaleana* (Linnaeus), ♂. (Bexley, Kent)

12 *Croesia forsskaleana* (Linnaeus), ♀. (Bexley, Kent)

13 *Croesia holmiana* (Linnaeus), ♂. (Charmouth, Dorset)

14 *Acleris latifasciana* (Haworth), ♀. --

15 *Acleris latifasciana* (Haworth), ♀. (Corfe Castle, Dorset)

16 *Acleris latifasciana* (Haworth), ♀. (Co. Armagh)

17 *Acleris latifasciana* (Haworth), ♀. (Co. Armagh)

Plate 39

Plate 40

TORTRICIDAE: TORTRICINAE

Figures 1–18

1 *Acleris comariana* (Lienig & Zeller), ♀. (Formby, Lancs)

2 *Acleris comariana* (Lienig & Zeller), ♂. (Formby, Lancs)

3 *Acleris comariana* (Lienig & Zeller), ♂. (Wisbech, Cambs)

4 *Acleris comariana* (Lienig & Zeller), ♂. (Folkestone, Kent)

5 *Acleris comariana* (Lienig & Zeller), ♂. (Folkestone, Kent)

6 *Acleris caledoniana* (Stephens), ♂. (Morridge, Staffs)

7 *Acleris caledoniana* (Stephens), ♀. (Rannoch, Perthshire)

8 *Acleris caledoniana* (Stephens), ♂. (Rannoch, Perthshire)

9 *Acleris sparsana* (Denis & Schiffermüller), ♂. (Bolton-le-Sands, Lancs)

10 *Acleris sparsana* (Denis & Schiffermüller), ♂. (New Forest, Hants)

11 *Acleris sparsana* (Denis & Schiffermüller), ♂. (New Forest, Hants)

12 *Acleris rhombana* (Denis & Schiffermüller), ♂. (Folkestone, Kent)

13 *Acleris rhombana* (Denis & Schiffermüller), ♀. (Isle of Wight)

14 *Acleris rhombana* (Denis & Schiffermüller), ♂. (Isle of Wight)

15 *Acleris rhombana* (Denis & Schiffermüller), ♂. (Isle of Wight)

16 *Acleris rhombana* (Denis & Schiffermüller), ♀. (Redhill, Surrey)

17 *Acleris aspersana* (Hübner), ♂. (Baddesley Common, Hants)

18 *Acleris aspersana* (Hübner), ♀. (Baddesley Common, Hants)

Plate 41

TORTRICIDAE: TORTRICINAE

Figures 1–16

PLATE 41

1 *Acleris tripunctana* (Hübner), ♀. (Rannoch, Perthshire)

2 *Acleris tripunctana* (Hübner), ♂. --

3 *Acleris tripunctana* (Hübner), ♂. (Wisley Common, Surrey)

4 *Acleris tripunctana* (Hübner), ♀. (Bexley, Kent)

5 *Acleris tripunctana* (Hübner), ♀. (Aviemore, Inverness-shire)

6 *Acleris tripunctana* (Hübner), ♂. (Bexley, Kent)

7 *Acleris ferrugana* (Denis & Schiffermüller), ♂. (Isle of Wight)

8 *Acleris ferrugana* (Denis & Schiffermüller), ♂. (Isle of Wight)

9 *Acleris ferrugana* (Denis & Schiffermüller), ♂. (Isle of Wight)

10 *Acleris ferrugana* (Denis & Schiffermüller), ♂. (Ashtead, Surrey)

11 *Acleris shepherdana* (Stephens), ♂. (Wicken Fen, Cambs)

12 *Acleris shepherdana* (Stephens), ♀. (Wicken Fen, Cambs)

13 *Acleris schalleriana* (Linnaeus), ♂. (New Forest, Hants)

14 *Acleris schalleriana* (Linnaeus), ♂. (Isle of Wight)

15 *Acleris schalleriana* (Linnaeus), ♂. (Darenth, Kent)

16 *Acleris schalleriana* (Linnaeus), ♀. (Burren, Co. Clare)

1

2

3

4

5

6

7

8

9

10

11

12

13

14

15

16

Tortricidae: Tortricinae

Plate 41

Plate 42

TORTRICIDAE: TORTRICINAE

Figures 1–14

1 *Acleris variegana* (Denis & Schiffermüller), ♂.
(Bare, Lancs)

2 *Acleris variegana* (Denis & Schiffermüller), ♂. --

3 *Acleris variegana* (Denis & Schiffermüller), ♂.
(Middlesbrough, Yorks)

4 *Acleris variegana* (Denis & Schiffermüller), ♂.
(Wilmington, Kent)

5 *Acleris variegana* (Denis & Schiffermüller), ♂.
(Bare, Lancs)

6 *Acleris permutana* (Duponchel), ♂. (Falmer, Sussex)

7 *Acleris boscana* (Fabricius), ♂. (Isle of Wight)

8 *Acleris boscana* (Fabricius), ♂. (Ashtead, Surrey)

9 *Acleris boscana* (Fabricius), ♂. (Isle of Wight)

10 *Acleris logiana* (Clerck), ♂. (Aviemore, Inverness-shire)

11 *Acleris logiana* (Clerck), ♀. (Aviemore, Inverness-shire)

12 *Acleris logiana* (Clerck), ♂. (Glen Tilt, Perthshire)

13 *Acleris umbrana* (Hübner), ♂. (New Forest, Hants)

14 *Acleris umbrana* (Hübner), ♀. (West Dorset)

1

2

3

4

5

6

7

8

9

10

11

12

13

14

Plate 42

Plate 43

TORTRICIDAE: TORTRICINAE

Figures 1–22

PLATE 43

1   *Acleris hastiana* (Linnaeus), ♂. (South Wales)

2   *Acleris hastiana* (Linnaeus), ♀. (Loch Inver, Sutherland)

3   *Acleris hastiana* (Linnaeus), ♂. (South Wales)

4   *Acleris hastiana* (Linnaeus), ♂. (South Wales)

5   *Acleris hastiana* (Linnaeus), ♀. (South Wales)

6   *Acleris hastiana* (Linnaeus), ♀. (Deal, Kent)

7   *Acleris hastiana* (Linnaeus), ♀. (South Wales)

8   *Acleris hastiana* (Linnaeus), ♂. (South Wales)

9   *Acleris hastiana* (Linnaeus), ♀. (Wicken, Cambs)

10 *Acleris hastiana* (Linnaeus), ♂. (South Wales)

11 *Acleris hastiana* (Linnaeus), ♂. (Wicken, Cambs)

12 *Acleris hastiana* (Linnaeus), ♂. (Soyea I., Sutherland)

13 *Acleris hastiana* (Linnaeus), ♂. (Wicken, Cambs)

14 *Acleris hastiana* (Linnaeus), ♂. (Wicken, Cambs)

15 *Acleris hastiana* (Linnaeus), ♀. (Rannoch, Perthshire)

16 *Acleris hastiana* (Linnaeus), ♂. (Isle of Wight)

17 *Acleris hastiana* (Linnaeus), ♂. (Co. Sligo)

18 *Acleris hastiana* (Linnaeus), ♂. (South Wales)

19 *Acleris hastiana* (Linnaeus), ♂. (South Wales)

20 *Acleris hastiana* (Linnaeus), ♂. (Loch Inver, Sutherland)

21 *Acleris hastiana* (Linnaeus), ♂. (South Wales)

22 *Acleris hastiana* (Linnaeus), ♀. (South Wales)

Plate 44

TORTRICIDAE: TORTRICINAE

Figures 1–17

PLATE 44

1 *Acleris cristana* (Fabricius), ♂. (Isle of Wight)

2 *Acleris cristana* (Fabricius), ♂. (Ashtead, Surrey)

3 *Acleris cristana* (Fabricius), ♂. (Ashtead, Surrey)

4 *Acleris cristana* (Fabricius), ♂. (New Forest, Hants)

5 *Acleris cristana* (Fabricius), ♂. (New Forest, Hants)

6 *Acleris cristana* (Fabricius), ♂. (Ashtead, Surrey)

7 *Acleris cristana* (Fabricius), ♂. (Dunsfold, Surrey)

8 *Acleris cristana* (Fabricius), ♂. (Sussex)

9 *Acleris cristana* (Fabricius), ♂. (Ashtead, Surrey)

10 *Acleris cristana* (Fabricius), ♀. (Ashtead, Surrey)

11 *Acleris cristana* (Fabricius), ♀. (Ashtead, Surrey)

12 *Acleris cristana* (Fabricius), ♀. (Isle of Wight)

13 *Acleris cristana* (Fabricius), ♂. (New Forest, Hants)

14 *Acleris cristana* (Fabricius), ♂.(Steeres Common, Surrey)

15 *Acleris cristana* (Fabricius), ♂. (Dunsfold, Surrey)

16 *Acleris cristana* (Fabricius), ♀. (Maidenhead, Berks)

17 *Acleris cristana* (Fabricius), ♂. (Nettlebed, Oxford)

Plate 44

Plate 45

TORTRICIDAE: TORTRICINAE

Figures 1–15

1 *Acleris hyemana* (Haworth), ♂. (Broadwater Forest, Kent)

2 *Acleris hyemana* (Haworth), ♀. (Corfe, Dorset)

3 *Acleris hyemana* (Haworth), ♀. (Studland, Dorset)

4 *Acleris hyemana* (Haworth), ♂. (Corfe, Dorset)

5 *Acleris lipsiana* (Denis & Schiffermüller), ♂. (Foulshaw Moss, Westmorland)

6 *Acleris lipsiana* (Denis & Schiffermüller), ♂. (Westmorland)

7 *Acleris lipsiana* (Denis & Schiffermüller), ♀. (Rannoch, Perthshire)

8 *Acleris rufana* (Denis & Schiffermüller), ♂. --

9 *Acleris rufana* (Denis & Schiffermüller), ♀. (Foulshaw Moss, Westmorland)

10 *Acleris rufana* (Denis & Schiffermüller), ♂. (Corfe, Dorset)

11 *Acleris rufana* (Denis & Schiffermüller), ♀. (New Forest, Hants)

12 *Acleris rufana* (Denis & Schiffermüller), ♂. (Foulshaw Moss, Westmorland)

13 *Acleris lorquiniana* (Duponchel), ♀. (Wicken Fen, Cambs)

14 *Acleris lorquiniana* (Duponchel), ♀. (Wicken Fen, Cambs)

15 *Acleris lorquiniana* (Duponchel), ♀. --

Plate 46

TORTRICIDAE: TORTRICINAE

Figures 1–11

PLATE 46

1 *Acleris abietana* (Hübner), ♂. (Germany)

2 *Acleris abietana* (Hübner), ♀. (Europe)

3 *Acleris abietana* (Hübner), ♀. (Aberfoyle, Perthshire)

4 *Acleris maccana* (Treitschke), ♂. (Rannoch, Perthshire)

5 *Acleris maccana* (Treitschke), ♂. (Rannoch, Perthshire)

6 *Acleris maccana* (Treitschke), ♂. (Rannoch, Perthshire)

7 *Acleris emargana* (Fabricius), ♂. (Corfe, Dorset)

8 *Acleris emargana* (Fabricius), ♀. (Isle of Wight)

9 *Acleris emargana* (Fabricius), ♀. (Isle of Wight)

10 *Acleris emargana* (Fabricius), ♂. (Dunsfold, Surrey)

11 *Acleris emargana* (Fabricius), ♀. (Braemar, Aberdeenshire)

1   2

3

4   5

6

7   8

9

10   11

Plate 47

TORTRICIDAE: TORTRICINAE

Figures 1–20

PLATE 47

1 *Acleris literana* (Linnaeus), ♀. (New Forest, Hants)

2 *Acleris literana* (Linnaeus), ♂. (New Forest, Hants)

3 *Acleris literana* (Linnaeus), ♂. (New Forest, Hants)

4 *Acleris literana* (Linnaeus), ♀. (Isle of Wight)

5 *Acleris literana* (Linnaeus), ♀. --

6 *Acleris literana* (Linnaeus), ♂. (Isle of Wight)

7 *Acleris literana* (Linnaeus), ♀. (New Forest, Hants)

8 *Acleris literana* (Linnaeus), ♀. (Isle of Wight)

9 *Acleris literana* (Linnaeus), ♀. (Isle of Wight)

10 *Acleris literana* (Linnaeus), ♀. (New Forest, Hants)

11 *Acleris literana* (Linnaeus), ♀. (New Forest, Hants)

12 *Acleris literana* (Linnaeus), ♂. (New Forest, Hants)

13 *Acleris literana* (Linnaeus), ♀. (New Forest, Hants)

14 *Acleris literana* (Linnaeus), ♂. (New Forest, Hants)

15 *Acleris literana* (Linnaeus), ♀. (New Forest, Hants)

16 *Acleris literana* (Linnaeus), ♂. --

17 *Acleris literana* (Linnaeus), ♀. (New Forest, Hants)

18 *Acleris literana* (Linnaeus), ♀. --

19 *Acleris literana* (Linnaeus), ♀. (Isle of Wight)

20 *Acleris literana* (Linnaeus), ♀. --

1

3

2

4

6

5

7

9

8

10

12

11

13

15

14

16

18

17

19

20

LARVAL FOODPLANTS

The botanical nomenclature of the British wild flowers in this list is based on *The Concise British Flora in Colour* by W. Keble Martin (1965), that of cultivated plants on the *Dictionary of Gardening* published by the Royal Horticultural Society (1951–69), and that of coniferous trees on *A Handbook of Coniferae and Ginkgoaceae* by Dallimore & Jackson (1966).

Abies spp.
  Argyrotaenia pulchellana; Archips xylosteana; Aphelia viburnana; Ptycholoma lecheana; Lozotaenia forsterana; Acleris abietana

Abies alba (Silver Fir)
  Pandemis cinnamomeana; Archips oporana

Acer spp.
  Pandemis cerasana, P. cinnamomeana; Archips xylosteana; Clepsis rurinana; Ptycholoma lecheana; Aleimma loeflingiana; Tortrix viridana

Acer campestre (Common Maple)
  Croesia forsskaleana; Acleris literana

Acer pseudoplatanus (Sycamore)
  Croesia forsskaleana; Acleris sparsana

Achillea millefolium (Yarrow)
  Aethes smeathmanniana, A. margaritana; Eupoecilia angustana; Cochylidia implicitana; Choristoneura diversana; Cnephasia pasiuana

Agropyron repens (Creeping Twitch, Couch)
  Aphelia paleana

Alchemilla spp.
  Acleris shepherdana

Alchemilla alpina (Alpine Lady's Mantle)
  Acleris caledoniana

Alchemilla vulgaris agg.
  Acleris aspersana

Alder, see *Alnus glutinosa*

Alder Buckthorn, see *Frangula alnus*

Alexanders, see *Smyrnium olusatrum*

Alisma spp.
  Phalonidia alismana; Aphelia viburnana

Alisma plantago-aquatica (Water Plantain)
  Phalonidia manniana, P. permixtana, P. alismana

Allium spp.
  Aphelia unitana

Alnus glutinosa (Alder)
  Pandemis cerasana; Adoxophyes orana; Capua vulgana; Eulia ministrana; Acleris tripunctana, A. umbrana, A. emargana

Alpine Lady's Mantle, see *Alchemilla alpina*

Anaphalis margaritacea (Pearly Everlasting)
  Cnephasia pasiuana

Andromeda polifolia (Marsh Andromeda)
  Aphelia viburnana

Anemone spp.
  Cnephasia gueneana

Angelica sylvestris (Wild Angelica)
  Aphelia unitana

Anthemis spp. (Chamomiles)
  Cnephasia longana, C. pasiuana

Anthemis arvensis (Corn Chamomile)
  Aethes smeathmanniana

Anthemis cotula (Stinking Chamomile)
  Cochylidia implicitana

Anthriscus spp.
  Exapate congelatella

Anthriscus cerefolium (Garden Chervil)
Clepsis rurinana

Anthyllis vulneraria (Kidney Vetch)
Paraclepsis cinctana

Antirrhinum majus (Snapdragon)
Falseuncaria ruficiliana; *Cochylis roseana*

Apple, see *Malus sylvestris*

Apricot, see *Prunus armeniaca*

Aquilegia vulgaris (Columbine)
Olindia schumacherana

Arctium spp. (Burdocks)
Aethes rubigana

Arctium lappa (Great Burdock)
Aethes rubigana; *Cochylis dubitana*

Arctium minus (Lesser Burdock)
Aethes rubigana

Arctostaphylos uva-ursi (Bear Berry)
Philedonides lunana

Armeria maritima (Sea Daisy, Thrift)
Philedone gerningana; *Cnephasia longana*;
Eana penziana

Arrhenatherum elatius (False Oat Grass)
Aphelia paleana

Artemisia spp.
Aphelia viburnana

Artemisia campestris (Field
Southernwood)
Hysterosia inopiana; *Cochylidia implicitana*;
Paraclepsis cinctana; *Cnephasia pasiuana*

Artemisia maritima (Sea Wormwood)
Aphelia viburnana; *Clepsis spectrana*

Ash, see *Fraxinus excelsior*

Aspen, see *Populus tremula*

Aster tripolium (Sea Aster)
Phalonidia affinitana; *Cochylidia implicitana*;
Falseuncaria ruficiliana; *Cochylis roseana*;
Argyrotaenia pulchellana; *Aphelia viburnana*;
Clepsis spectrana; *Cnephasia longana*,
C. pasiuana

Autumn Felwort, see *Gentianella
amarella*

Azalea spp.
Acleris comariana

Baccharis spp.
Epiphyas postvittana

Barberry, see *Berberis vulgaris*

Bay, see *Laurus nobilis*

Beach Pine, see *Pinus contorta*

Bean, see *Vicia* spp.

Bear Berry, see *Arctostaphylos uva-ursi*

Beech, see *Fagus sylvatica*

Bellis perennis (Common Daisy)
Falseuncaria ruficiliana

Berberis vulgaris (Barberry)
Pseudargyrotoza conwagana; *Exapate
congelatella*; *Acleris variegana*

Betula spp. (Birches)
Cochylis nana; *Pandemis cerasana*,
P. cinnamomeana, *P. heparana*; *Argyrotaenia
pulchellana*; *Archips betulana*; *Choristoneura
diversana*, *C. hebenstreitella*; *Syndemis
musculana*; *Adoxophyes orana*; *Eulia ministrana*;
Tortricodes alternella; *Neosphaleroptera
nubilana*; *Acleris sparsana*, *A. tripunctana*,
A. logiana, *A. cristana*, *A. lipsiana*, *A. literana*,
A. emargana

Bilberry, see *Vaccinium* spp.;
Vaccinium myrtillus

Birch, see *Betula* spp.

Bird Cherry, see *Prunus padus*

Bird's-eye Primrose, see *Primula farinosa*

Blackcurrant, see *Ribes* spp.

Black Poplar, see *Populus nigra*

Blackthorn, see *Prunus* spp.; *Prunus
spinosa*

Blood-red Geranium, see *Geranium
sanguineum*

Bluebell, see *Endymion non-scriptus*

Blue Fleabane, see *Erigeron acer*

Boehmeria nivea
  Choristoneura lafauryana

Bog Asphodel, see *Narthecium ossifragum*

Bogbean, see *Menyanthes trifoliata*

Bog Myrtle, see *Myrica gale*

Bog Whortleberry, see *Vaccinium uliginosum*

Box, see *Buxus sempervirens*

Bramble, see *Rubus* spp.

Bristly Ox-tongue, see *Picris echioides*

Bryonia dioica (White Bryony)
  Phtheochroa rugosana

Broom, see *Sarothamnus scoparius*

Buckthorn, see *Rhamnus catharticus*

Buddleia davidii
  Epiphyas postvittana

Burdock, see *Arctium* spp.

Burnet Rose, see *Rosa pimpinellifolia*

Butomus umbellatus (Flowering Rush)
  Phalonidia permixtana

Buttercup, see *Ranunculus* spp.

Buxus sempervirens (Box)
  Ditula angustiorana

Calabrian Pine, see *Pinus halepensis* var. *brutia*

Calluna vulgaris (Ling)
  Eupoecilia angustana; Argyrotaenia pulchellana; Philedonides lunana; Exapate congelatella; Acleris hyemana

Caltha palustris (Marsh Marigold)
  Aphelia paleana

Camellia spp.
  Homona menciana

Canadian Fleabane, see *Conyza canadensis*

Carduus spp. (Thistles)
  Agapeta hamana

Carduus acanthoides (Welted Thistle)
  Cochylis dubitana

Carlina vulgaris (Carline Thistle)
  Cnephasia stephensiana

Carline Thistle, see *Carlina vulgaris*

Carnation, see *Dianthus* spp.

Carpinus betulus (Hornbeam)
  Clepsis consimilana; Capua vulgana; Tortricodes alternella; Aleimma loeflingiana; Tortrix viridana; Acleris sparsana, A. umbrana, A. cristana

Carrot, see *Daucus carota*

Cat's-ear, see *Hypochoeris* spp.

Centaurea spp. (Knapweeds)
  Agapeta zoegana; Pandemis dumetana; Argyrotaenia pulchellana; Aphelia viburnana, A. paleana; Sparganothis pilleriana; Cnephasia stephensiana

Centaurea jacea
  Cochylis dubitana

Centaurea nigra (Common Knapweed, Hardheads)
  Stenodes straminea; Agapeta zoegana; Aethes smeathmanniana

Centaurea scabiosa (Greater Knapweed)
  Stenodes alternana; Cnephasia pasiuana

Centranthus ruber (Red Valerian)
  Epiphyas postvittana

Chamomile, see *Anthemis* spp.

Chelidonium majus (Greater Celandine)
  Clepsis rurinana

Chenopodium spp.
  Phalonidia gilvicomana; Cnephasia stephensiana

Cherry, see *Prunus* spp.

Chincherinchee, see *Ornithogalum thyrsoides*

Chondrilla juncea
  Hysterophora maculosana

Chrysanthemum spp.
  Cochylidia implicitana; Epiphyas postvittana;
  Cnephasia longana, C. communana,
  C. conspersana, C. stephensiana, C. interjectana,
  C. pasiuana, C. genitalana, C. incertana

Chrysanthemum leucanthemum (Ox-eye Daisy)
  Cnephasia conspersana, C. stephensiana,
  C. interjectana, C. pasiuana; Eana incanana

Chrysanthemum vulgare (Tansy)
  Aethes margaritana

Chrysosplenium spp. (Golden Saxifrages)
  Olindia schumacherana

Cinquefoil, see *Potentilla* spp.

Cirsium spp. (Thistles)
  Agapeta hamana; Cnephasia stephensiana,
  C. pasiuana

Cirsium oleraceum
  Aethes cnicana, A. rubigana

Cirsium palustre (Marsh Thistle)
  Aethes cnicana

Cirsium vulgare (Spear Thistle)
  Aethes cnicana, A. rubigana; Cochylis dubitana

Clematis vitalba (Traveller's Joy)
  Sparganothis pilleriana

Cloud Berry, see *Rubus chamaemorus*

Clover, see *Trifolium* spp.

Cocksfoot, see *Dactylis glomerata*

Colchicum autumnale (Meadow Saffron)
  Cnephasia pasiuana

Coltsfoot, see *Tussilago farfara*

Columbine, see *Aquilegia vulgaris*

Comfrey, see *Symphytum officinale*

Common Daisy, see *Bellis perennis*

Common Knapweed, see *Centaurea nigra*

Common Laurel, see *Prunus laurocerasus*

Common Maple, see *Acer campestre*

Common Reed, see *Phragmites communis*

Common Rockrose, see *Helianthemum chamaecistus*

Common Toadflax, see *Linaria vulgaris*

Conifers, see Pinaceae

Conium maculatum (Hemlock)
  Aethes beatricella

Convallaria majalis (Lily-of-the-Valley)
  Clepsis senecionana

Convolvulus arvensis (Lesser Bindweed)
  Clepsis rurinana

Conyza canadensis (Canadian Fleabane)
  Cochylidia heydeniana

Corn Chamomile, see *Anthemis arvensis*

Coronilla spp.
  Aphelia viburnana

Corylus avellana (Hazel)
  Pandemis corylana, P. cerasana; Archips
  betulana, A. xylosteana, A. rosana; Choristoneura
  hebenstreitella; Adoxophyes orana; Capua
  vulgana; Eulia ministrana; Tortricodes
  alternella; Acleris rhombana, A. variegana,
  A. cristana, A. emargana

Couch, see *Agropyron repens*

Cowberry, see *Vaccinium vitis-idaea*

Cow Parsnip, see *Heracleum sphondylium*

Cowslip, see *Primula veris*

Crack Willow, see *Salix fragilis*

Crataegus spp. (Hawthorns)
  Clepsis consimilana; Epiphyas postvittana;
  Epagoge grotiana; Isotrias rectifasciana;
  Tortricodes alternella; Exapate congelatella;
  Neosphaleroptera nubilana; Croesia holmiana;
  Acleris latifasciana, A. rhombana, A. variegana,
  A. umbrana, A. cristana

Creeping Twitch, see *Agropyron repens*

Creeping Willow, see *Salix repens*

Crepis spp. (Hawk's Beards)
   Aethes tesserana; *Cochylis dubitana*,
   C. hybridella

Crinitaria linosyris (Goldilocks)
   Cochylidia rupicola

Crocosmia crocosmi-flora (Montbretia)
   Epiphyas postvittana

Cudweed, see *Gnaphalium* spp.

Currant, see *Ribes* spp.

Cyclamen spp.
   Clepsis spectrana

Cydonia spp.
   Croesia holmiana

Dactylis glomerata (Cocksfoot)
   Aphelia paleana

Dandelion, see *Taraxacum* spp.

Daucus carota (Wild Carrot)
   Aethes williana, *A. francillana*

Devil's Bit Scabious, see *Succisa pratensis*

Dianthus spp. (Carnations; Pinks)
   Cacoecimorpha pronubana; *Epichoristodes acerbella*

Dipsacus fullonum (Teasel)
   Cochylis roseana

Dock, see *Rumex* spp.

Dog's Mercury, see *Mercurialis perennis*

Dogwood, see *Thelycrania sanguinea*

Dorycnium spp.
   Clepsis senecionana

Drooping Star of Bethlehem, see *Ornithogalum nutans*

Dryas octopetala (Mountain Avens)
   Argyrotaenia pulchellana; *Aphelia viburnana*;
   Clepsis conspersana; *Acleris aspersana*

Dyer's Greenweed, see *Genista tinctoria*

Eared Willow, see *Salix aurita*

Echium vulgare (Viper's Bugloss)
   Cnephasia pasiuana

Elder, see *Sambucus nigra*

Elm, see *Ulmus* spp.

Endymion non-scriptus (Bluebell)
   Hysterophora maculosana; *Eana incanana*

Epilobium spp. (Willow Herbs)
   Clepsis spectrana; *Eulia ministrana*

Erica spp. (Heathers)
   Argyrotaenia pulchellana; *Aphelia viburnana*;
   Philedonides lunana; *Exapate congelatella*;
   Acleris hyemana

Erigeron acer (Blue Fleabane)
   Cochylidia heydeniana; *Cnephasia pasiuana*

Eryngium campestre
   Aethes williana

Eryngium maritimum (Sea Holly)
   Aethes margarotana, *A. williana*

Escallonia spp.
   Epiphyas postvittana

Euonymus japonicus (Japanese Spindle Tree)
   Archips podana; *Cacoecimorpha pronubana*;
   Epiphyas postvittana

Eupatorium cannabinum (Hemp Agrimony)
   Cochylidia rupicola

Euphorbia spp. (Spurges)
   Clepsis rurinana

Euphorbia amygdaloides (Wood Spurge)
   Cacoecimorpha pronubana

Euphrasia spp. (Eyebrights)
   Phalonidia permixtana

Eyebright, see *Euphrasia* spp.

Fagus sylvatica (Beech)
   Aphelia paleana; *Clepsis rurinana*; *Eulia
   ministrana*; *Tortrix viridana*; *Acleris sparsana*,
   A. tripunctana, *A. ferrugana*, *A. cristana*,
   A. literana

False Acacia, see *Robinia pseudacacia*

False Oat Grass, see *Arrhenatherum elatius*

Ferula communis
Aethes williana

Festuca ovina (Sheep's Fescue)
Eana penziana

Field Scabious, see *Knautia arvensis*

Field Southernwood, see *Artemisia campestris*

Figwort, see *Scrophularia* spp.

Filipendula ulmaria (Meadow Sweet)
Aphelia viburnana, A. paleana; Clepsis
spectrana; Acleris latifasciana, A. aspersana,
A. shepherdana, A. rufana

Flax, see *Linum vulgare*

Fleabane, see *Pulicaria dysenterica*

Flowering Rush, see *Butomus umbellatus*

Forsythia spp.
Choristoneura lafauryana

Fragaria spp. (Strawberries)
Pandemis dumetana; Choristoneura lafauryana;
Cacoecimorpha pronubana; Clepsis spectrana;
Lozotaenia forsterana; Cnephasia longana,
C. interjectana, C. incertana; Acleris comariana

Fragaria vesca (Wild Strawberry)
Acleris comariana, A. aspersana

Frangula alnus (Alder Buckthorn)
Hysterophora sodaliana; Eupoecilia ambiguella;
Spatalistis bifasciana; Croesia bergmanniana

Fraxinus excelsior (Ash)
Pandemis corylana; Archips crataegana,
A. xylosteana; Pseudargyrotoza conwagana;
Eulia ministrana

Fuchsia spp.
Cacoecimorpha pronubana

Garden Chervil, see *Anthriscus cerefolium*

Genista tinctoria (Dyer's Greenweed)
Paraclepsis cinctana

Gentian, see *Gentiana* spp.

Gentiana spp. (Gentians)
Falseuncaria ruficiliana

Gentiana lutea
Phalonidia permixtana

Gentiana pneumonanthe (Marsh Gentian)
Argyrotaenia pulchellana

Gentianella amarella (Autumn Felwort)
Clepsis senecionana

Geranium sanguineum (Blood-red Geranium)
Aphelia viburnana

Germander, see *Teucrium* spp.

Geum spp.
Acleris aspersana

Geum rivale (Water Avens)
Acleris comariana

Gipsy-wort, see *Lycopus europaeus*

Gnaphalium spp. (Cudweeds)
Aethes williana; Cochylidia implicitana

Golden Rod, see *Solidago virgaurea*

Golden Saxifrage, see *Chrysosplenium* spp.

Goldilocks, see *Crinitaria linosyris*

Gooseberry, see *Ribes* spp.

Gramineae (Grasses—various species)
Paramesia gnomana; Eana argentana, E. osseana

Grape Vine, see *Vitis vinifera*

Grasses, see Gramineae

Great Burdock, see *Arctium lappa*

Great Burnet, see *Sanguisorba officinalis*

Great Fen Ragwort, see *Senecio paludosus*

Great Sallow, see *Salix caprea*

Greater Celandine, see *Chelidonium majus*

Greater Knapweed, see *Centaurea scabiosa*

Guelder Rose, see *Viburnum opulus*

Hardheads, see *Centaurea nigra*

Hawk's Beard, see *Crepis* spp.

Hawkweed, see *Hieracium* spp.

Hawkweed Ox-tongue, see *Picris hieracioides*

Hawthorn, see *Crataegus* spp.

Hazel, see *Corylus avellana*

Heather, see *Erica* spp.

Hebe spp.
  Epiphyas postvittana

Hedera helix (Ivy)
  Eupoecilia ambiguella; Pandemis dumetana; Choristoneura hebenstreitella; Clepsis consimilana; Epiphyas postvittana; Lozotaenia forsterana; Ditula angustiorana

Helianthemum spp.
  Aphelia viburnana; Philedone gerningana; Cnephasia conspersana

Helianthemum chamaecistus (Common Rockrose)
  Acleris aspersana

Helichrysum stoechas
  Aethes williana

Hemlock, see *Conium maculatum*

Hemp Agrimony, see *Eupatorium cannabinum*

Heracleum spp.
  Aphelia unitana; Cnephasia stephensiana

Heracleum sphondylium (Cow Parsnip, Hogweed)
  Aethes dilucidana

Hieracium spp. (Hawkweeds)
  Aethes tesserana; Cochylis dubitana; Cnephasia conspersana, C. stephensiana, C. genitalana

Hippophae rhamnoides (Sea Buckthorn)
  Cacoecimorpha pronubana; Ditula angustiorana

Hogweed, see *Heracleum sphondylium*

Holly, see *Ilex aquifolium*

Honeysuckle, see *Lonicera periclymenum*

Hop, see *Humulus lupulus*

Hornbeam, see *Carpinus betulus*

Houseleek, see *Sempervivum tectorum*

Humulus lupulus (Hop)
  Clepsis spectrana; Adoxophyes orana; Sparganothis pilleriana

Hypericum spp. (St. John's Worts)
  Archips xylosteana; Epiphyas postvittana

Hypochoeris spp. (Cat's-ears)
  Cnephasia longana, C. conspersana, C. pasiuana

Ilex aquifolium (Holly)
  Ditula angustiorana

Inula spp.
  Falseuncaria ruficiliana; Cnephasia stephensiana

Inula conyza (Ploughman's Spikenard)
  Aethes tesserana

Iris pseudacorus (Yellow Flag)
  Clepsis spectrana; Paramesia gnomana; Sparganothis pilleriana

Ivy, see *Hedera helix*

Japanese Spindle Tree, see *Euonymus japonicus*

Jasione montana (Sheep's-bit)
  Cochylis pallidana

Jasminum spp.
  Epiphyas postvittana

Jonquil, see *Narcissus jonquilla*

Juniper, see *Juniperus communis*

Juniperus communis (Juniper)
Aethes rutilana; *Archips oporana*; *Ditula angustiorana*

Kidney Vetch, see *Anthyllis vulneraria*

Knapweed, see *Centaurea* spp.

Knautia arvensis (Field Scabious)
Agapeta zoegana; *Aethes hartmanniana*; *Cochylis flaviciliana*

Knotgrass, see *Polygonum* spp.

Lapsana communis (Nipple-wort)
Phalonidia gilvicomana

Larch, see *Larix decidua*

Larix decidua (Larch)
Pandemis cinnamomeana; *Argyrotaenia pulchellana*; *Archips oporana*; *Syndemis musculana*; *Ptycholomoides aeriferanus*; *Clepsis senecionana*; *Ptycholoma lecheana*; *Lozotaenia forsterana*; *Ditula angustiorana*

Lathyrus spp. (Peas)
Cnephasia stephensiana, *C. interjectana*, *C. incertana*

Lathyrus palustris (Marsh Pea)
Pandemis dumetana

Laurus nobilis (Bay)
Cacoecimorpha pronubana; *Ditula angustiorana*

Lavandula spp. (Lavender)
Epiphyas postvittana

Lavender, see *Lavandula* spp.

Ledum palustre
Aphelia viburnana

Leontodon hispidus (Rough Hawkbit)
Cnephasia conspersana

Lesser Bindweed, see *Convolvulus arvensis*

Lesser Burdock, see *Arctium minus*

Lesser Celandine, see *Ranunculus ficaria*

Ligularia spp.
Cnephasia longana

Ligustrum spp. (Privets)
Cacoecimorpha pronubana; *Clepsis consimilana*; *Epiphyas postvittana*; *Lozotaenia forsterana*; *Pseudargyrotoza conwagana*; *Exapate congelatella*

Lilac, see *Syringa* spp.

Lilium spp. (Lilies)
Clepsis rurinana

Lily, see *Lilium* spp.

Lily-of-the-Valley, see *Convallaria majalis*

Lime, see *Tilia* spp.

Limonium vulgare (Sea Lavender)
Clepsis spectrana; *Sparganothis pilleriana*

Linaria vulgaris (Common Toadflax)
Falseuncaria degreyana

Ling, see *Calluna vulgaris*

Linum vulgare (Flax)
Falseuncaria ruficiliana; *Cnephasia longana*

Lonicera periclymenum (Honeysuckle)
Eupoecilia ambiguella; *Pandemis heparana*; *Archips xylosteana*; *Choristoneura diversana*; *Aphelia viburnana*; *Clepsis rurinana*, *C. consimilana*; *Adoxophyes orana*; *Lozotaenia forsterana*

Lotus spp. (Trefoils)
Aphelia viburnana; *Clepsis senecionana*; *Paraclepsis cinctana*; *Philedone gerningana*; *Cnephasia communana*, *C. incertana*

Lousewort, see *Pedicularis sylvatica*

Lucerne, see *Medicago sativa*

Lychnis spp.
Cnephasia longana

Lycopus europaeus (Gipsy-wort)
Phalonidia manniana; *Cochylidia rupicola*

Lycopersicum esculentum (Tomato)
Cacoecimorpha pronubana

Lysimachia spp.
Pandemis cerasana, *P. dumetana*; *Aphelia viburnana*; *Clepsis senecionana*

Lythrum spp.
    Aphelia viburnana

Lythrum salicaria (Purple Loosestrife)
    Acleris lorquiniana

Malus sylvestris (Apple)
    Pandemis cerasana, P. heparana;
    Argyrotaenia pulchellana; Archips podana,
    A. rosana; Choristoneura diversana,
    C. hebenstreitella, C. lafauryana; Clepsis
    consimilana; Epiphyas postvittana; Adoxophyes
    orana; Ptycholoma lecheana; Ditula
    angustiorana; Cnephasia incertana;
    Neosphaleroptera nubilana; Croesia holmiana;
    Acleris rhombana, A. variegana, A. cristana,
    A. lipsiana, A. literana

Marjoram, see *Origanum vulgare*

Marsh Andromeda, see *Andromeda polifolia*

Marsh Arrow-grass, see *Triglochin palustris*

Marsh Cinquefoil, see *Potentilla palustris*

Marsh Gentian, see *Gentiana pneumonanthe*

Marsh Marigold, see *Caltha palustris*

Marsh Pea, see *Lathyrus palustris*

Marsh Thistle, see *Cirsium palustre*

Matricaria recutita (Wild Chamomile)
    Phalonidia luridana; Aethes margaritana;
    Cochylidia implicitana

Meadow Saffron, see *Colchicum autumnale*

Meadow Sweet, see *Filipendula ulmaria*

Medicago sativa (Lucerne)
    Cnephasia incertana

Mentha spp. (Mints)
    Pandemis dumetana; Epiphyas postvittana;
    Philedonides lunana

Mentha aquatica (Water Mint)
    Phalonidia manniana

Menyanthes trifoliata (Bogbean)
    Phalonidia minimana

Mercurialis perennis (Dog's Mercury)
    Olindia schumacherana

Mesembryanthemum spp.
    Epiphyas postvittana

Milk Parsley, see *Selinum carvifolia*

Mint, see *Mentha* spp.

Mistletoe, see *Viscum album*

Montbretia, see *Crocosmia crocosmi-flora*

Mosses, see Musci

Mountain Ash, see *Sorbus* spp.; *Sorbus aucuparia*

Mountain Avens, see *Dryas octopetala*

Musci (Mosses—various species)
    Eana argentana, E. osseana

Mycelis muralis (Wall Lettuce)
    Phalonidia gilvicomana

Myrica gale (Bog Myrtle)
    Pandemis heparana; Argyrotaenia pulchellana;
    Archips betulana; Choristoneura hebenstreitella,
    C. lafauryana; Aphelia viburnana; Clepsis
    senecionana; Philedonides lunana; Acleris
    caledoniana, A. tripunctana, A. lipsiana,
    A. rufana, A. maccana

Narcissus jonquilla (Jonquil)
    Cnephasia gueneana

Narthecium ossifragum (Bog Asphodel)
    Sparganothis pilleriana

Nettle, see *Urtica* spp.

Nipple-wort, see *Lapsana communis*

Oak, see *Quercus* spp.

Onobrychis viciifolia (Sainfoin)
    Clepsis senecionana

Origanum vulgare (Marjoram)
    Eupoecilia angustana; Sparganothis pilleriana

Ornithogalum nutans (Drooping Star of Bethlehem)
    Eana incanana

Ornithogalum thyrsoides (Chincherinchee)
  Epichoristodes acerbella

Ox-eye Daisy, see *Chrysanthemum leucanthemum*

Ox-tongue, see *Picris* spp.

Pastinaca sativa (Wild Parsnip)
  Aethes dilucidana; *Aphelia viburnana*

Pea, see *Lathyrus* spp.

Pear, see *Pyrus* spp.; *Pyrus communis*

Pearly Everlasting, see *Anaphalis margaritacea*

Pedicularis spp.
  Phalonidia permixtana

Pedicularis palustris (Red Rattle)
  Phalonidia minimana

Pedicularis sylvatica (Lousewort)
  Falseuncaria ruficiliana

Pelargonium spp.
  Clepsis spectrana

Petasites albus (White Butterbur)
  Epagoge grotiana

Peucedanum spp.
  Philedone gerningana

Peucedanum officinale (Sea Hog's Fennel)
  Aethes francillana

Phleum pratense (Timothy Grass)
  Aphelia paleana

Phragmites communis (Common Reed)
  Aphelia paleana

Picea spp. (Spruces)
  Argyrotaenia pulchellana; *Syndemis musculana*;
  Clepsis senecionana; *Ptycholoma lecheana*;
  Lozotaeniodes formosanus; *Lozotaenia forsterana*;
  Eana argentana; *Acleris abietana*

Picea glauca (White Spruce)
  Archips oporana

Picea sitchensis (Sitka Spruce)
  Eupoecilia angustana; *Pandemis cinnamomeana*;
  Aphelia paleana; *Philedone gerningana*;
  Philedonides lunana

Picris spp. (Ox-tongues)
  Cochylis dubitana

Picris echioides (Bristly Ox-tongue)
  Aethes tesserana; *Cochylis hybridella*

Picris hieracioides (Hawkweed Ox-tongue)
  Aethes tesserana; *Cochylis hybridella*

Pinaceae (Conifers—various species)
  Archips rosana; *Cnephasia incertana*

Pine, see *Pinus* spp.

Pinks, see *Dianthus* spp.

Pinus spp. (Pines)
  Argyrotaenia pulchellana; *Aphelia viburnana*;
  Clepsis senecionana; *Lozotaenia forsterana*;
  Acleris abietana

Pinus halepensis var. *brutia* (Calabrian Pine)
  Sparganothis pilleriana

Pinus contorta (Beach Pine)
  Philedonides lunana; *Acleris caledoniana*,
  A. hyemana

Pinus pinea (Stone Pine)
  Sparganothis pilleriana

Pinus sylvestris (Scots Pine)
  Archips oporana; *Lozotaeniodes formosanus*;
  Ditula angustiorana

Plantago spp. (Plantains)
  Eupoecilia angustana; *Aphelia paleana*;
  Paramesia gnomana; *Philedone gerningana*;
  Sparganothis pilleriana; *Olindia schumacherana*;
  Cnephasia communana, *C. stephensiana*,
  C. interjectana, *C. incertana*

Plantago lanceolata (Ribwort Plantain)
  Falseuncaria degreyana

Plantago maritima (Sea Plantain)
  Eana penziana

Plantain, see *Plantago* spp.

Ploughman's Spikenard, see *Inula conyza*

Plum, see *Prunus* spp.; *Prunus domestica*

Polygonatum spp. (Solomon's Seal)
  Clepsis senecionana

Polygonum spp. (Knotgrasses)
  Clepsis consimilana

Poplar, see *Populus* spp.

Populus spp. (Poplars)
  Adoxophyes orana; *Ptycholoma lecheana*;
  Tortrix viridana; *Acleris latifasciana*,
  A. sparsana, *A. emargana*

Populus alba (White Poplar)
  Acleris hastiana, *A. rufana*

Populus nigra (Black Poplar)
  Hysterosia schreibersiana

Populus tremula (Aspen)
  Acleris tripunctana

Potentilla spp. (Cinquefoils)
  Aphelia viburnana; *Clepsis senecionana*,
  C. spectrana; *Philedone gerningana*;
  Philedonides lunana; *Exapate congelatella*;
  Acleris caledoniana, *A. aspersana*

Potentilla palustris (Marsh Cinquefoil)
  Acleris comariana

Primrose, see *Primula vulgaris*

Primula farinosa (Bird's-eye Primrose)
  Falseuncaria ruficiliana

Primula veris (Cowslip)
  Falseuncaria ruficiliana; *Cnephasia stephensiana*

Primula vulgaris (Primrose)
  Falseuncaria ruficiliana

Privet, see *Ligustrum* spp.

Prunus spp. (Blackthorn; Cherries; Plums)
  Eupoecilia ambiguella; *Pandemis corylana*,
  P. cerasana, *P. cinnamomeana*, *P. heparana*;
  Archips rosana; *Argyrotaenia pulchellana*;
  Choristoneura diversana, *C. hebenstreitella*;
  Adoxophyes orana; *Ptycholoma
  lecheana*; *Ditula angustiorana*; *Eulia
  ministrana*; *Neosphaleroptera nubilana*;
  Croesia holmiana; *Acleris latifasciana*,
  A. rhombana, *A. variegana*, *A. hastiana*,
  A. cristana

Prunus armeniaca (Apricot)
  Neosphaleroptera nubilana

Prunus cerasus (Sour Cherry)
  Adoxophyes orana; *Acleris tripunctana*

Prunus domestica (Wild Plum)
  Neosphaleroptera nubilana

Prunus laurocerasus (Common Laurel)
  Acleris rhombana

Prunus padus (Bird Cherry)
  Hysterosia schreibersiana; *Acleris umbrana*

Prunus spinosa (Blackthorn)
  Tortricodes alternella; *Exapate congelatella*;
  Acleris variegana, *A. permutana*, *A. cristana*

Pulicaria spp.
  Clepsis rurinana; *Epiphyas postvittana*

Pulicaria dysenterica (Fleabane)
  Hysterosia inopiana

Purple Loosestrife, see *Lythrum salicaria*

Pyracantha spp.
  Epiphyas postvittana

Pyrus spp. (Pears)
  Pandemis cerasana, *P. cinnamomeana*,
  P. heparana; *Archips rosana*;
  Choristoneura diversana, *C. hebenstreitella*,
  C. lafauryana; *Adoxophyes orana*; *Ditula
  angustiorana*; *Sparganothis pilleriana*; *Exapate
  congelatella*; *Neosphaleroptera nubilana*;
  Croesia holmiana; *Acleris rhombana*,
  A. variegana

Pyrus communis (Pear)
  Acleris tripunctana

Quercus spp. (Oaks)
  Pandemis corylana, *P. cerasana*,
  P. cinnamomeana, *P. dumetana*; *Archips
  crataegana*, *A. xylosteana*; *Choristoneura
  diversana*, *C. hebenstreitella*; *Syndemis
  musculana*; *Aphelia paleana*; *Clepsis rurinana*;
  Ptycholoma lecheana; *Epagoge grotiana*;
  Capua vulgana; *Ditula angustiorana*; *Eulia
  ministrana*; *Tortricodes alternella*; *Aleimma
  loeflingiana*; *Tortrix viridana*; *Acleris sparsana*,
  A. rhombana, *A. tripunctana*, *A. ferrugana*,
  A. cristana, *A. literana*

Ragwort, see *Senecio* spp.; *Senecio jacobaea*

Ranunculus spp. (Buttercups)
    Cnephasia longana, C. stephensiana,
    C. interjectana, C. pasiuana, C. genitalana,
    C. incertana

Ranunculus ficaria (Lesser Celandine)
    Olindia schumacherana

Raspberry, see *Rubus idaeus*

Red Rattle, see *Pedicularis palustris*

Red Valerian, see *Centranthus ruber*

Rhamnus catharticus (Buckthorn)
    Hysterosia sodaliana; Choristoneura diversana;
    Eulia ministrana; Exapate congelatella;
    Spatalistis bifasciana; Croesia bergmanniana

Rhinanthus spp.
    Phalonidia permixtana

Rhododendron spp.
    Ditula angustiorana; Acleris comariana

Ribes spp. (Currants; Gooseberry)
    Eupoecilia ambiguella; Pandemis heparana;
    Archips rosana; Choristoneura lafauryana;
    Adoxophyes orana; Lozotaenia forsterana;
    Exapate congelatella

Ribwort Plantain, see *Plantago lanceolata*

Robinia pseudacacia (False Acacia)
    Cacoecimorpha pronubana

Rosa spp. (Roses)
    Clepsis rurinana; Adoxophyes orana; Epagoge
    grotiana; Eulia ministrana; Croesia
    bergmanniana, C. holmiana; Acleris
    latifasciana, A. rhombana, A. variegana,
    A. permutana

Rosa pimpinellifolia (Burnet Rose)
    Croesia bergmanniana; Acleris permutana

Rose, see *Rosa* spp.

Rough Hawkbit, see *Leontodon hispidus*

Rowan, see *Sorbus aucuparia*

Rubus spp. (Brambles)
    Pandemis corylana, P. dumetana; Archips
    xylosteana; Syndemis musculana; Aphelia
    unitana; Epiphyas postvittana; Adoxophyes
    orana; Epagoge grotiana; Capua vulgana;
    Exapate congelatella; Croesia holmiana;
    Acleris latifasciana, A. sparsana, A. aspersana,
    A. variegana

Rubus chamaemorus (Cloud Berry)
    Eulia ministrana; Acleris caledoniana

Rubus idaeus (Raspberry)
    Archips rosana; Cnephasia interjectana;
    Acleris tripunctana, A. rufana

Rue, see *Thalictrum* spp.

Rumex spp. (Docks; Sorrels)
    Cnephasia communana, C. stephensiana,
    C. interjectana, C. incertana

Rush, see *Scirpus* spp.

Sainfoin, see *Onobrychis viciifolia*

St. John's Wort, see *Hypericum* spp.

Salad Burnet, see *Sanguisorba minor*

Salix spp. (Sallows; Willows)
    Pandemis cerasana, P. heparana; Archips
    crataegana, A. xylosteana; Choristoneura
    diversana, C. lafauryana; Aphelia viburnana;
    Adoxophyes orana; Ptycholoma lecheana; Exapate
    congelatella; Eana argentana; Tortrix viridana;
    Acleris latifasciana, A. aspersana, A. umbrana,
    A. hastiana, A. emargana

Salix aurita (Eared Willow)
    Acleris hastiana

Salix caprea (Great Sallow)
    Choristoneura hebenstreitella; Acleris ferrugana,
    A. rufana, A. emargana

Salix fragilis (Crack Willow)
    Acleris hastiana

Salix repens (Creeping Willow)
    Acleris hastiana

Sallow, see *Salix* spp.

Sambucus nigra (Elder)
    Choristoneura hebenstreitella

Sanguisorba minor (Salad Burnet)
    Acleris aspersana, A. variegana

Sanguisorba officinalis (Great Burnet)
    Aphelia viburnana; Acleris aspersana,
    A. shepherdana

Sarothamnus scoparius (Broom)
    Paraclepsis cinctana

Saw-wort, see *Serratula* spp.; *Serratula tinctoria*

Saxifraga spp. (Saxifrages)
  Cnephasia incertana

Saxifrage, see *Saxifraga* spp.

Scabiosa spp. (Scabiouses)
  Aphelia paleana

Scabiosa columbaria (Small Scabious)
  Agapeta zoegana; Aethes hartmanniana; Philedone gerningana

Scabious, see *Scabiosa* spp.

Scentless Chamomile,
see *Tripleurospermum maritimum* ssp. *inodorum*

Scirpus spp. (Rushes)
  Clepsis spectrana

Scots Pine, see *Pinus sylvestris*

Scrophularia spp. (Figworts)
  Aphelia viburnana

Sea Arrow-grass, see *Triglochin maritima*

Sea Aster, see *Aster tripolium*

Sea Buckthorn, see *Hippophae rhamnoides*

Sea Campion, see *Silene maritima*

Sea Daisy, see *Armeria maritima*

Sea Hog's Fennel, see *Peucedanum officinale*

Sea Holly, see *Eryngium maritimum*

Sea Lavender, see *Limonium vulgare*

Sea Plantain, see *Plantago maritima*

Sea Wormwood, see *Artemisia maritima*

Selinum carvifolia (Milk Parsley)
  Cnephasia longana

Sempervivum tectorum (Houseleek)
  Cnephasia longana

Senecio spp. (Ragworts)
  Cochylis dubitana; Cnephasia conspersana, C. genitalana

Senecio jacobaea (Ragwort)
  Commophila aeneana; Cochylis atricapitana

Senecio paludosus (Great Fen Ragwort)
  Commophila aeneana

Serratula spp. (Saw-worts)
  Cnephasia stephensiana

Serratula salicina
  Agapeta zoegana

Serratula tinctoria (Saw-wort)
  Agapeta hamana

Sheep's-bit, see *Jasione montana*

Sheep's Fescue, see *Festuca ovina*

Silene maritima (Sea Campion)
  Cnephasia conspersana

Silver Fir, see *Abies alba*

Sitka Spruce, see *Picea sitchensis*

Small Scabious, see *Scabiosa columbaria*

Smyrnium olusatrum (Alexanders)
  Aethes beatricella; Philedonides lunana

Snapdragon, see *Antirrhinum majus*

Solanum dulcamara (Woody Nightshade)
  Adoxophyes orana

Solidago virgaurea (Golden Rod)
  Phalonidia curvistrigana; Eupoecilia angustana; Cochylidia implicitana, C. subroseana; Falseuncaria ruficiliana; Cochylis dubitana

Solomon's Seal, see *Polygonatum* spp.

Sonchus spp. (Sowthistles)
  Cnephasia stephensiana

Sorbus spp. (Mountain Ash; White Beams)
  Pandemis cerasana; Choristoneura hebenstreitella; Capua vulgana; Eulia ministrana; Acleris latifasciana, A. sparsana, A. rhombana, A. umbrana

Sorbus aria agg. (White Beams)
  Acleris cristana

Sorbus aucuparia (Mountain Ash, Rowan)
  Pandemis cinnamomeana

Sorrel, see *Rumex* spp.

Sour Cherry, see *Prunus cerasus*

Sowthistle, see *Sonchus* spp.

Spear Thistle, see *Cirsium vulgare*

Spruce, see *Picea* spp.

Spurge, see *Euphorbia* spp.

Stachys spp. (Woundworts)
  Lozotaenia forsterana; *Paramesia gnomana*;
  Sparganothis pilleriana

Stinking Chamomile, see *Anthemis cotula*

Stone Pine, see *Pinus pinea*

Strawberry, see *Fragaria* spp.

Succisa pratensis (Devil's Bit Scabious)
  Aethes piercei

Sycamore, see *Acer pseudoplatanus*

Symphytum officinale (Comfrey)
  Acleris latifasciana

Syringa spp. (Lilacs)
  Choristoneura diversana; *Clepsis consimilana*;
  Pseudargyrotoza conwagana; *Exapate
  congelatella*

Tansy, see *Chrysanthemum vulgare*

Taraxacum spp. (Dandelions)
  Paramesia gnomana; *Cnephasia conspersana*,
  C. stephensiana

Taxus baccata (Yew)
  Ditula angustiorana

Teasel, see *Dipsacus fullonum*

Teucrium spp. (Germanders)
  Aphelia viburnana; *Cnephasia conspersana*,
  C. interjectana, *C. genitalana*

Thalictrum spp. (Rues)
  Pandemis dumetana; *Cnephasia pasiuana*

Thelycrania sanguinea (Dogwood)
  Eupoecilia ambiguella; *Pandemis corylana*;
  Spatalistis bifasciana; *Acleris umbrana*

Thistle, see *Carduus* spp.; *Cirsium* spp.

Thrift, see *Armeria maritima*

Thyme, see *Thymus* spp.

Thymus spp. (Thymes)
  Eupoecilia angustana

Tilia spp. (Limes)
  Pandemis cerasana, *P. heparana*; *Archips
  crataegana*, *A. xylosteana*; *Eulia ministrana*;
  Tortricodes alternella; *Acleris cristana*,
  A. literana

Timothy Grass, see *Phleum pratense*

Tomato, see *Lycopersicum esculentum*

Traveller's Joy, see *Clematis vitalba*

Trefoil, see *Trifolium* spp.

Trifolium spp. (Clovers; Trefoils)
  Choristoneura diversana; *Cnephasia incertana*

Triglochin maritima (Sea Arrow-grass)
  Phalonidia vectisana

Triglochin palustris (Marsh Arrow-grass)
  Phalonidia vectisana

Tripleurospermum maritimum ssp. *inodorum*
(Scentless Chamomile)
  Cochylidia implicitana

Tussilago farfara (Coltsfoot)
  Aphelia paleana; *Cnephasia stephensiana*

Ulmus spp. (Elms)
  Hysterosia schreibersiana; *Pandemis cerasana*;
  Archips crataegana, *A. xylosteana*;
  Choristoneura hebenstreitella; *Exapate
  congelatella*; *Acleris variegana*, *A. boscana*,
  A. cristana

Urtica spp. (Nettles)
  Pandemis dumetana; *Clepsis rurinana*,
  C. spectrana; *Epiphyas postvittana*; *Tortrix
  viridana*

Vaccinium spp. (Bilberries)
  Pandemis cerasana, P. cinnamomeana,
  P. heparana; Argyrotaenia pulchellana;
  Archips betulana; Choristoneura hebenstreitella;
  Aphelia viburnana; Lozotaenia forsterana;
  Epagoge grotiana; Capua vulgana; Philedone
  gerningana; Philedonides lunana;
  Olindia schumacherana; Exapate congelatella;
  Eana incanana; Tortrix viridana;
  Acleris latifasciana, A. variegana, A. maccana

Vaccinium myrtillus (Bilberry)
  Aphelia viburnana; Clepsis senecionana;
  Lozotaenia forsterana; Epagoge gnomana;
  Eulia ministrana; Spatalistis bifasciana;
  Acleris caledoniana, A. lipsiana, A. maccana

Vaccinium uliginosum (Bog Whortleberry)
  Spatalistis bifasciana

Vaccinium vitis-idaea (Cowberry)
  Aphelia viburnana; Acleris caledoniana,
  A. lipsiana, A. maccana

Valeriana spp.
  Philedonides lunana

Vetch, see *Vicia* spp.

Viburnum spp.
  Aphelia viburnana

Viburnum lantana (Wayfaring Tree)
  Acleris schalleriana

Viburnum opulus (Guelder Rose)
  Acleris schalleriana

Vicia spp. (Vetches)
  Cnephasia longana, C. stephensiana,
  C. interjectana, C. incertana

Viper's Bugloss, see *Echium vulgare*

Viscum album (Mistletoe)
  Ditula angustiorana

Vitis vinifera (Grape Vine)
  Eupoecilia ambiguella; Argyrotaenia pulchellana;
  Ditula angustiorana; Sparganothis pilleriana;
  Cnephasia incertana

Wall Lettuce, see *Mycelis muralis*

Water Avens, see *Geum rivale*

Water Plantain, see *Alisma plantago-aquatica*

Wayfaring Tree, see *Viburnum lantana*

Welted Thistle, see *Carduus acanthoides*

White Beam, see *Sorbus* spp.; *Sorbus aria*

White Bryony, see *Bryonia dioica*

White Butterbur, see *Petasites albus*

White Poplar, see *Populus alba*

White Spruce, see *Picea glauca*

Wild Angelica, see *Angelica sylvestris*

Wild Carrot, see *Daucus carota*

Wild Chamomile, see *Matricaria recutita*

Wild Parsnip, see *Pastinaca sativa*

Wild Plum, see *Prunus domestica*

Wild Strawberry, see *Fragaria vesca*

Willow, see *Salix* spp.

Willow Herb, see *Epilobium* spp.

Wood Spurge, see *Euphorbia amygdaloides*

Woody Nightshade, see *Solanum dulcamara*

Woundwort, see *Stachys* spp.

Yarrow, see *Achillea millefolium*

Yellow Flag, see *Iris pseudacorus*

Yew, see *Taxus baccata*

BIBLIOGRAPHY

Adkin, R., 1888. *Tortrix piceana* L. in Surrey. *Entomologist's mon. Mag.*, **25**: 160

Adkin, R., 1906. *Tortrix pronubana* Hb. reared from British larvae. *Entomologist's mon. Mag.*, **42**: 274

Adkin, R., 1908. Life-history of *Tortrix pronubana*. *Entomologist*, **41**: 49–51, pl. 2, figs. 1–7

Adkin, R., 1924. Apple fruit attacked by the larva of *Tortrix heparana*. *Entomologist*, **57**: 188–189

Adkin, R., 1932. Additions to the lepidopterous fauna of the county of Sussex since the publication of the Victoria History List, 1905. *Entomologist*, **65**: 28–33

Allen, A. A., 1959. A note on certain apple-feeding microlepidoptera at Blackheath. *Entomologist's Rec. J. Var.*, **71**: 151–153

Aston, A. 1966. Additions to the Suffolk microlepidoptera. *Trans. Suffolk Nat. Soc.*, **13**: 159–162

Atmore, E. A., 1881a. *Tortrix lafauryana* Ragonot, a species new to Britain. *Entomologist's mon. Mag.*, **18**: 17

Atmore, E. A., 1881b. *Tortrix lafauryana* Ragonot, a species new to the British fauna. *Entomologist*, **14**: 153–154

Atmore, E. A., 1889. Notes on *Tortrix decretana* Tr., with a description of its larva. *Entomologist's mon. Mag.*, **25**: 243–245

Baker C. R. B., 1964. *Homona menciana* (Walker) (Lep., Tortricidae) imported into Britain. *Entomologist*, **97**: 275

Baker, C. R. B., 1968. Notes on *Epiphyas* (=*Austrotortrix*) *postvittana* (Walker) (Lep., Tortricidae). *Entomologist's Gaz.*, **19**: 167–172, fig. 1 (distribution map)

Balachowsky, A. S. et al., 1966. *Entomologie appliquée à l'agriculture*. Lépidoptères. **2**(1): 1–1057, figs. 1–373. Paris

Balfour, A. B., 1930. Butterflies and moths found in East Lothian. *Trans. E. Loth. Antiq. Fld Nat. Soc.*, **1**(5): 169–184

Bankes, E. R., 1899. Notes on the life-history of *Phalonia vectisana* Westw., with descriptions of the larva and pupa. *Entomologist's mon. Mag.*, **35**: 178–180

Bankes, E. R., 1909. Notes on the life-history of *Clepsis rusticana* Tr., with descriptions of the larva and pupa. *Entomologist's mon. Mag.*, **45**: 151–154

Barrett, C. G., 1861. Irish captures. *Entomologist's wkly Intell.*, **10**: 11–14

Barrett, C. G., 1872. Notes on British Tortrices, with descriptions of two new species. *Entomologist's mon. Mag.*, **9**: 124–130

Barrett, C. G., 1874. Notes on British Tortrices. *Entomologist's mon. Mag.*, **11**: 28–31

Barrett, C. G., 1880. Notes on British Tortrices. *Entomologist's mon. Mag.*, **17**: 35–38

Barrett, C. G., 1881. Notes on British Tortrices. *Entomologist's mon. Mag.*, **18**: 152–154

Barrett, C. G., 1883. Notes on British Tortrices. *Entomologist's mon. Mag.*, **20**: 132–135

Barrett, C. G., 1884. Notes on British Tortrices. *Entomologist's mon. Mag.*, **20**: 236–244

Barrett, C. G., 1887. Notes on British Tortrices. *Entomologist's mon. Mag.*, **24**: 34–36

Barrett, C. G., 1904–05. *The lepidoptera of the British Islands*, **10**: 1–381, pls. 425–469. London

Barrett, C. G., 1905–07. *The lepidoptera of the British Islands*, **11**: i–lxxv, 1–293, pls. 470–504. London

Barritt, N. W., 1952. The Tortrix menace. *Fruit Grower*, no. 2928: 235–236

Beale, S. C. Tress, 1858. Captures near Alkham. *Entomologist's wkly Intell.*, **4**: 35

Beirne, B. P., 1941. A list of the microlepidoptera of Ireland. *Proc. R. Ir. Acad.*, **47**(B)4: 53–147

Benander, P., 1929. Zur Biologie einiger Kleinschmetterlinge. *Ent. Tidskr.*, **50**: 123–145, figs. 1–12

Benander, P., 1950. Tortricina. *Svensk Insektfauna*, no. 39: 1–173, pls. 1–9

Benander, P., 1965. Notes on larvae of Swedish micro-lepidoptera. II. *Opusc. ent.*, **30**: 1–23

Bentinck, G. A. Graaf & Diakonoff, A., 1968. De Nederlandse Bladrollers (Tortricidae). *Monogr. Ned. ent. Ver.*, no. 3: 1–201, pls. 1–99

Birkett, N. L., 1955. *Phalonia implicitana* Wocke in North Lancashire. *Entomologist's Rec. J. Var.*, **67**: 331

Blair, K. G., 1925. The lepidoptera of the Scilly Isles. *Entomologist*, 58: 3–10

Bolam, G., 1929. The lepidoptera of Northumberland and the Eastern Borders. *Hist. Berwicksh. Nat. Club*, 27: 115–142

Bradley, J. D., 1952a. *Adoxophyes orana* (Fischer von Roeslerstamm, 1834) (Lepidoptera: Tortricidae). *Entomologist*, 85: 1–4, text figs. 1–4, pl. 1, figs. 1, 2

Bradley, J. D., 1952b. *Proc. S. Lond. ent. nat. Hist. Soc.*, 1951/52: 16–17

Bradley, J. D., 1953. Microlepidoptera from the Bantry-Glengarriff area, West Cork. *Ir. Nat. J.*, 11: 16–18

Bradley, J. D., 1958. Microlepidoptera from the Islands of Canna and Sanday, Inner Hebrides. *Entomologist*, 91: 9–14

Bradley, J. D., 1959. An illustrated list of the British Tortricidae. Part II: Olethreutinae. *Entomologist's Gaz.*, 10: 60–80, text figs. 1–12, pls. 1–19

Bradley, J. D., 1962. On *Acleris literana* (Linnaeus) (Lep., Tortricidae), with descriptions of new aberrations from the L. T. Ford collection. *Entomologist's Gaz.*, 13: 117–128, pls. 1, 2

Bradley, J. D., 1964. Lepidoptera in Ireland, May–June, 1962. Part III. *Entomologist's Gaz.*, 15: 74–82, pl. 3, figs. 1–6, text figs. 1–14

Bradley, J. D. & Martin, E. L., 1956. An illustrated list of the British Tortricidae. Part I: Tortricinae and Sparganothinae. *Entomologist's Gaz.*, 7: 151–156, pls. 1–10

Bradley, J. D. & Mere, R. M., 1964. Natural history of the garden of Buckingham Palace. Lepidoptera. *Proc. S. Lond. ent. nat. Hist. Soc.*, 1963(2): 55–74, pls. 7, 8

Bradley, J. D. & Pelham-Clinton, E. C., 1967. The lepidoptera of the Burren, Co. Clare, W. Ireland. *Entomologist's Gaz.*, 18: 115–153, pl. 5, text fig. 1 (map)

Brown, S. C. S., 1953. *Eulia formosana* in Hants and Dorset. *Entomologist*, 86: 83

Chalmers-Hunt, J. M., 1951. *Eulia formosana* in Kent in 1951. *Entomologist*, 84: 272

Chalmers-Hunt, J. M., 1968a. Manx entomology. *Entomologist's Rec. J. Var.*, 80: 41–48

Chalmers-Hunt, J. M., 1968b. A previously unpublished record of *Adoxophyes orana* F. v. R. *Entomologist's Rec. J. Var.*, 81: 179

Chalmers-Hunt, J. M., 1969. Breconshire and Monmouthshire entomology. *Entomologist's Rec. J. Var.*, 81: 39–46

Chapman, T. A., 1907. *Tortrix semialbana* Gn. in Argyleshire [sic]. *Entomologist's mon. Mag.*, 43: 258

Chipperfield, H. E., 1967. Suffolk lepidoptera, 1966. *Trans. Suffolk Nat. Soc.*, 13: 318–320

Chipperfield, H. E., 1968. Suffolk lepidoptera in 1967. *Trans. Suffolk Nat. Soc.*, 14: 60–63

Chipperfield, H. E., 1970. *Adoxophyes orana* F. v. R. in Suffolk. *Entomologist's Rec. J. Var.*, 81: 87

Chrétien, P., 1897. Les premiers états de la *Tortrix grotiana*. *Naturaliste*, 19: 258–260

Common, I. F. B., 1970. *In* Mackerras, I. M., *The insects of Australia*, pp. xiii, 1029, col. pls., text figs. Melbourne

Conquest, G. H., 1911. Six days at Glen Tilt, Perthshire. *Entomologist*, 44: 155–157

Cottam, A., 1903. *Aphelia argentana* in Norfolk. *Entomologist's mon. Mag.*, 39: 226–227

Cox, W. E., 1940a. *Cacoecia pronubana* Hübn. in Glamorgan. *Entomologist*, 73: 236

Cox, W. E., 1940b. *Cnephasia genitalana* Pierce in Glamorgan. *Entomologist*, 73: 236

Cram, W. T. & Tonks, N. V., 1959. Note on occurrence in British Columbia of the omnivorous leaf tier, *Cnephasia longana* (Haw.) (Lepidoptera: Tortricidae), as a pest of strawberry. *Can. Ent.*, 91: 155–156

Dale, C. W., 1886. *The lepidoptera of Dorsetshire or a catalogue of butterflies and moths found in the county of Dorset*, pp. xiv, 90. Dorchester

Dale, C. W., 1891. *The lepidoptera of Dorsetshire: or a catalogue of butterflies and moths found in the county of Dorset*, pp. 76 (second edition). Dorchester

Dallimore, W. & Jackson, A. B., 1966. *A handbook of Coniferae and Ginkgoaceae*, pp. 729, 46 pls., 131 text figs. London

De Joannis, J., 1900. Note sur quelques lépidoptères observés par H. Lhotte aux environs de Paris. *Bull. Soc. ent. Fr.*, 1900: 298–300

De Lucca, C., 1951. Notes on the biology of *Cnephasia gueneana* Duponchel (Lepidoptera: Tortricidae). *Entomologist*, 84: 205–207

De Worms, C. G. M., 1952. *Tortrix formosana* near Woking. *Entomologist*, 85: 215

Diakonoff, A., 1952. Microlepidoptera of New Guinea. *Verh. K. ned. Akad. Wet.*, (2)49(1): 1–167, figs. 1–208, pl. 1

Diakonoff, A., 1953. *Ibidem. Verh. K. ned. Akad. Wet.*, (2)49(3): 1–166, figs. 1–372

Döring, E., 1954. Ein Beitrag zur Kenntnis der Kremaster von Microlepidopteren. *Dt. ent. Z.*, (N.F.)1: 23–32

Donovan, C., 1902. A list of the lepidoptera of County Cork. *Entomologist*, 35: 10–14

Doubleday, H., 1847. Remarks on the introduction of exotic insects into collections professedly British. *Zoologist*, 5: 1728–1730

Doubleday, H., 1873. *Synonymic list of British lepidoptera, third supplement*, pp. [1]–[4]. London

Dunn, T. C., 1969. Collecting in the Aran Islands, July 28th–August 10th, 1968. *Entomologist's Gaz.*, 20: 271–278

Edwards, T. G. & Wakely, S., 1958. A fortnight at Dungeness, 1957. *Entomologist's Rec. J. Var.*, 70: 92–94

Edwards, W. D. & Mote, D. C., 1936. Omnivorous leaf tier, *Cnephasia longana* Haw. *J. econ. Ent.*, 29: 1118–1123

Elisha, G., 1891. Early stages of *Argyrolepia maritimana* Guen. *Entomologist*, 24: 277–278, pl. 5, figs. 1–7

Emmet, A. M., 1968. Lepidoptera in West Galway. *Entomologist's Gaz.*, 19: 45–58

Emmet, A. M., 1969. *Choristoneura diversana* Hübn. in Somerset. *Entomologist's Rec. J. Var.*, 81: 96

Emmet, A. M., 1971. More lepidoptera in West Galway. *Entomologist's Gaz.*, 22: 3–18

Evans, W., 1897. Notes on lepidoptera collected in the Edinburgh district. *Ann. Scot. nat. Hist.*, 1897: 89–110

Evans, W., 1905. Lepidoptera from the Edinburgh (or Forth) district. *Ann. Scot. nat. Hist.*, 1905: 153–160

Fairclough, R., 1955. *Cacoecia aeriferana* H.-S. in Sussex. *Entomologist's Rec. J. Var.*, 67: 34

Fairclough, R., 1961. Collecting lepidoptera in 1960. *Entomologist's Rec. J. Var.*, 73: 11–17

Fassnidge, W., 1936. *Phalonia flaviciliana* Wilk. in Hants. *Entomologist's Rec. J. Var.*, 48: 123–124

Fenn, C., 1890. A concise life-history of *Tortrix diversana=transitana. Entomologist's mon. Mag.*, 26: 216–217

Fisher, R. C., 1924. The life-history and habits of *Tortrix pronubana* Hb., with special reference to the larval and pupal stages. *Ann. appl. Biol.*, 11: 395–447, figs. 1–16, pl. 15

Fletcher, T. Bainbrigge, 1933. An Australian tortricid (*Tortrix postvittana* Walker) in England. *Entomologist's Rec. J. Var.*, 45: 165–166

Fletcher, T. Bainbrigge, 1945. A new locality for *Phalonia gilvicomana* (Lep., Phaloniadae). *Entomologist's Rec. J. Var.*, 57: 21

Fletcher, W. H. B., 1905. *Tortrix pronubana* Hb.: a species new to the British list, in Sussex. *Entomologist's mon. Mag.*, 41: 276

Ford, L. T., 1936a. Micro-collecting—mid May to mid June. *Entomologist's Rec. J. Var.*, 48: 61–62

Ford, L. T., 1936b. Micro collecting. Mid-July to mid-August. *Entomologist's Rec. J. Var.*, 48: 84–85

Ford, L. T., 1945. Collecting Micros. *Amat. Ent.*, 8: 26–29, fig. 26

Ford, L. T., 1949. *A guide to the smaller British lepidoptera*, pp. 230. London

Ford, L. T., 1958. *Supplement to a guide to the smaller British lepidoptera*, pp. 16. London

Fracker, S. B., 1915. The classification of lepidopterous larvae. *Illinois biol. Monogr.*, 2(1): 1–169, pls. 1–10

Frampton, R. E. E., 1936. Notes on some *Tortrix* larvae, May and early June. *Entomologist's Rec. J. Var.*, 48: 62–63

Freeman, T. N., 1958. The Archipinae of North America (Lepidoptera: Tortricidae). *Can. Ent.*, Suppl., 7: 1–89

Freeman, T. N., 1967. Annotated keys to some nearctic leaf-mining lepidoptera on conifers. *Can. Ent.*, 99: 419–435, text figs. 1–31c

Fryer, J. C. F., 1926. Miscellaneous observations, 1925. *Entomologist*, 59: 114

Fryer, J. C. F., 1928. Polymorphism in the moth *Acalla comariana* Zeller. *J. Genet.*, **20**: 157–178, pl. 2, figs. 1–7

Fryer, J. C. F., 1931. Further notes on the tortricid moth *Acalla comariana* Zeller. *J. Genet.*, **24**: 195–202

Fryer, J. C. F., 1934a. *Tortrix postvittana* Walker in England. *Entomologist's Rec. J. Var.*, **46**: 7

Fryer, J. C. F., 1934b. *Cacoecia pronubana* Hb. *Entomologist's Rec. J. Var.*, **46**: 7–8

Gair, R., 1959. A tortricid caterpillar affecting Timothy seed crops. *Pl. Path.*, **8**: 95–96, pl. 4, figs. 1–5

Gasow, H., 1925. Der grüne Eichenwickler (*Tortrix viridana* Linné) als Forstschädling. *Arb. biol. BundAnst. Land-u. Forstw.*, **12**: 355–508, pls. 1–8

Gerasimov, A. M., 1952. *Fauna SSSR*, Lepid., **1**(2): 1–338, figs. 1–140 [in Russian]

Gregson, C. S., 1870. Description of *Argyrolepia luridana* Gregson, a Tortrix new to science. *Entomologist*, **5**: 80

Gregson, C. S., 1873a. Remarks on Mr. Barrett's "Notes on British *Tortrices*". *Entomologist's mon. Mag.*, **9**: 176–178

Gregson, C. S., 1873b. Description of the larva of *Sciaphila penziana*. *Entomologist*, **6**: 360–361

Gregson, C. S., 1880a. Life-history of *Peronea permutana*. *Entomologist*, **13**: 45–46

Gregson, C. S., 1880b. Life-history of *Amphysa prodromana* Hüb.=*walkeri* Curtis. *Entomologist*, **13**: 90

Groves, J. R., 1951. *Adoxophyes orana* F. R. (Lep., Tortricidae), a moth new to Britain. *Entomologist's mon. Mag.*, **87**: 259

Groves, J. R., 1952. A preliminary account of the summer fruit tortricid, *Adoxophyes orana* (F.R.), in Great Britain. *Rep. E. Malling Res. Stn*, **1951**: 152–154, figs. 1–4

Guthrie, W. G., 1897. Lepidoptera of the Hawick district. *Hist. Berwicksh. Nat. Club*, **15**: 332–345

Hannemann, H. J., 1961. *Die Tierwelt Deutschlands*. Part 48. Kleinschmetterlinge oder Microlepidoptera. I. Die Wickler (s. str.) (Tortricidae). pp. i–xi, 233, 457 text figs., 22 pls. Jena

Hannemann, H. J., 1964. *Die Tierwelt Deutschlands*. Part 50. Kleinschmetterlinge oder Microlepidoptera. II. Die Wickler (s.l.) (Cochylidae und Carposinidae). Die Zünslerartigen (Pyraloidea). pp. viii, 401, text figs., 22 pls. Jena

Hardman, J. A., 1953. The summer fruit tortrix moth, *Adoxophyes orana* (Fisch. v. Roesl.), in Britain: a review of available information. *Entomologist*, **86**: 264–272

Haworth, A. H., 1803–28. *Lepidoptera Britannica*, pp. xxxvi, 609. London

Hayward, A. R., 1927. *Peronea umbrana* in Somerset. *Entomologist*, **60**: 44

Hayward, H. C., 1929a. *Peronea umbrana* Hb. in Dorset. *Entomologist*, **62**: 19

Hayward, H. C., 1929b. Some notes on lepidoptera in the Isle of Man. *Entomologist*, **62**: 49–51

Heddergott, H., 1957. *Cnephasia argentana* Cl. (Lep., Tortricidae) als Schädling an Fichtenkulteren. *Z. angew. Ent.*, **40**: 332–342, figs. 1–7

Heinrich, C., 1923. Revision of the North American moths of the subfamily Eucosminae of the family Olethreutidae. *Bull. U.S. natn. Mus.*, **123**: 1–298, pls. 1–59

Heinrich, C., 1926. Revision of the North American moths of the subfamilies Laspeyresiinae and Olethreutinae. *Bull. U.S. natn. Mus.*, **132**: 1–216, pls. 1–76

Hering, E. M., 1951. *Biology of leaf miners*, pp. iv, 420, 180 text figs. 's-Gravenhage

Hering, E. M., 1957. *Blattminen von Europa*, **1**: 1–648; **2**: 1–1185; **3**: 1–221, pls. 1–81, figs. 1–725

Hey, G. L. & Massee, A. M., 1934. *Tortrix* investigations in 1933. *Rep. E. Malling Res. Stn*, **1933**: 228–230, table 1

Hinton, H. E., 1943. The larvae of the lepidoptera associated with stored products. *Bull. ent. Res.*, **34**: 163–212, figs. 1–128

Hinton, H. E., 1946. On the homology and nomenclature of the setae of lepidopterous larvae, with some notes on the phylogeny of the lepidoptera. *Trans. R. ent. Soc. Lond.*, **97**: 1–37, figs. 1–24

Hinton, H. E., 1948. The dorsal cranial areas of caterpillars. *Ann. Mag. nat. Hist.*, (11)14: 843–852, text figs. 1–6

Hinton, H. E., 1956. The larvae of the species of Tineidae of economic importance. *Bull. ent. Res.*, 47: 251–346, text figs. 1–216

Hodgkinson, J. B., 1869. *Peronea umbrana* in Westmoreland [sic]. *Entomologist's mon. Mag.*, 5: 224

Hoffmann, A., 1886. The lepidoptera of the Shetland Islands. *Scott. Nat.*, 8: 165–179, 244–250

Holst, P. L., 1962. *Tortrix unitana* Hübner, a distinct species (Lepidoptera, Tortricidae). *Ent. Meddr.*, 31: 303–310, figs. 1–24

Huggins, H. C., 1923. Notes on Tortrices observed in 1922. *Entomologist*, 56: 15–16

Huggins, H. C., 1924. A few notes on the habits of *Peronea hastiana*. *Entomologist*, 57: 281–282

Huggins, H. C., 1928. The habits of *Peronea fissurana* Pierce. *Entomologist*, 61: 17–18

Huggins, H. C., 1932. The possible occurrence of *Capua* (*Dichelia*) *gnomana* Clerck in the British Isles. *Entomologist*, 65: 104–105

Huggins, H. C., 1933. *Peronea fissurana* Pierce and Metcalfe in Essex. *Entomologist*, 66: 39

Huggins, H. C., 1951. Obituary. Percival Frederic Harris. *Entomologist*, 84: 120

Huggins, H. C., 1953a. *Adoxophyes orana* (v. Roesl.) (Lep., Tortricidae) in south-east Essex. *Entomologist*, 86: 189

Huggins, H. C., 1953b. The lepidoptera of Glengarriff, Co. Cork (Part II). *Entomologist*, 86: 242–254

Huggins, H. C., 1954. *Adoxophyes orana* (Fisch. v. Roesl.) in Essex. *Entomologist*, 87: 128

Huggins, H. C., 1958a. Notes on microlepidoptera. *Entomologist's Rec. J. Var.*, 70: 53–55

Huggins, H. C., 1958b. Notes on microlepidoptera. *Entomologist's Rec. J. Var.*, 70: 107–108

Huggins, H. C., 1958c. *Eulia formosana* Huebner in Essex. *Entomologist*, 91: 269

Huggins, H. C., 1959. Notes on the microlepidoptera. *Entomologist's Rec. J. Var.*, 71: 121

Huggins, H. C., 1961a. Notes on the microlepidoptera. *Entomologist's Rec. J. Var.*, 73: 149–152

Huggins, H. C., 1961b. Notes on the microlepidoptera. *Entomologist's Rec. J. Var.*, 73: 181

Huggins, H. C., 1962. Insect movements in 1961. *Entomologist's Rec. J. Var.*, 74: 40–41

Huggins, H. C., 1964. Notes on the microlepidoptera. *Entomologist's Rec. J. Var.*, 76: 158–159

Huggins, H. C., 1964. Notes on the microlepidoptera. *Entomologist's Rec. J. Var.*, 76: 230–231

Hulme, D. C., 1962. Derbyshire lepidoptera—first supplement. *Entomologist's Rec. J. Var.*, 74: 77–84

Humphreys, H. N. & Westwood, J. O., 1845. *British moths and their transformations*, 2: i–xix, 1–268, pls. 57–124. London

Image, S., 1906. *Tortrix pronubana* Hb. at Eastbourne. *Entomologist's mon. Mag.*, 42: 13–14

Imms, A. D., 1960. *A general textbook of Entomology*, pp. x, 886, 609 text figs. London

Jacobs, S. N. A., 1931. *Peronea ferrugana* in January. *Entomologist*, 64: 41–42

Jacobs, S. N. A., 1933. *Proc. S. Lond. nat. Hist. Soc.*, 1932–33: 70

Jacobs, S. N. A., 1952. *Adoxophyes orana* in Kent. *Entomologist's Rec. J. Var.*, 64: 86–87

King, J. J. F. X., 1887. *Proc. nat. Hist. Soc. Glasgow*, (N.S.)1: xlix

Kloet, G. S. & Hincks, W. D., 1972. A check list of British Insects, second edition (completely revised), part 2. *Handbk Ident. Br. Insects*, 11(2): i–viii, 1–153

Klots, A. B., 1941. Two European Tortricidae (Lepidoptera) not hitherto recorded from North America. *Bull. Brooklyn ent. Soc.*, 36: 126–127

Klots, A. B., 1970. *In* Tuxen, S. L., *Taxonomist's glossary of genitalia in insects*, pp. 359, 248 text figs. Copenhagen

Mackay, M. R., 1959. Larvae of the North American Olethreutidae (Lepidoptera). *Can. Ent.*, Suppl., 10: 1–338, figs. 1–161

Mackay, M. R., 1962. Larvae of the North American Tortricinae (Lepidoptera: Tortricidae). *Can. Ent.*, Suppl., 28: 1–182, figs. 1–86

Mackay, M. R., 1963. Evolution and adaptation of larval characters in the Tortricidae. *Can. Ent.*, **95**: 1321–1344, figs. 1–26

Maling, W., 1877. Notes on the occurrence of lepidoptera in Northumberland and Durham in 1875. *Trans. nat. Hist. Soc. Northumb.*, **5**: 277–282

Manley, W. B. L., 1961. A tortricid new to Britain. *Entomologist's Gaz.*, **12**: 18

Manley, W. B. L., 1973. A guide to *Acleris cristana* [Denis & Schiffermüller, 1775] (Lepidoptera: Tortricidae) in Britain. *Entomologist's Gaz.*, **24**: 89–206, pls. 4–7

Mansbridge, W., 1905. Tortrices in the Liverpool district. *Entomologist*, **38**: 115–116

Mansbridge, W., 1914. *Tortrix costana* F. vars. *liverana* and *intermedia*. *Rep. Lancs. Chesh. ent. Soc.*, **1913**: 18–19, pl., figs.

Mansbridge, W., 1918. Hibernation of *Peronea hastiana*. *Entomologist*, **51**: 140

Martin, E. L., 1962. Keys to the identification of the families of lepidopterous larvae [English translation from Gerasimov, 1952]. *Coridon*, (A)**5**: 1–[16]

Martin, W. Keble, 1965. *The concise British flora in colour*, pp. 231, 100 pls. London

Massee, A. M., 1946. *The pests of fruit and hops*, pp. 284, 26 pls. London

Matthey, W., 1967. The natural history of three oak-feeding tortricids: *Ptycholoma lecheana* (L.), *Pandemis cerasana* (Hb.) and *Batodes angustiorana* (Haw.). *Entomologist*, **100**: 115–120, tables 1–4

Meldola, R., 1915. Hibernation of *Peronea sponsana* Fab. *Entomologist*, **48**: 198

Mere, R. M., 1959. The Isles of Scilly in 1958. *Entomologist's Gaz.*, **10**: 107–110

Mere, R. M. & Pelham-Clinton, E. C., 1966. Lepidoptera in Ireland, 1963, 1964 and 1965. *Entomologist's Gaz.*, **17**: 163–182

Messenger, J. L., 1951. Light trap captures at Weybridge. *Entomologist*, **84**: 272

Meyrick, E., 1895. *A handbook of British lepidoptera*, pp. vi, 843. London

Meyrick, E., 1913. Lepidoptera Heterocera Fam. Tortricidae. *Genera Insect.*, **149**: 1–81, pls. 1–5

Meyrick, E., 1928. *A revised handbook of British lepidoptera*, pp. 914. London

Meyrick, E., 1937. *Tortrix postvittana* Walk. (Microlepidoptera), a species new to Britain. *Entomologist*, **70**: 256

Michaelis, H. N., 1953. Microlepidoptera in Cheshire and South Lancashire in 1952. *Entomologist's Rec. J. Var.*, **65**: 74–76

Michaelis, H. N., 1954a. *In* Roberts, J. E. et al., The butterflies and moths found in the county of Radnorshire . . . Lepidoptera Tortricoidea. *Proc. Chester Soc. nat. Sci.*, **5**: 55–80

Michaelis, H. N., 1954b. Lepidoptera, 1952/53. *30th Rep. Lancs. Chesh. Fauna Comm.*: 53–56

Michaelis, H. N., 1956. The foodplants of *Peronea caledoniana* Steph. *Entomologist's Rec. J. Var.*, **68**: 235–237

Michaelis, H. N., 1958. Microlepidoptera in Lancashire and Cheshire, 1955–57. *Entomologist's Rec. J. Var.*, **70**: 122–124

Michaelis, H. N., 1965. *Aphelia unitana* Hübner (Lep., Tortricidae) in Yorkshire. *Entomologist's Gaz.*, **16**: 16

Michaelis, H. N., 1966. Records of lepidoptera from Lancashire, Cheshire and Wales. *Proc. Lancs. Chesh. ent. Soc.*, **1966**: 106–111

Michaelis, H. N., 1969a. Microlepidoptera found near the estuary of the River Conway, North Wales, 1964/68. *Entomologist's Rec. J. Var.*, **81**: 1–6

Michaelis, H. N., 1969b. Lepidoptera from the mountains of North Wales. *Nature Wales*, **11**: 183–191

Mourikis, P. A. & Vassilaina-Alexopoulou, P., 1972. Observations on the laboratory rearing and biology of *Cacoecimorpha pronubana* (Hbn.). *Entomologist*, **105**: 209–216, text fig. 1, tables 1–3

Newton, J., 1961. Microlepidoptera in Gloucestershire. *Entomologist's Rec. J. Var.*, **73**: 86–88

Obraztsov, N., 1952. Über einige palaearktische *Aethes* Billb. Arten (Lepidoptera, Agapetidae = Phaloniidae). *Ent. Z.*, **61**: 157–160, 164–168, 174–176, 181–182

Obraztsov, N. S., 1954–57. Die Gattungen der palaearktischen Tortricidae. *Tijdschr. Ent.*, **97**: 141–231, text figs. 1–248; *ibidem*, **98**: 147–228, text figs. 249–366; *ibidem*, **99**: 107–154; *ibidem*, **100**: 309–347

Obraztsov, N. S., 1963. Some North American moths of the genus *Acleris* (Lepidoptera: Tortricidae). *Proc. U.S. natn. Mus.*, **114**: 213–270, pls. 1–18

Opheim, M., 1961. Revision of some Norwegian species of the genus *Phalonia* Hb. (Lep., Phaloniidae). *Ent. Tidskr.*, **11**: 250–254

Opheim, M., 1965. Notes on the Norwegian Tortricidae III (Lepidoptera). *Norsk ent. Tidsskr.*, **13**: 23–30, figs. 1–15

Parfitt, R. W., 1947. *Eulia formosana* Geyer (Lep., Tortricidae): a species new to the British list. *Entomologist*, **80**: 225–227

Patočka, J., 1958. Bemerkungen zur Morphologie der Puppen und Bionomie einiger Wickler aus dem Tribus Archipsini (Lepidoptera, Tortricidae). *Cas. čsl. Spol. ent.*, **55**: 185–197, figs. 1–17

Patočka, J., 1960. *Die Tannenschmetterlinge der Slowakei*, pp. 214, 470 figs. Bratislava

Peers, J., 1864. Description of the larva of *Peronia* [sic] *caledoniana*. *Entomologist*, **2**: 63

Peers, J., 1865. Description of the egg and larva of *Tortrix ministrana*. *Entomologist*, **2**: 250–251

Pelham-Clinton, E. C., 1959. Microlepidoptera new to Scotland. *Entomologist's Rec. J. Var.*, **71**: 68–72

Pelham-Clinton, E. C., 1967. *Acleris abietana* (Hübner, 1819–22), a tortricid moth new to the British Isles. *Entomologist's Rec. J. Var.*, **79**: 151–152, pl. 8, figs. 1–3

Pelham-Clinton, E. C., 1969. *Epichoristodes acerbella* (Walker) (Lep., Tortricidae) imported alive into Britain. *Entomologist's Gaz.*, **20**: 72

Perring, F. H. & Walters, S. M., 1962. *Atlas of the British Flora*, pp. xxiv, 432. London

Petherbridge, F. R., 1920. The life history of the strawberry tortrix, *Oxygrapha comariana* (Zeller). *Ann. appl. Biol.*, **7**: 6–10, pl. 1, figs. 1–3

Pierce, F. N., 1909. *The genitalia of the group Noctuidae of the lepidoptera of the British Islands*, pp. xii, 88, 32 pls. Liverpool

Pierce, F. N., 1914. *The genitalia of the group Geometridae of the lepidoptera of the British Islands*, pp. xxix, 88, 48 pls. Liverpool

Pierce, F. N. & Metcalfe, J. W., 1915. An easy method of identifying the species of the genus *Cnephasia*=*Sciaphila* (Tortricidae). *Entomologist's Rec. J. Var.*, **27**: 99–102, pl. 3

Pierce, F. N. & Metcalfe, J. W., 1922. *The genitalia of the group Tortricidae of the lepidoptera of the British Islands*, pp. xxii, 101, 34 pls. Oundle

Pierce, F. N. & Metcalfe, J. W., 1935. *The genitalia of the Tineid families of the lepidoptera of the British Islands*, pp. xxii, 116, 68 pls. Oundle

Pierce, F. N. & Metcalfe, J. W., 1960. *The genitalia of the group Tortricidae of the lepidoptera of the British Islands*, pp. xxii, 101, 34 pls. Facsimile edition, Hampton

Povolný, D., 1951. *Tortrix* (*Cnephasia* Curt.) *nubilana* Hbn. škůdcem meruňky na jižní Moravé. *Ent. Listy*, **14**: 188–196, figs., pl. 1, figs. 1–22

Powell, J. A., 1964. Biological and taxonomic studies on tortricine moths, with reference to the species in California. *Univ. Calif. Publs Ent.*, **32**: 1–317, figs. 1–108, pls. 1–8, maps

Prest, W., 1877. Food of *Tortrix viburnana*. *Entomologist*, **10**: 49

Purdey, W., 1899a. Notes on the habits of *Lozopera beatricella* Wlsm. *Entomologist's mon. Mag.*, **35**: 289

Purdey, W., 1899b. Note on *Lozopera beatricella*. *Entomologist*, **32**: 306

Raynor, G. H., 1882. *Argyrolepia schreibersiana* re-discovered. *Entomologist's mon. Mag.*, **19**: 44

Razowski, J., 1959. Europejskie gatunki Cnephasiini (Lepidoptera, Tortricidae). *Acta zool. cracov.*, **4**: 179–423, pls. 17–67

Razowski, J., 1965. The palaearctic Cnephasiini (Lepidoptera, Tortricidae). *Acta zool. cracov.*, **10**: 199–343, pls. 12–26, text figs. 1–154

Razowski, J., 1966. *World fauna of the Tortricini* (*Lepidoptera, Tortricidae*), pp. 576, 832 text figs., 41 pls. Kraków

Razowski, J., 1969. *Klucze Oznacz. Owad. Pol.*, **27**, 41b Tortricidae: Tortricinae i Sparganothinae, pp. 131, 522 text figs.

Razowski, J., 1970. *In* Amsel, H. G., Gregor, F. & Reisser, H., *Microlepid. palaearct.*, **3**: i–xiv, 1–528, pls. 1–27 (colour), 1–161 (line drawings). Wien

Reid, J. A., 1941. Mating and oviposition in *Cnephasia chrysantheana* (Dup.) (Lepidopt., Tortricidae). *Proc. R. ent. Soc. Lond.*, (A)**16**: 24–28, tables 1–3

Reid, W., 1893a. List of the lepidoptera of Aberdeenshire and Kincardineshire. *Br. Nat.*, **3**: 8–10

Reid, W., 1893b. *List of the lepidoptera of Aberdeenshire and Kincardineshire*, pp. 35

Richardson, N. M., 1885. Habits and description of the larva of *Tortricodes hyemana*. *Entomologist's mon. Mag.*, 21: 252–253

Richardson, N. M., 1890. Notes on *Eupoecilia notulana* and *Halonota cirsiana*. *Entomologist's mon. Mag.*, 26: 299–300

Richardson, N. M., 1891. The larva of *Eupoecilia geyeriana*. *Entomologist's mon. Mag.*, 27: 239–240

Richardson, N. M., 1892. Notes on Dorset lepidoptera in 1891. *Proc. Dorset nat. Hist. antiq. Fld Club*, 13: 168–177, pl., figs.

Richardson, A. & Mere, R. M., 1958. Some preliminary observations on the lepidoptera of the Isles of Scilly with particular reference to Tresco. *Entomologist's Gaz.*, 9: 115–147, pls. 8, 9

Robinson, G. S., 1967. Lepidoptera at Heversham. *Entomologist's Rec. J. Var.*, 79: 271–278

Rosenstiel, R. G., 1941. Oviposition of the omnivorous leaf-tier. *J. econ. Ent.*, 34: 255

Rosenstiel, R. G., Ferguson, G. R. & Mote, D. C., 1944. Some ecological relationships of *Cnephasia longana*. *J. econ. Ent.*, 37: 814–817, figs. 1, 2

Royal Horticultural Society, 1951–69. *Dictionary of Gardening*. Vols. 1–4; 2 Suppls. Oxford

Sadler, E., 1967. *Austrotortrix postvittana* Walk. in Hampshire. *Entomologist's Rec. J. Var.*, 79: 87

Sang, J., 1885. Occurrence of *Sciaphila abrasana*. *Entomologist's mon. Mag.*, 21: 192

Scott, G., 1952. *Cacoecia aeriferana* Herrich-Schäffer in England. *Entomologist*, 85: 170

Scott, G., 1953. On *Cacoecia aeriferana* H.-S. in Kent. *Entomologist*, 86: 139

Sheldon, W. G., 1888. Notes on some British Tortrices. *Entomologist*, 21: 102–104

Sheldon, W. G., 1891. Notes on *Eupoecilia sodaliana* Haw. (*amandana* H.-S.), with description of the larva. *Entomologist's mon. Mag.*, 27: 301

Sheldon, W. G., 1917. *Peronea cristana*: its life-history, habits of the imago, distribution of the various named forms, and some speculations on the present trend of its variation. *Entomologist*, 50: 217–222, 245–250, 268–273

Sheldon, W. G., 1918. *Ibidem. Entomologist*, 51: 10–14

Sheldon, W. G., 1919. The earlier stages of *Peronea maccana* Tr., *P. lipsiana* Schiff., *P. rufana* Schiff. and *P. schalleriana* L. *Entomologist*, 52: 252–255, 271–274

Sheldon, W. G., 1920a. The life-cycle of *Cacoecia unifasciana* Duponchel. *Entomologist*, 53: 49–52

Sheldon, W. G., 1920b. Notes on the variation of *Peronea cristana* Fab., with descriptions of six new forms, and the reasons for sinking the names at present in use of six others. *Entomologist*, 53: 265–271

Sheldon, W. G., 1921a. *Ibidem. Entomologist*, 54: 12–16, 35–39, 64–67

Sheldon, W. G., 1921b. *Oxigrapha literana* L.: its life-cycle, distribution, and variation. *Entomologist*, 54: 129–135, 157–161, 187–190, 209–210, pl. 1, figs. 1–10

Sheldon, W. G., 1922. On the earlier stages of *Cacoecia crataegana* Hübn. *Entomologist*, 55: 194–195

Sheldon, W. G., 1923. *Peronea hastiana* L.: its distribution, habits, life-cycle and variation. *Entomologist*, 56: 75–81, 100–104, 128–131, 149–153, 173–178, 197–202, 221–226, 248–252, 269–271, pl. 2, figs. 1–24

Sheldon, W. G., 1925. *Peronea comariana* Zeller, and its variation. *Entomologist*, 58: 281–285

Sheldon, W. G., 1930. Notes on the nomenclature and variation of British species of the *Peronea* group of the Tortricidae. *Entomologist*, 63: 121–124, 148–151, 175–178, 193–198, 222–225, 242–246, 273–277, pl. 4, figs. 1–18

Sheldon, W. G., 1931. *Ibidem. Entomologist*, 64: 2–6, 30–34, 60–64, 77–82, 99–103, 124–127

Sheldon, W. G., 1937. *Phalonia gilvicomana* Zeller; its history as a British species, with notes on its foodplant and larva. *Entomologist*, 70: 197–199

Sich, A., 1914. *Tortrix pronubana* Hb., in Chiswick. *Entomologist's mon. Mag.*, **50**: 250

Sich, A., 1916. Life cycle of *Tortrix viridana* L. *Proc. S. Lond. ent. nat. Hist. Soc.*, 1915/16: 15–20

Sich, A., 1918. Field notes from Bath, with special reference to *Teras contaminana* Hb. *Entomologist's Rec. J. Var.*, **30**: 69–71

Smart, H. D., 1917. Note on egg-laying of *Tortrix pronubana*. *Entomologist*, **50**: 280

South, R., 1893. Lepidoptera of the Shetland Islands. *Entomologist*, **26**: 98–102

South, R., 1897. *Peronea permutana* in Sussex. *Entomologist*, **30**: 220

South, R., 1898. Tortrices occurring in the vicinity of the Chesham line. *Entomologist*, **31**: 90–94, 116–119, 133–136

Stainton, G. H., 1845. Capture of lepidopterous insects in Scotland in 1845. *Zoologist*, **3**: 1090–1091

Stainton, H. T., 1858. Lepidoptera. New British species in 1857. *Entomologist's Annu.* 1858: 85–98

Stainton, H. T., 1859. *A manual of British butterflies and moths*, pp. xi, 480, text figs. London

Standish, F. O., 1879. Capture of *Argyrolepia schreibersiana*. *Entomologist*, **12**: 205–206

Stephens, J. F., 1834–35. *Illustrations of British entomology*. Haustellata, **4**: 1–[436]. London

Styles, J. H., 1955. *Cacoecia aeriferana* Herrich-Schäffer (Lep., Tortricidae) in Norfolk. *Entomologist*, **88**: 82–83, fig. 1

Styles, J. H., 1959. Notes on some microlepidoptera. *Entomologist's Gaz.*, **10**: 43–44

Styles, J. H., 1960. *Syndemis musculana* Hübner (Lep., Tortricidae) in conifer plantations and forest nurseries in the British Isles. *Entomologist's Gaz.*, **11**: 144–148, fig. 1, pl. 5, figs. 1, 2

Swatschek, B., 1958. Die Larvalsystematik der Wickler (Tortricidae und Carposinidae). *Abh. Larvalsyst. Insekt.*, no. 3: 1–269, figs. 1–276

Thomas, D. C. & Waterston, A. R., 1936. *In* Forrest, J. E., Waterston, A. R. & Watson, E. V., The natural history of Barra, Outer Hebrides. *Proc. R. phys. Soc. Edinb.*, **22**: 241–296

Thurnall, A., 1902. A list of Tortrices taken in South Essex between 1885 and 1901. *Entomologist*, **35**: 129–134

Thurnall, A., 1911. *Loxopera beatricella* Wlsm. in Essex. *Entomologist's mon. Mag.*, **47**: 260

Thurnall, A., 1920. *Cacoecia unifasciana* Dup. *Entomologist*, **53**: 93

Tremewan, W. G., 1957. *Austrotortrix postvittana* in Cornwall. *Entomologist*, **90**: 76

Turner, A. H., 1955. *Lepidoptera of Somerset*, pp. 188, 1 map. Taunton

Turner, J. R. G., 1968. The ecological genetics of *Acleris comariana* (Zeller) (Lepidoptera: Tortricidae), a pest of strawberry. *J. Anim. Ecol.*, **37**: 489–520, figs. 1–8

Tutt, J. W., 1890. Habits of *Tortrix forsterana* (*adjunctana*). *Entomologist's Rec. J. Var.*, **1**: 65

Tutt, J. W., 1897. *Peronea permutana* at Barnes? *Entomologist's Rec. J. Var.*, **9**: 209–210

Van Deurs, W., 1956. Sommerfugle VIII. Viklere. *Danm. Fauna*, **61**: 1–292, pls. 1–31

Vernon, J. D. R., 1971. Observations on the biology and control of tortricid larvae on strawberries. *Pl. Path.*, **20**: 73–80, pls. 2, 3

Wakely, S., 1936. Notes on lepidoptera collected during 1935. *Entomologist*, **69**: 197–199

Wakely, S., 1937. Further notes on *Phalonia gilvicomana* Zeller. *Entomologist*, **70**: 225–227

Wakely, S., 1953. *Cacoecia aeriferana* Herrich-Schäffer (Lep., Tortricidae): further Kent records. *Entomologist*, **86**: 302

Wakely, S., 1955. *Cacoecia aeriferana* H.-S. in Britain. *Entomologist*, **88**: 141

Wakely, S., 1965. Entomological notes, 1964. *Entomologist's Rec. J. Var.*, **77**: 91–94

Wakely, S., 1966. Entomological notes of captures and observations in 1965. *Entomologist's Rec. J. Var.*, **78**: 99–103

Wakely, S., 1968a. Entomological notes for 1967. *Entomologist's Rec. J. Var.*, **80**: 164–169

Wakely, S., 1968b. Holiday at Thorpeness, Suffolk, 1968. *Entomologist's Rec. J. Var.*, **80**: 312–314

Wakely, S., 1969. Occurrence of *Adoxophyes orana* F. v. R. at Camberwell, London. *Entomologist's Rec. J. Var.*, **81**: 95

Walsingham, Lord, 1893. *Conchylis degreyana* McL.: an enigma. *Entomologist's mon. Mag.*, **29**: 202–203

Walsingham, Lord, 1898. *Lozopera francillonana* F. compared with its allies. *Entomologist's mon. Mag.*, 34: 70–76

Warren, W., 1883. On the habits of the larva of *Eupoecilia rupicola. Entomologist's mon. Mag.*, 20: 17

Warren, W., 1887a. Occurrence of *Lozotaenia* (*Cacoecia* Hb.) *decretana* Tr. in Norfolk. *Entomologist's mon. Mag.*, 24: 125–126

Warren, W., 1887b. Description of the larva of *Eupoecilia degreyana. Entomologist's mon. Mag.*, 24: 134

Watts, C. W., 1894. Lepidoptera taken in the Belfast district. *Proc. Belf. Nat. Fld Club*, 2 (Appendix 4): 115–131

Weir, J. J., 1881. Notes on the lepidoptera of the Outer Hebrides. *Entomologist*, 14: 218–223

Weir, J. J., 1882. Notes on the lepidoptera of the Orkney Islands. *Entomologist*, 15: 1–5

West, W., 1877. *Leptogramma scabrana* bred from the eggs of *L. boscana. Entomologist*, 10: 303–304

West, W., 1906. The lepidoptera of the south-eastern district of London. *Entomologist's Rec. J. Var.*, 18: 229–236

White, F. Buchanan, 1875. A new British Tortrix—*Ablabia argentana* Cl. *Entomologist's mon. Mag.*, 12: 85–86

White, J. S., 1880. *Amphysa prodromana. Entomologist*, 13: 114–116

Whittingham, A. G., 1931. *Peronea cristana*: breeding experiments. *Entomologist*, 64: 237–239

Whittingham, A. G., 1934. Further notes on *Peronea cristana. Entomologist*, 67: 90–91

Wickham, A. P., 1927. Notes on Sutherland and Inverness lepidoptera. *Entomologist*, 60: 42–44

Wilkinson, S. J., 1859. *The British Tortrices*, pp. viii, 328, 4 pls. London

Williams, B. S., 1915. Hibernation of *Peronea sponsana* Fab. *Entomologist*, 48: 220

Wood, J. H., 1878. Description of the larva of *Eupoecilia maculosana*, and its habits. *Entomologist's mon. Mag.*, 15: 149–150

Wood, W., 1839. *Index entomologicus*, pp. xii, 266, 54 pls. London

Wormell, P., 1969. *In* Steel, W. O. & Woodroffe, G. E., The entomology of the Isle of Rhum National Nature Reserve. *Trans. Soc. Br. Ent.*, 18: 91–167, 1 map

Wright, A. E., 1935. Notes on breeding *Eupoecilia affinitana* Douglas. *Entomologist*, 68: 263–264

Wright, A. E., 1947. Note on the foodplant of *Phalonia luridana* Gregson. *Entomologist's Rec. J. Var.*, 59: 69–70

Zangheri, S., 1965. La *Choristoneura* (=*Cornicacoecia*) *lafauryana* Rag. osservazioni sulla sua comparsa e diffusione nella Pianura Veneta (Lep., Tortricidae). *Memorie Soc. ent. ital.*, 44: 5–11, fig. 1

ADDENDUM

Fletcher, T. Bainbrigge & Clutterbuck, C. Granville, 1939. Microlepidoptera of Gloucestershire. *Proc. Cotteswold Nat. Fld Club*, 26: 298–317; 1940, *ibidem*, 27: 24–46; 1945, *ibidem*, 28: 65–66

INDEX TO GENERA, SPECIES
AND FORMS

(Synonyms and infrasubspecific names, i.e. names of forms, are placed *in italics*)